T0212042

Bedford Cultural Editions

GERTRUDE STEIN
Three Lives

EDITED BY
Linda Wagner-Martin
University of North Carolina — Chapel Hill

BEDFORD/ST. MARTIN'S   BOSTON ♦ NEW YORK

ISBN 978-1-349-62513-0   ISBN 978-1-137-07021-0 (eBook)
DOI 10.1007/978-1-137-07021-0

For Bedford/St. Martin's
Developmental Editor: Jennifer L. Rush
Project Management: Stratford Publishing Services, Inc.
Production Supervisor: Joe Ford
Marketing Manager: Karen Melton
Cover Design: Zenobia Rivetna
Cover Photos: Top and bottom photos (details): *Laundresses, Cooks, Parlor Maids, and Scullery Girls, Black River Falls, WI, ca. 1890.* Photo by Charles Van Schaick, reproduced with permission of the State Historical Society of Wisconsin Visual Material Archives. Middle photo (detail): Untitled, St. Paul, MN, circa 1910. Reproduced with permission of CORBIS/Minnesota Historical Society.
Composition: Stratford Publishing Services, Inc.
Printing and Binding: Haddon Craftsmen, an R. R. Donnelley & Sons Company

President: Charles H. Christensen
Editorial Director: Joan E. Feinberg
Director of Editing, Design, and Production: Marcia Cohen
Manager, Publishing Services: Emily Berleth

Library of Congress Catalog Card Number: 99-63687

5   4   3   2   1   0
f   e   d   c   b   a

For information, write: Bedford/St. Martin's, 75 Arlington Street, Boston, MA 02116 (617-399-4000)

ISBN: 978-0-312-18356-1 (paperback)

Acknowledgments
    Excerpts from Anna Julia Haywood Cooper's "The Higher Education of Women" from *A Voice from the South,* reprinted with the permission of The Schomburg Library of Nineteenth-Century Black Women Writers, 1988, Oxford Press.

Acknowledgments and copyrights are continued at the back of the book on page 387, which constitutes an extension of the copyright page.

Transferred to Digital Printing 2013

About the Series

The need to "historicize" literary texts — and even more to analyze the historical and cultural issues all texts embody — is now embraced by almost all teachers, scholars, critics, and theoreticians. But the question of how to teach such issues in the undergraduate classroom is still a difficult one. Teachers do not always have the historical information they need for a given text, and contextual documents and sources are not always readily available in the library — even if the teacher has the expertise (and students have the energy) to ferret them out. The Bedford Cultural Editions represent an effort to make available for the classroom the kinds of facts and documents that will enable teachers to use the latest historical approaches to textual analysis and cultural criticism. The best scholarly and theoretical work has for many years gone well beyond the "new critical" practices of formalist analysis and close reading, and we offer here a practical classroom model of the ways that many different kinds of issues can be engaged when texts are not thought of as islands unto themselves.

The impetus for the recent cultural and historical emphasis has come from many directions: the so-called new historicism of the late 1980s, the dominant historical versions of both feminism and Marxism, the cultural studies movement, and a sharply changed focus in older movements such as reader response, structuralism, deconstruction, and psychoanalytic theory. Emphases differ, of course, among schools and individuals, but what these movements and approaches have in common is a commitment to explore — and to have students in the classroom study interactively — texts in their full historical and cultural dimensions. The aim is to discover how older texts (and those from other traditions) differ from our own assumptions and expectations, and thus the focus in teaching falls on cultural and historical difference rather than on similarity or continuity.

The most striking feature of the Bedford Cultural Editions — and the one most likely to promote creative classroom discussion — is the inclusion of a generous selection of historical documents that contextualize the main text in a variety of ways. Each volume contains works (or passages from works) that are contemporary with the main text: legal and social documents, journalistic and autobiographical accounts, histories, sections from conduct books, travel books, poems, novels, and other historical sources. These materials have several uses. Often they provide information beyond what the main text offers. They provide, too, different perspectives on a particular theme, issue, or event central to the text, suggesting the range of opinions contemporary readers would have brought to their reading and allowing students to experience for themselves the details of cultural disagreement and debate. The documents are organized in thematic units — each with an introduction by the volume editor that historicizes a particular issue and suggests the ways in which individual selections work to contextualize the main text.

Each volume also contains a general introduction that provides students with information concerning the political, social, and intellectual context for the work as well as information concerning the material aspects of the text's creation, production, and distribution. There are also relevant illustrations, a chronology of important events, and, when helpful, an account of the reception history of the text. Finally, both the main work and its accompanying documents are carefully annotated in order to enable students to grasp the significance of historical references, literary allusions, and unfamiliar terms. Everywhere we have tried to keep the special needs of the modern student — especially the culturally conscious student of the turn of the millennium — in mind.

For each title, the volume editor has chosen the best teaching text of the main work and explained his or her choice. Old spellings and capitalizations have been preserved (except that the long "s" has been regularized to the modern "s") — the overwhelming preference of the two hundred teacher-scholars we surveyed in preparing the series.

Original habits of punctuation have also been kept, except for occasional places where the unusual usage would obscure the syntax for modern readers. Whenever possible, the supplementary texts and documents are reprinted from the first edition or the one most relevant to the issue at hand. We have thus meant to preserve — rather than counter — for modern students the sense of "strangeness" in older texts, expecting that the oddness will help students to see where older texts are *not* like modern ones, and expecting too that today's historically informed teachers will find their own creative ways to make something of such historical and cultural differences.

In developing this series, our goal has been to foreground the kinds of issues that typically engage teachers and students of literature and history now. We have not tried to move readers toward a particular ideological, political, or social position or to be exhaustive in our choice of contextual materials. Rather, our aim has been to be provocative — to enable teachers and students of literature to raise the most pressing political, economic, social, religious, intellectual, and artistic issues on a larger field than any single text can offer.

J. Paul Hunter, University of Chicago
William E. Cain, Wellesley College
Series Editors

About This Volume

Here at the turn of the twenty-first century, Gertrude Stein has become popular once more. During the past two decades, she has frequently been studied as an inventive precursor of postmodernism, a woman writer who claimed the riches of the American language for her own sometimes idiosyncratic uses. Recently, attention has fallen most often on *Tender Buttons,* her 1914 poem collection; the 1934 *Four Saints in Three Acts,* the opera libretto she wrote for her friend Virgil Thomson's music; the last of her three autobiographies, *Wars I Have Seen* (1946); and the truly difficult novels *Lucy Church Amiably* (1930), *A Novel of Thank You* (1925–26, 1958), and *Ida* (1941). The poems and novels have also been studied as pieces illustrative of Stein's conflicts with her lesbianism.

Before this most recent round of enthusiastic response, Gertrude Stein was regularly accorded the distinction of being one of America's premier innovative modernists. Stein's writing was taught alongside that of John Dos Passos, Ernest Hemingway, T. S. Eliot, and William Faulkner. Critical focus fell on works such as her long novel, *The Making of Americans* (1925), and the first of her memoirs, *The Autobiography of Alice B. Toklas* (1933), the title of which was a pun intended to remind readers that the couple, Stein and Toklas, were literally indistinguishable. In circles of literary criticism, Stein's three autobiographies were also regularly studied as examples of that most American of literary forms, "self-writing," which began with Benjamin Franklin's *Autobiography* and which has dominated much of late-twentieth-century publishing.

The literary world has never dismissed Gertrude Stein, but her works have sometimes been treated with less respect than they merit. This edition of her first published book, *Three Lives,* written in 1905–06 and published in 1909, is meant to bring attention to the origins of Stein's unique art. It is important to remember that the young Gertrude Stein was not going to college to become a writer. She was majoring in philosophy and psychology, and she then completed four years of medical school. The fact that she was in her late twenties before she decided to write makes her journey to the publication of *Three Lives* an interesting study.

The documents of cultural and literary history assembled here are meant to give the reader insight into the world of literature and professionalism as Gertrude Stein would have experienced it; the sections strive to chart her own assimilation of information, information that would ultimately lead to the publication of *Three Lives.* For example, because Stein was the youngest daughter of her family, she watched her older sister assume the responsibilities of the household after the death of their mother; for this reason, Gertrude became interested in the progress (or lack thereof) of women's rights. Chapter 1 of the Cultural Contexts segment of the book reprints excerpts from such famous American women activists as Elizabeth Cady Stanton, Anna Julia Haywood Cooper, and Charlotte Perkins Stetson Gilman, whose important book *Women and Economics* Stein quoted in a speech she gave to an audience of Baltimore women when she was in medical school at Johns Hopkins University. Also included are selections pertinent to the state of "women's medicine" at the turn of the century, excerpted from the works of three well-known doctors, S. Weir Mitchell, William Osler, and J. Whitridge Williams; these documents enable the reader to sense how different Stein's views of women as medical subjects were from the opinions of male physicians.

Chapter 2 of the documents includes material relating to gender, sexuality, and race and ethnicity (African American and Jewish). Beginning with Oscar Wilde's poem, "The Ballad of Reading Gaol," which openly professes his homosexuality, the section also includes a recently published memoir by a lesbian writer, Margaret C. Anderson, as well as key treatises by Sigmund Freud, Havelock Ellis, Otto Weininger, Ida B. Wells-Barnett, and Stein herself. In Chapter 3, writings by the professors Stein found extremely influential at Radcliffe and Harvard — William James and George Santayana — are juxtaposed with pieces by philosophers who were influencing world culture, Henri Bergson and Alfred North Whitehead. These texts are of primary importance for any understanding of modernism in the Western world.

The literary background immediately impacting on Stein appears in Chapter 4, with excerpts from the work of Henry James, Hamlin Garland, Stephen Crane, Frank Norris, and W. E. B. Du Bois. This American configuration of "realism," which greatly influenced what Stein created in her narratives of Anna, Lena, and Melanctha in *Three Lives,* is foregrounded against the European expression of that movement, evident in Gustave Flaubert's letter to Émile Zola. To re-create the aesthetic movement that grew up alongside *Three Lives,* Chapter 5 presents letters from Paul Cézanne and Gertrude's older brother, the art collector Leo Stein. Also included are two essays, written soon after the publication of *Three Lives,* by Carl Van Vechten and Mabel Dodge Luhan, both of whom were instrumental in Stein's achieving her reputation in the United States as early as 1914.

ACKNOWLEDGMENTS

My personal thanks for the assistance of Jane Degenhart (for work with permissions and the headnote to W. E. B. Du Bois) and Kathleen Drowne (for various other kinds of help). The edition has benefited tremendously from the suggestions of Professors Ellen Berry, Shari Benstock, William Cain, Kurt Curnutt, Marianne DeKoven, and Wendy Steiner. Errors of fact and judgment are my own.

The staff at Bedford/St. Martin's, especially Kathy Retan and Jennifer Rush, is exemplary. Special thanks also goes to Emily Berleth, and to Linda DeMasi and Leslie S. Connor of Stratford Publishing Services.

Linda Wagner-Martin
Chapel Hill, North Carolina

Contents

Illustrations

Part One

Three Lives
The Complete Text

Gertrude Stein as a student at Radcliffe College (the Harvard Annex), circa 1897. Reproduced by permission of the Beinecke Library American Literature Collection, Yale University.

Introduction:
Cultural and
Historical Background

William James called *Three Lives* "a fine new kind of realism" (Gallup 50). Sherwood Anderson claimed it had helped make him into a writer. And what reviews saw print were favorable (see Part Two, Chapter 5). Avant-garde readers saw Gertrude Stein's 1909 fiction as an impressive experiment, describing three lower-class women characters in strangely repetitious language, so that the good Anna, the gentle Lena, and the controversial Melanctha left indelible impressions. For those readers who ridiculed the very notion of the avant-garde, Stein was a strange Parisian, the "Mama of Dada," a person to be avoided rather than read. As a proponent of the dadaism movement, Stein joined other artistic minds and challenged established canons of art and literature, thought and morality.

As a classically formalist experiment published in the midst of realistic American fiction, *Three Lives* illustrated the new. As a set of scientifically objective portraits, the characterizations of women in *Three Lives* puzzled readers. Who *were* these women? Why were they valued enough to become the subjects of literary narrative? Where did their heroism, their nobility, lie? What was Gertrude Stein's relationship to these characters? Why was Stein — an educated woman who lived on proceeds from a family trust fund — writing so carefully about two German immigrant women and the black, lower-middle-class Melanctha? And why — despite the book's title of *Three Lives* — did each story end with the death of the character?

GENDER AND STEIN AS WRITER

> Melanctha now really was beginning as a woman. . . . Girls who are brought up with care and watching can always find moments to escape into the world, where they may learn the ways that lead to wisdom. (p. 95)

The role of the woman who wants to write has often been complicated by the infringement of expected social duties on her writing time; there is much reading and thinking, living and acting necessary before she can even dare to imagine becoming a writer. For Gertrude Stein, the youngest child of a traditional German Jewish family, being a daughter was both a role and a discipline. Daughters (like Gertrude and her older sister Bertha) had different roles in a family than did sons (like Michael, Simon, and Leo). Sons had tutors and learned difficult materials; daughters had governesses and learned to be sociable: they played the piano, drew, visited, and wore velvet dresses. Because the Steins traveled and lived in Europe for much of Gertrude's childhood, differences within the Austrian and French cultures made gender and sex differences even more noticeable.

Born in 1874, Gertrude lived in both Baltimore and the San Francisco area during the height of American "separate spheres" socialization, a social code pertinent to white middle- and upper-class society. Men earned good livings and maintained control over their wives and families; wives usurped what power they could from managing the household and the children. Men went out into the world and women lived sheltered lives at home, supposedly innocent of the brutality of capitalistic endeavors. Sons of wealthy enough families were sent away for educations; daughters helped at home. As the youngest of the five Stein children, Gertrude quickly saw that she wanted to lead a son's life rather than a daughter's. Her intellectual propensity coupled with her tomboyish nature made her more like her brothers than like her sister, Bertha. After the lingering illness and eventual death of their mother, Millie Keyser Stein, Gertrude (still called "Baby" by her siblings) realized that Bertha had become their mother, tending to all the household needs, the cooking, and what organization she could maintain. To avoid helping Bertha in the mother role, Gertrude stopped having anything to do with her sister, and with the chores.

Note: Page references are to the work reprinted in this volume.

Much to her often irate father's dismay, Gertrude shaped herself into a student. When she was not in school, she read and studied at one of San Francisco's large libraries. For entertainment, she and Leo frequented used book stores and saved their money to buy drawings, books, and other curiosities. The oldest son, Mike, was venerated in the family because, after starting college at University of California at Berkeley, he transferred to and graduated from Johns Hopkins. Then he did a year of postgraduate work. Leo's courses of study took longer, and moved him from the West Coast to the East — to both Harvard and Johns Hopkins — and eventually Gertrude followed him to Massachusetts. When she enrolled at the Harvard Annex (later to become Radcliffe College) as a special student (because she had never graduated from secondary school), she was probably the first woman of her family to attend a university.

Stein's personal fight for gender equality and her right to an education mirrored the efforts of many women late in the nineteenth century. Included in this text are cultural documents that show the almost continuous battle for not only women's suffrage (the right to vote), but for women's opportunity to become formally educated (see Part Two, Chapter 1). Elizabeth Cady Stanton's cry for suffrage foreshadows the eloquent arguments of Anna Julia Haywood Cooper, a black North Carolina teacher, and Charlotte Perkins Stetson Gilman, an economist, that women be both educated and valued for the real worth of their work. Without training, women were ineffectual in the labor market, and the essays by Fanny Fern and the anonymous farmer's wife included here describe those extremely limited lives (see pp. 219–46).

In a complex way, Stein's realization that her art could differ from that of men because of her gender grew slowly, modulated by her successful experiences at Radcliffe, where she did psychological research with Harvard graduate students, and at Johns Hopkins, where she chose not to attend seminars in gynecology that she found offensive. Just as her graduate work could be defined in part by her sex and her gendered interests, so could her writing. That she chose women characters as her subjects, grounding them in the very dailiness of the lives they would have had to cope with, is one marker of her consciousness about the restrictions of separate spheres. Her gender informed much more than her fictional pursuits, as was illustrated when she wrote so forcefully to her critic/friend, Henri Pierre Roche, "You are a man and I am a woman but I have a much more constructive mind than you have. I am a genuinely creative artist and being such my personality determines my art just as Matisse's or Picasso's or Wagner's. . . . Now you if I were a man would not write me such a letter because you would respect the *inevitable* character of my art. . . . . But being a man and believing that a man's business is to be constructive you forget the much greater constructive power of my mind and the absolute nature of my art which if I were a man you would respect. . . ." (see p. 246). Had it not been for the decades of women writing and speaking to gain equality — in education, in the workplace, and in society at large — Stein would have been unable to base her argument on any recognizable logic. She used the concept of women's rights effectively in her fiction because she understood what it meant to be recognized as a woman artist.

THE STORY OF THE STORY

In 1903, approaching the age of thirty, Gertrude Stein had decided she would become a writer. A magna cum laude graduate in philosophy and psychology from Radcliffe College, she had gone to classes for the four years of medical school at the Johns Hopkins University. But she had not graduated. Stein had intended to specialize in "women's medicine," but her beliefs about the female body and the care that was appropriate for it conflicted with medical practices at the turn of the century (see Part Two, Chapter 1). One of the professors of gynecology with whom she argued made the motion during a meeting of the medical school faculty that Miss Stein not be graduated with her class. The intention of the faculty was that she attend summer school, do the remainder of her required work, and then graduate. But in Stein's recollection years later, in the first of her memoirs, *The Autobiography of Alice B. Toklas* (1933), she said that she was relieved not to have become a physician. She had been much more bored than she knew, and saw the delay in her graduation as a way to avoid that continued boredom (82).

Stein spent more than a year after medical school doing independent research on the human brain under the direction of Professor Llewellys F. Barker; he admired her work and urged the leading journal in the field to publish her long essay, with its quantity of drawings. For political reasons connected with the medical school, the work was never published (Wagner-Martin). Stein had, however, already published papers in the *Harvard Psychological Review*: one was coauthored; the other was her senior thesis, based on research into students' ability to pay attention when they were tired. Stein was fascinated with the powers of the human mind, and her undergraduate work with Harvard's brilliant William James, one of the founders of the discipline of psychology, shaped her interests for the rest of her life (see Part Two, Chapter 3).

Gertrude Stein knew she could write scientific papers. But what she decided she wanted to write, after traveling abroad in the months between her scientific training and her moving to Paris to live with her art collecting brother, Leo, was fiction.

Three Lives was not her first effort. During 1903 and 1904, while she was still living in Baltimore and doing research on the brain, Stein was also writing narratives about love triangles. Both *Q.E.D. (Quod Erat Demonstrandum)* and *Fernhurst* were short works of fiction that dealt with lovers who were not yet fully committed to each other and were involved with several people. In one story, the affair was heterosexual; in the other, it was lesbian. Because the only person to whom Stein showed her writing was her brother Leo, and because he was offended by its intimate subject matter, she did not attempt to publish these fictions until decades later. But her efforts in writing them gave her confidence.

After her move to Paris, Stein immersed herself in reading. She had always read a great deal, and continued throughout her life to read a book a day. She knew all the British and the European novels, and she also knew that realistic writing was moving to the forefront of literary taste, particularly in the United States. Having been a student of William James, she was well aware of the work of his younger brother, Henry James, one of the most prominent of America's realists. As a beginning writer, she avidly followed William Dean Howells's pronouncements, and she knew who Stephen Crane, Hamlin Garland, and Frank Norris were (see Part Two, Chapter 4). She admired all these writers for experimenting with *realism* — they used common people as characters, they told their stories in everyday language, and they relied on the concrete details of place and time to convey life's circumstances.

Despite her interest in living abroad (for its advantages both economically, being inordinately cheaper, and socially, being much less restrictive about friendships and sexual liaisons), Stein enjoyed the fact that she was an American. She wanted to be known as an American writer, an American person, and the work she considered her greatest early effort — *The Making of Americans* — testified to that enthusiasm. She remained stridently American throughout her life, even though she lived the last forty years of that life in France. Yet because of her brother's negative reaction to materials he could identify as personally reflective, she understood that she might be better off writing fiction that could not be identified as autobiographical. So, Stein looked around her for narratives she could write about other people.

When realism, then, became the mode of preferred writing in the United States, with characters often drawn from the lower echelons of society, Stein felt that she understood both the literary and the human motives: she could tell stories about something other than the elite themes and characters of traditional literature, about something other than her own relatively wealthy, well-educated culture. With information gleaned from a lifelong penchant for observing others, she could follow the adventurous trends of current realist/naturalist writing in Émile Zola's *Nana,* Gustave Flaubert's *Madame Bovary,* and Stephen Crane's *Maggie: A Girl of the Streets.* In these tales of hungering women, the latter character a prostitute, Stein found subject matter that — as a woman herself — she was interested in. She also found value in describing and probing cultural restrictions on women's lives: the Victorian period put women in spheres separate from men, "protecting" them from life's realities, keeping them, literally, virginal and chaste. As exemplified by Stein, herself, that period was about to end (see Part Two, Chapter 1).

As an intellectual Radcliffe graduate and a medical student, Stein was already an anomaly. Few upper-middle-class women received formal educations; most were tutored at home or sent to finishing schools for a brief time. With the prospect of analyzing women's sexual lives as subjects for her fiction, Stein would define herself even further outside the realm of Victorian womanhood. Sexually experienced women, in particular, were dangerous subjects for a woman to write about; the reader would be bound to wonder where the author had gotten her information. In American realism, Stein found a touchstone of technique: she learned that with careful nondisclosive language, the writer could portray subjects that even the most sheltered readers would not find offensive. Style, language choices, and tone might be made to be more important to the writing and the reader than the more obvious plot and theme of the work. Style might, in essence, eclipse the story itself.

It is not accidental that Stein felt great confidence in her ability to become an American realist. One of her great enthusiasms in life was what she saw as her ability to understand the average person. Later in life, when she was asked to teach an informal Great Books course at

the University of Chicago, she prided herself on the fact that undergraduate students would talk with her, whereas they were much less
responsive with the regular professors, Mortimer Adler and Robert
Hutchins, men who had founded the program. Her hobby in life was
walking the streets of Paris, talking in her idiosyncratic French with
people she met. She had behaved similarly in California as an adolescent, as well as in Baltimore and Boston and London — wherever she
traveled. While in Baltimore, she considered the high point of her medical school training being able to work in a medical clinic and make
house calls, attending pregnant women who had no other medical
attention. Many of these women were black. Such shared intimacy
was rare at the turn-of-the-century for there were few social situations,
especially in a southern city, where white and black races met. For
Stein, this experience held great importance. A decade later, when the
European art world discovered African sculpture and painting, forcing
acknowledgment of what had previously been considered mere primitive art, Stein felt herself, once again, on intimate terms with the black
race, comfortable with the knowledge she had acquired about the
African American women she had known in urban Baltimore.
Whether or not her experience gave her the right to assume such an
intimacy is questionable; but she felt that her knowledge about black
culture was more valid and true than her friends' less personal information. In Stein's mind, then, it was not a difficult choice to write
about Melanctha.

The catalyst for *Three Lives* (which Stein originally titled *Three
Histories*) was, however, less the knowledge she had accumulated than
it was an exercise her brother suggested to her during a period of
writer's block: that she translate, from French into English, Flaubert's
story *"Un coeur simple"* (a simple heart), taken from his book *Trois
contes*. The translation, Leo said, would both improve her language
skills and, perhaps, give her a model for realistic writing (see Wineapple). While Stein wrote away on this practice, she absorbed Flaubert's
story of Félicité, the young serving girl, and her mind moved quickly to
her own servant, the German woman who had kept house for her and
Leo when they were students living in their own rented house in Baltimore. Lena Lebender, loving and self-effacing, whose devotion to
Gertrude was well known, became the model for Stein's first character, that of the ostensibly "good" Anna Federner. Complex and somewhat older than either Lena or the Flaubertian model, the Anna in
Stein's story lived discretely and amiably — she was, after all, despite
her "arduous and troubled life," characterized as "good" — but she

managed in the midst of her duties to carve out a lesbian relationship for herself.

Stein reserved the name "Lena" for what was to be the third story in the collection, the short tale of the young German maidservant who let herself be directed into an unsuitable and unwanted marriage — which led to her eventual death in childbirth. Stein wrote this story to criticize the well-intentioned, first-generation immigrant relatives in the States who saw no alternative but marriage for the wistful young German cousin brought to America as a serving girl. "The Gentle Lena" was Stein's attempt to discredit at least some of the assimilationist attitudes in the States at the turn of the century (see Dearborn). The implicit question at the end of Lena's literal and figurative journey is: Why must all women marry? A more basic, gendered question is: Why do women who are themselves settled into unwanted marriages try to coerce other women into the same patterns? Is there no choice available in a woman's existence?

By the time she had completed both "The Good Anna" and "The Gentle Lena," Stein had her prose method — of insistence, not mere repetition — firmly in hand (see Part Two, Chapter 3). Her fiction was unlike any she had ever read, in any language that she knew; it was her attempt to make words fresh. The reader who began the narrative of the work-obsessed Anna would long remember the dutiful woman, even if her life was a repetitious maze of largely self-inflicted painful service:

> Anna led an arduous and troubled life.
> Anna managed the whole little house for Miss Mathilda. It was a funny little house, one of a whole row of all the same kind that made a close pile like a row of dominoes that a child knocks over, for they were built along a street which at this point came down a steep hill. They were funny little houses, two stories high, with red brick fronts and long white steps. (p. 37)

Strange as the description may at first appear, the Baltimore row house emerges accurately from the simple language, shaded sentence by sentence. The house is simply compared, in the "row of dominoes" image. Nothing about this paragraph would be off-putting to the reader, nor would the language send him or her to a dictionary. There is also a hint of observant humor in the description, attained through the hyperbolic adjectives "arduous" and "troubled" and the almost childlike emphasis, "the *whole* little house."

As Stein herself expressed, her technique of repeating the same general information by varying it slightly, appropriately, makes the reader pay attention. She preferred to think of it as insistence rather than repetition. What may be most surprising is that after nearly a century, her descriptions — even though written in very common language — still work. For instance, this is a paragraph that describes the changes in Anna's voice, depending on her audience:

> Her voice was a pleasant one, when she told the histories of bad Peter and of Baby and of little Rags [her dogs]. Her voice was a high and piercing one when she called to the teamsters and to the other wicked men, what she wanted that should come to them, when she saw them beat a horse or kick a dog. She did not belong to any society that could stop them and she told them so most frankly, but her strained voice and her glittering eyes, and her queer piercing german english first made them afraid and then ashamed. . . . (p. 39)

What is repeated here is less the language than it is the structures of Stein's phrases. Taken word by word, her description is remarkable — "piercing" (repeated), "strained," "glittering," "german english" written as a noncapitalized melding of the two spoken languages. Anna's denigration of her own social position — or lack of it — as she tries to coerce the teamsters (men who drive a wagon with a team of horses) out of their habitual cruelty to animals is effectively used. There is no question that Stein is dealing with a lower-class saga, focused on the frustrated German servant who is so powerless she cannot change anyone's behavior. Anna becomes a classic realist figure and as she does, Stein becomes a pioneer of twentieth-century realism.

Even with Stein's use of repetition, there is nothing mechanical in the narrations composed for *Three Lives*. As this passage from "The Gentle Lena" shows, as Stein conveys information that has a higher emotional content, her use of insistence becomes more noticeable, but not stagnant and fixed. For example, the following passage reveals that Lena, the shy German girl, has no language for extreme and debilitating fear, the emotion she had first felt on the stormy passage over to America, and which she feels again at the prospect of birthing a child:

> And so things went on, the same way, a little longer. Poor Lena was not feeling any joy to have a baby. She was scared the way she had been when she was so sick on the water. She was scared now every time when anything would hurt her. She was scared and still and lifeless, and sure that every minute she would die. Lena had no power to be strong in this

kind of trouble, she could only sit still and be scared, and dull, and life-less, and sure that every minute she would die. (p. 213)

Stein's language truly insists on the powerlessness of the fearful young woman. Actual repetition of words combines with the shortness of the paced phrases to create in the reader the reality of Lena's fear, allowing the rawness of Lena's pain to be fully absorbed. Drawing on her experiences with the ill-informed pregnant women she had treated in Baltimore, Stein manages to write an objective and unsentimental portrait of Lena's paralysis.

The portraits in *Three Lives* reflect the emergence of irony in modern literature, requiring readers to look beneath the surface of the characters to find their hidden stories. Stein took great pride in her self-perceived radicalism — her ability to work on multiple levels, to describe a life convincingly enough to convey depth while still maintaining the utmost in simple, accurate language (see Farber). Fatalistic trends were also creeping into modern literature, as evinced by the feelings of doomed inevitability so often expressed by Stein's characters. For all the "goodness" and "gentleness" of Anna and Lena (qualities that should have brought them happiness in the Victorian world) both women die miserable, their only happiness coming from friendships with other women. In the case of Lena, the women she trusts as mentors place her into the irrevocable marriage that leads to her death; the friendships that can save her are those with other lower-class servants — the Irish and Italian girls she is dissuaded from seeing. In the case of Anna, the breaking of her relationship with her lover leads to her spiritual and physical wearing away. Even the epigraph to *Three Lives*, Jules Laforgue's line *"Donc je suis un malheureux et ce n'est ni ma faute ni celle de la vie"* (Thus I am unhappy and this is neither my fault nor that of life) exemplifies this burgeoning fatalism in modern realism.

Once Stein had honed her new realistic style, she began the most difficult challenge of the three narratives, that of the bisexual, black Melanctha Herbert. It is this middle story, the longest of *Three Lives*, that made Stein comparatively famous. At the time, very little fiction about black characters was available in the United States. What writing did exist was either autobiographical (slave or postslave narratives or tales of pastoral or southern culture) or "uplift" literature, fiction and poetry intended to reach out with a strong moral purpose. Motivated by both Christian and heterosexual belief systems, writers of uplift literature wanted to lead readers into acceptable life paths. The dominance of uplift literature tended to encourage women to lead exemplary lives: if culture itself did not coerce women to behave in virtuous ways, "the polemics [arguments] of nineteenth century 'racial uplift' fiction" did (Blackmer, 232). Stein's creation of a sexually experienced yet not socially "fallen" woman — one treated in a nonjudgmental manner — was unique to the time and culture. As Blackmer notes, in Stein's writing "racial and sexual taboos inevitably intersect and function to contain the desires of nonconformist and independent women to define themselves" (234).

NARRATIVE METHOD AND STYLE IN "MELANCTHA"

It is important to emphasize that the American Gertrude Stein, living in Paris, was sitting for a portrait Pablo Picasso was painting of her before she began writing the story of Melanctha during the winter of 1905–06. During her more than eighty sittings at his humble studio, Stein listened to Picasso's mistress read and tell stories; she gained a new appreciation for the spoken narrative. Not only was she surrounded by the avant-garde in Picasso's studio, when she returned home to 27 rue de Fleurus, she wrote, daily, sitting under Paul Cézanne's large and striking "Portrait of Mme Cézanne." A few blocks away, her brother Michael and his wife, Sarah, were collecting the most recent work of Henri Matisse as fast as their finances would allow. In "Melanctha," then, in light of all these influences, Stein's style changes: her use of repetition intensifies, her syntax grows more complex, and she seems to see for the first time what the qualities of impressionist and modernist painting might have to do with writing, especially with her own innovative brand of realism (see Part Two, Chapter 5).

Stein's presentation of the black protagonist of this novella is both more complicated, and more sympathetic, than her characterizations of Anna and Lena; her narration continues to be innovative, but it is also risky. The style of incremental repetition that attracts readers' attention can also repel, and in this fiction, that repetitive style has been elongated. Not only are phrases and sentences repetitive, but the prose circles back on itself: whole sections of the text seem to be repeating other sections. Blackmer comments that in "Melanctha," Stein may have been trying to create a three-dimensional effect. The fact that Melanctha is clearly bisexual (unlike the very tentative portrayal of Anna's sexuality) might also drive readers away. It is clear that Melanctha has had heterosexual relationships during her adolescence; the reader is told that she spent much time "wandering" at the docks (p. 96). But later she becomes intimate with the alcoholic Jane Harden who, at twenty-three, is sexually adept. Stein says clearly, "It was not from the men that Melanctha learned her wisdom. It was always Jane Harden herself who was making Melanctha begin to understand" (p. 101). The two years of the women's relationship pass quietly, Melanctha spending "long hours with Jane in her room," a description that echoes lovemaking scenes from Stein's earlier novella *Q.E.D.* (p. 101) Melanctha's later liaison with Rose, which is the story that opens the novella, adds to the lesbian strand of the narrative and suggests Melanctha's double injury when Rose betrays her — first by marrying Sam and then by disguising their lesbian affair.

After thirty-five pages of Melanctha's varied bisexual history, Gertrude introduces Jeff Campbell, the black doctor who grows to love Melanctha while he tends her dying mother. The story then becomes an extended dialogue between the arbitrarily rational Dr. Campbell and the purposefully inarticulate Melanctha, a tour de force of voiced dialogue unlike anything yet published in American literature. During the lengthy Jeff-Melanctha interchanges, Stein defines Jeff as the rational speaker who wants permanence, exclusivity, security. His polemical insistence is shown to be absurd, however, when contrasted with Melanctha's meaningful silences. She loves through acts; she gives Jeff what she has to give and does not talk about it. While he accepts her love, he verbalizes all parts of their relationship and forces her into language that becomes destructive. Whatever Melanctha says, Jeff argues with. By the end of the fifty-page dialogue, the reader sees that Stein has constructed a classic philosophical discourse between reason and emotion. Because the language sounds so much like actual speech — circular, repetitious, boring — its classic pattern is often overlooked.

Stein's fiction furthered what was becoming her life process, melding the knowledge she had acquired from her studies of philosophy and psychology at Radcliffe College and her studies of brain anatomy and medicine at Johns Hopkins with her understanding of literature and painting. Her main interest was presenting the person; her fascination with the "portrait" — a form in which she worked for the next twenty years — was a culmination of years of formal study as well as the result of the contemporary artistic excitement over Cézanne, Picasso, and Matisse as they worked to change the nature of painting, particularly through *their* portraits. In "Melanctha," Gertrude not only creates the title character, she also creates a fictional portrait of herself as a deeply divided individual — a double portrait, if you will. Although the long dialogue between Jeff and Melanctha has been described as typical of conversations Stein and her female lover often had, with Jeff Campbell representing Stein, the author also reveals aspects of herself in the character of Melanctha. Born of very different — and irreconcilable — parents and later isolated from her family, the maturing Melanctha — like Stein — tried to escape her feelings of difference and looked to sexual love for self-knowledge. Jeff and Melanctha's impasse mirrors Stein's own conflicted emotional loyalties to different facets of herself.

Reading "Melanctha" as a double portrait of Stein herself — with Jeff Campbell representing the logical, argumentative, and philosophically trained consciousness and Melanctha standing for the intuitive and warmly sexual being, the persona who did not want to talk about her emotions — creates three critical problems. First is the issue of gender. How can Stein as author identify with both the male speaker and the female? How can she avoid stereotyping male and female response to sexuality? If Stein was comfortable with her own sexuality, which — despite her tentative romances with men in Cambridge — appeared to be lesbian, how did she translate that homoerotic response into a heterosexual world that fiction expected? The second concern is the issue of class, for despite what Stein saw as her comparative poverty, she was wealthy enough to live abroad, to collect paintings (however inexpensive those paintings were), and to travel widely. Her life was as different from the lives she described in *Three Lives* as possible — none of those characters had freedom of choice. Third, and probably most troubling to today's reader, is the issue of racial identity. Despite the account of Richard Wright's reading "Melanctha" aloud to a group of African American workers, to their enthusiastic response (see Weiss), Stein's story of a sexually promiscuous and aberrant black woman appears to rely on racial stereotyping — although in some ways, Stein's personal identification with Melanctha may counter what appears to be her reliance on obvious stereotypes.

In her unpublished notebooks, Stein identified herself as a sexual, and therefore — in society's eyes — an objectionable, woman. She several times repeated that her nature was sensual, even "dirty": "the Rabelaisian, nigger abandonment . . . daddy side, bitter taste fond of it" (Wagner-Martin 77–80). By likening herself to Rabelais, a sixteenth-century French satirist known for his coarse and obscene humor, Stein locates herself in the camp of the sexual, and uses the stereotype of the sexual black woman as a kind of disguised self-portrait; the racial difference between author and character is a protective device. Further, aligning Melanctha with the paternal line of her family — "her black brute" of a father — instead of with her better-born (and lighter-skinned) mother is a means of justifying Stein's own alliance with her father, and her own grudging acceptance of sexuality — though readers of today must still deplore her choice of language (14).

It is Stein's willingness to take on these three controversial issues that makes *Three Lives* such a conundrum of social attitudes. Readers respond today to the radicalism in all three narratives, but particularly to "Melanctha," and to Stein's choice of characters who are lower-class, bisexual (in two of the three cases), and ethnically identified (and, therefore, categorized as beneath polite society). That her protagonists were women also marked *Three Lives* as an unusual, and challenging, set of literary texts (see Part Two, Chapter 2).

THE PLACE OF PHILOSOPHY IN STEIN'S WRITING

Much of the assumed difficulty of Stein's fiction stems from the fact that her undergraduate training was in philosophy, a discipline that was based on logical argument and abstract correlation. The diction of philosophy was marked with a great many prepositional phrases; arguments tended to be carefully reasoned and often expressed in cumbersome language; and abstractions obscured concrete nouns and verbs. In short, reading philosophy could be time-consuming and less than precise, and writing philosophy shared the same stylistic markers. Some of Stein's language in her fiction seems to be drawn almost directly from philosophical discourse. Her language was often highly abstract, with comparatively little concrete detail. For instance, when Stein describes the deadening of Anna's passion for the great love of her life, Mrs. Lehntman, she uses only four sentences:

> Mrs. Lehntman she saw very rarely. It is hard to build up new on an old friendship when in that friendship there has been bitter disillusion. They did their best, both these women, to be friends, but they were never able to again touch one another nearly. There were too many things between them that they could not speak of, things that had never been explained nor yet forgiven. (p. 84)

Rather than cause the reader to feel sorrow, Stein's descriptive passage remains remote, distant. Of what does the women's "disillusion" consist? Why can't they "touch one another nearly"? The reader remains faced with the same dilemma as the characters, trying to understand "things that had never been explained."

This reliance on the abstract is a way of disguising the women's emotions, but it also reflects the nature of philosophical discourse. As Steven Meyer says in his introduction to Stein's *A Novel of Thank You,* "Stein is concerned to *characterize* thought — not thought generally, but thought as it enters into the composition" (xxi). There is a great amount of "talk about thought" in Stein's fiction (xxii).

Best illustrated in "Melanctha," Stein's "talk about thought" is the basis of the relationship between Jeff Campbell and Melanctha:

> Always now Jeff wondered did Melanctha love him. Always now he was wondering, was Melanctha right when she said, it was he had made all their beginning. Was Melanctha right when she said, it was he had the real responsibility for all the trouble they had and still were having now between them. If she was right, what a brute he always had been in his acting. If she was right, how good she had been to endure the pain he had made so bad so often for her. But no, surely she had made herself to bear it, for her own sake, not for his to make him happy. Surely he was not so twisted in all his long thinking. Surely he could remember right what it was had happened every day in their long loving. Surely he was not so poor a coward as Melanctha always seemed to be thinking. Surely, surely, and then the torment would get worse every minute in him. (p. 149)

Jeff Campbell's torment when he is alone is doubled when he is with Melanctha. In contrast to his effervescent rhetoric, she is likely to punctuate his talk with a comment like "What you mean Jeff by your talking?" But the impasse Stein draws remains — language does not serve as a bridge over the emotional chasm, and the affair, reluctantly, dies.

Although it seems at variance with Meyer's comment, Margaret Dickie points out that there is a great deal of description in Stein's early writing — Stein used her scientific training "to classify and describe what she saw" (*Gendered* 6). She did so, however, says Dickie, with a difference. Although "science is continuously busy with the complete description of something . . . with ultimately the complete description of everything," Stein eventually outgrew that notion of the usefulness of description (*Lyrics* 64). But in *Three Lives,* page after page describes, often to the point of reiteration if not exact repetition. Carefully arranged on the page, Stein's descriptions comprise much of the narrative of each story. For instance, after Lena's marriage to Herman, Stein's description creates the seemingly persistent march

of events related to her sexual role — pregnancy, childbirth, the deadly
doldrums of her life.

> Before very long, Lena had her baby. He was a good, healthy little
> boy, the baby. Herman cared very much to have the baby. When Lena
> was a little stronger he took a house next door to the old couple, so he
> and his own family could eat and sleep and do the way they wanted.
> This did not seem to make much change now for Lena. She was just the
> same as when she was waiting with her baby. She just dragged around
> and was careless with her clothes and all lifeless, and she acted always
> and lived on just as if she had no feeling. (p. 213)

Stein's detailed prose suggests the quasi-scientific objectivity of a
removed observer, and, through that tone, the reader avoids thinking
author and persona have anything in common.

For an understanding of what Stein's philosophical training con-
tributed to her writing, essays by William James, George Santayana,
Henri Bergson, and Alfred North Whitehead provide both content
and stylistic models (see Part Two, Chapter Three). The student of
both Santayana and James, Stein wrote in their mode so well that each
professor considered her a star pupil (except for her last term with
James, when she did not take his final exam). She read Bergson, as did
all the modernists in England and France; and she spent the first
months of World War I with the family of Alfred North Whitehead, a
philosopher-mathematician whose work she had long admired. The
two of them spent their time on long walks, perfect for discussions.
Philosophical discourse was second nature to Gertrude Stein.

In keeping with this pervasive training, it makes sense that it was
less the details of Anna's and Lena's lives that interested Stein than it
was the larger questions about human behavior, women's behavior.
Rather than one telling detail, Stein wanted to map the complexity of
her characters. While she might focus on Anna's need to control people
(and, for comic effect, her dogs), Stein's real attention was less on that
dimension of Anna's character than it was on her basic psychological
need to take charge of everyone — often to her own detriment. In fact,
even after Anna has her own business and is responsible for charging
enough to make a profit from her arduous life, she fails to make ends
meet: she creates a martyr of herself, even though there is no reason
for that martyrdom. People would stay with her no matter what she
charged. Her inability to ask what she is worth is a grave character
flaw, and it leads — Stein implies — to her death. The warning for
readers found in all three of Stein's "lives" helps to place her work

into an instructive model that is as appropriate to philosophy as to literature. In this way too, Stein was writing out of the larger impulse to graph lives, and to make the narratives of those lives serve their readers.

QUESTIONS OF OTHERNESS — RACE, ETHNICITY, SEXUALITY

Despite Gertrude Stein's prestigious education, she — and her family — were never part of mainstream America. That they had immigrated from Austria midway through the nineteenth century, that they had worked ferociously and as tradespeople to haul themselves out of severe poverty, and — most of all, by the time of Gertrude's adolescence — that they were Jewish in a largely white, Protestant culture served to keep them marginalized. The latter category had taken on importance as the century moved to its end: the Russian pogroms (the organized massacre of Jews from 1880 to 1885), inscription in various national armies, the general milieu of persecution both subtle and blatant led thousands of Jews to migrate to the United States. But with increasing numbers of Jews in America, prejudice grew. The caricatures and cartoons that peppered the world's newspapers and magazines were continual reminders that to be a Jew was to be not only different, but objectionable.

Crystallizing much of what might have gone as unspoken anti-Semitism during the 1890s was the very visible Alfred Dreyfus affair. The innocent Alfred Dreyfus, a Jewish officer in the French army, was framed and convicted of treason and imprisoned for life on Devil's Island. Such blatant societal discrimination was paralleled, in a sense, by Irish writer Oscar Wilde's conviction and imprisonment for sodomy: whatever the difference, mainstream culture had the power to deem it criminal. The power of definition — and of life and death — rested with the majority.

In Stein's own lifetime, she saw her family's move to San Francisco take on a socially exclusionary drama once they had left their extended family, and their roots, in Baltimore. Well-placed in Jewish Baltimore, the Keysers, her mother's family, were confident about their standing; their friendships with other notable Jewish families — among them the wealthy Cones — kept the matrix of social life intact. But in California, where families of long-standing position and wealth controlled society, no such welcome existed. Daniel Stein's business skills were erratic, and his anger at not being well connected may have kept him from earning the money he desired. After the family moved out of the plush Tubbs' Hotel in Oakland, they had difficulty keeping friendships afloat — even though Daniel, for the first time, tried to ingratiate himself with the local rabbi.

In retrospect, Stein considered the Johns Hopkins Medical School to be anti-Semitic, but she felt little prejudice at Radcliffe or Harvard; perhaps she had missed some nuances while she was in Cambridge (see the 1900 *Harvard Lampoon* cover, p. 297). It is clear, though, that when Gertrude arrived as a freshman, she became part of her brother's social circle — Leo had already been at Harvard for a year — and the names of friends and visitors indicate that their social circles, both in Cambridge and during their later travels in Europe, particularly at their Paris salon, were largely Jewish. To be invited to either of the Stein salons — fashionable artistic receptions at Leo and Gertrude's or Michael and Sally's — meant having some Jewish connection, through birth, family, or friendship. Being so joined with others who feared the increasingly blatant anti-Semitism meant that no one needed to talk about the condition: what remains unsaid in all of Stein's memoirs is the threat of being found unacceptable because of her Jewishness.

Being of German descent tended, in part, to neutralize the Jewishness. To be German made one a part of the largest migrant group in the United States, a group that had already earned acceptance in many circles. Considered "'thrifty, frugal, and industrious'" by employers (Bodnar 69), Germans seemed content with modest employment — many were laborers, unskilled as well as skilled, and many German girls came to the States expressly to be paid housekeepers or nurses. The patterns Stein traces in both Anna's and Lena's stories were customary, though Anna's financial success was unusual. As the largest nineteenth-century migrant group, the Germans had settled throughout urban and Midwestern America, and the sizable population in Baltimore — where Millie Keyser Stein had grown up and Daniel's family had come when he was a youth — meant a secure society. Diffident and disdainful about competing with blacks for work, most Germans went no farther south than Washington, D.C., and Baltimore; had there been competition, households would have hired Germans before African Americans. According to a social worker in 1900, the black woman got "the job that the white girl does not want" (Woloch 228).

Throughout the early segments of "The Gentle Lena," Stein presents recognizable portraits of both Italian and Irish nursemaids. Again typed by national characteristics — more volatile in temperament, more prone to alcoholism, perhaps a rung below the Germans on the social scale of acceptability — these women parallel the German working class by being similarly disinterested in education. Women from all three immigrant groups would have been seen as religious (or at least pious), obedient, and heterosexual — marriage was their destiny. They were working women who worked only until marriage rescued them. The Catholicism that all three groups shared (although Germans were often Protestant) added to their perceived intellectual inferiority.

While the condition of being Jewish and German was less visible than being black, there were affiliations. For the most part, even into the 1920s when the Harlem Renaissance (the New York–based artistic movement that saw African Americans championing racial pride and cultural identity) made black society acceptable in New York, wealthy whites would not have known blacks (the servants in white households would have come from the poor white cultures of the Irish, the Italian, and the German). For all the post–Civil War migration to the North, most blacks lived in urban ghettos far from middle-class whites; indeed, those ghettos of the poorest housing were where the Irish and Italians might intermarry with African Americans or Hispanics. Class reconfirmed the separation that race seemed to mandate.

When Stein describes Melanctha in racially coded language (*blues, wisdom, desire, wandering, trouble*), she intends the descriptions to be exotic for most of her readers. As Blackmer points out, one of the constant metaphors in "Melanctha" is "the central role that racial visibility and invisibility [plays] in establishing gender roles and sexual identities for black and white women alike. . . . [the work employs] the mask as a mediating metaphor of concealment and revelation" (250, 233). Whether what is being disguised is Jewishness or blackness or lesbianism, the method and the principle of masking serves to intentionally mislead the reader.

As the materials included in Part Two, Chapter 2 suggest, a woman's "inversion" or "perversion" — the accepted terms for lesbianism at the turn of the century — was heinous, and was often discussed (when that discussion did occur) in combination with other "deviant" behaviors such as prostitution or racial degeneracy. Both racial difference and sexual difference constituted pathology, a diseased "otherness." Dependent on the so-called scientific observations of Havelock Ellis, George M. Beard, and Richard von Krafft-Ebing, theories of aberrant sexuality irreparably branded any person inclined to experimentation — with bisexuality as well as homosexuality.

Bearing weight on Stein's writing was the coupling of sexual aberrance with theories of scientific racism, which attempted to prove that one race (specifically, Caucasian) might be superior to another because of biological and racial differences. Unfortunately, Stein's studies in medical school had done much to overexpose her to such prejudicial information.

From this perspective, her clear intention to make the characters of at least Anna and Melanctha bisexual, if not singly lesbian, assumed a great risk. Sensitive to the body of knowledge that the literate world could claim, Stein knew that most of her readers would be unsympathetic to Anna's great love for her woman friend, just as they would wonder why Melanctha had such disappointing affairs with the men in her life. It is only with women that Stein's most significant woman characters are truly happy. *Three Lives* thus became a kind of "brave new world" of feminine portraiture — as well as a brave new kind of realism.

THE PLACE OF AESTHETICS IN STEIN'S WRITING

Changing taste in literature seemed harmless in the midst of the immense social disarray brought on by the close of the Victorian age. Literature was merely reflecting reality; changes in perceived gender roles, class distinctions, and issues involving propriety and sexuality were brewing. Within proscribed black writing, the move from a clearly mandated "uplift" aesthetic (writing geared to instruct readers both black and white in the morality of the day) to more organic and individualized creative forms (writing that might deviate from mandated content) became visible. As W. E. B. Du Bois wrote in his 1903 *The Souls of Black Folk,* the marginalized writer or artist, already an outsider, deals with "a world which yields him no true self-consciousness, but only lets him see himself through the revelation of the other world. It is a peculiar sensation, this double-consciousness, this sense of always looking at one's self through the eyes of others, of measuring one's soul by the tape of a world that looks on in amused contempt and pity. One ever feels his two-ness, — an American, a Negro; two souls, two thoughts, two unreconciled strivings . . ." (see p. 354; see North 72–75, on Stein's mask of language in "Melanctha").

Central to the evolution of American realism, Du Bois's statement of outsider vision linked the less racially directed comments of Frank Norris and Hamlin Garland. To be of "the folk," the common American person and experience, gave a different kind of validity to the burgeoning local color fiction. Garland's insistence on the writer's emotion as the most important quality of literary production dovetailed with William James's insistence that theoretical scholarship somehow isolate and quantify emotion: American realism had become philosophy. "The secret of every lasting success in art or literature lies, I believe, in a powerful, sincere, emotional concept of life first, and, second, in the acquired power to convey that concept to others" (see pp. 338–39). Garland spoke in a plainer idiom, perhaps a more nearly scientific one, than did Henry James, although the latter's pronouncements (particularly in the prefaces to the novels issued in the New York edition — books that Stein purchased in London) remain basic to any understanding of American realism.

Both Garland and James insisted that realism meant new forms, new models, new kinds of characters, and fresh styles. Norris, too, saw that style could only, legitimately, reflect the folk simplicity and character (see Norris on the difficulty of "the simple treatment," which "demands more of the artist," p. 350). Years later, when Gertrude Stein was reviewing her own writing career in the essays she wrote for her 1934 United States speaking tour, she reminisced about this earlier decade:

> [N]othing changes from generation to generation except the composition in which we live and the composition in which we live makes the art which we see and hear. I said in *Lucy Church Amiably* that women and children change, I said if men have not changed women and children have. But it really is of no importance even if this is true. The thing that is important is the way that portraits of men and women and children are written, by written I mean made. And by made I mean felt. (*Lectures* 165)

American realism, of course, was a coda to the dramatic break from traditional letters that Émile Zola had instigated. Through her admiration for Flaubert, who in turn championed Zola's art, Gertrude Stein identified herself and her writing with the Continental (European) as well as the United States method. While she understood that her choice of subject and characters for *Three Lives* put her squarely in the realist, even naturalist, line, she did not see her work as imitative in any way. She was writing for readers she herself admired. Accordingly, the stories went first of all to her sister-in-law Sally Stein. Once Sally (who was to become known as both the collector and the pupil of Henri Matisse, as well as the founder of his influential, if short-lived, school) had approved the work, Gertrude asked her friend Etta Cone to type the manuscript for her. *Three Lives* became the product — perhaps a subversive one — of women's friendship, even collusion.

Once the manuscript was ready, Stein sent it to the most influential of Leo's friends, the well-published Hutchins Hapgood, whose praise of the work did not disguise his warning that "Your stories are not easy reading. . . . They lack all the minor qualities of art, — construction, etc., etc. They often irritate me by the innumerable and often as it seems to me unnecessary repetitions; by your painstaking but often clumsy phraseology, by what seems sometimes almost an affectation of style." Despite his reservations about Stein's methods, Hapgood was clear about the fact that her stories were "excellent. . . . extremely good — full of reality, truth, unconventionality. I am struck with their deep humanity, and with the really remarkable way you have of getting deep into human psychology." Given that Hapgood was not a proponent of realism, he said nothing specifically about the Anna or Lena narratives. His attention fell several times on "Melanctha," and he praised its newness, its unique subject matter as well as style: "It is the very best thing on the subject of the Negro that I have ever read" (Gallup 32).

Whatever that praise actually meant, considering the state of culture in 1906, Hapgood failed to be the proper agent to place *Three Lives* (still titled *Three Histories,* a phrase that emphasized the objective style of the narratives: factual, uninvolved, even journalistic). No publishers were interested in the work. Several years went by: in desperation, Stein turned to Mabel Weeks, an American academic friend, who gave the manuscript and other of Stein's writings to May Bookstaver, the now-married woman who had been the great love of Gertrude's life in Baltimore. It was May Bookstaver Knoblauch who found the vanity publisher Grafton Press, which agreed to publish *Three Lives* for the sum of $660 — a very high fee for 1908, and nearly half of Stein's annual income.

It was also May who approached Alfred Steiglitz about using Stein's later "portraits" in his journal *Camera Work;* published as they were in 1912, in the company of reproductions of paintings by Picasso and Matisse, her portraits of the two painters made Stein a part of the avant-garde movement. That designation served her better than did that of American realist, because the 1913 Armory Show, an internationally acclaimed New York art exhibit, was bringing the surprise of French impressionism and cubism into United States life. Mabel Dodge Luhan's essay on Stein's writing (see p. 372) was published during the furor over the Armory show, which Dodge Luhan had helped to organize, and in the serendipitous manner that sometimes works far better than planning, the name *Gertrude Stein* became synonymous with all that was innovative in Europe.

Such a characterization was not entirely a misnomer; as Stein recalled in her last interview (1946):

> Everything I have done has been influenced by Flaubert and Cézanne, and this gave me a new feeling about composition. Up to that time composition had consisted of a central idea, to which everything else was an accompaniment and separate but was not an end in itself, and Cézanne conceived the idea that in composition one thing was as important as another thing. Each part is as important as the whole, and that impressed me enormously, and it impressed me so much that I began to write *Three Lives* under this influence and this idea of composition and I was more interested in composition at that moment, this background of word-system, which had come to me from this reading that I had done. I was obsessed by this idea of composition, and the Negro story was a quintessence of it. . . . [I]t was not solely the realism of the characters but the realism of the composition which was the important thing, the realism of the composition of my thoughts." (15) (See Part Two, Chapter 5.)

And with her customary accuracy about her own psyche and motives, Stein added the telling sentence, "After all, to me one human being is as important as another human being . . ." (16).

Gertrude Stein during her student days, holding her nephew Allen on a visit to California, circa 1899. Reproduced by permission of the Beinecke Library American Literature Collection, Yale University.

Chronology of Stein's Life and Times

1874

Gertrude Stein born in Allegheny, PA to Amelia (Millie) Keyser Stein (18??–1888), and Daniel Stein (18??–1891), the seventh child (four others living). The youngest, Gertrude would be known as "Baby" for much of her life.

1875–78

The Stein family lives in both Austria and France.

1879–80

The family returns to Baltimore, and then moves to Oakland, California, where Daniel works with the street railway system.

Fyodor Dostoevsky (1821–1881), *The Brothers Karamazov*; Henrik Ibsen (1828–1906), *A Doll's House*; Émile Zola (1840–1902), *Nana*.

1888

Amelia Stein dies after a several-year bout with cancer.

Edward Bellamy (1850–1898), *Looking Backward, 2000–1887*.

1889

The Stein family is in disarray; Michael, the oldest son, returns from studying at Johns Hopkins.

Montana, Washington, and Dakotas admitted to the Union.

Walt Whitman (1819–1892), *Leaves of Grass* (eighth edition).

Motion pictures introduced by W. K. L. Dickson.

1890

Gertrude drops out of secondary school after the building burns; she goes daily to libraries and reads.

Idaho and Wyoming admitted to the Union.

James G. Frazer (1854–1941), *The Golden Bough.*

William James (1842–1910), *Principles of Psychology.*

1891

Daniel Stein dies suddenly; Mike moves his four siblings to San Francisco where he supports them through his street railway post.

Famine in Russia.

International Copyright Act.

Thomas Hardy (1840–1928), *Tess of the D'Urbervilles;* Oscar Wilde (1854–1900), *The Picture of Dorian Gray;* Hamlin Garland (1860–1940), *Main-Travelled Roads.*

1892

Gertrude, with Leo and Bertha, goes to Baltimore to live with her mother's family. Leo enrolls at Harvard.

Ellis Island is opened as an immigration center.

Joel Chandler Harris (1848–1908), *Uncle Remus and His Friends;* Henry James (1843–1916), *The Lesson of the Master.*

1893

Gertrude enrolls as a special student (without secondary school diploma) at the Harvard Annex, soon to become Radcliffe College. Majoring in philosophy (which includes psychology), she and Leo live in the same boarding house.

World's Columbia Exhibition, with its famed Women's Building, opens in Chicago.

Anti-Saloon League formed.

Stephen Crane (1871–1900), *Maggie: A Girl of the Streets.*

1894

Gertrude works under Hugo Münsterberg and eventually William James in the Harvard Psychological Laboratory; does well in most of her classes, especially George Santayana's.

Economic troubles: Pullman Car Co. strike; "Coxey's Army" marches on Washington.

The Dreyfus Affair signals anti-Semitism.

George Bernard Shaw (1856–1950), *Arms and the Man;* George du Maurier (1834–1896), *Trilby;* George Santayana (1863–1952), *Sonnets and Other Verses.*

1895

Gertrude is secretary of the Philosophy Club; arranges speakers. Earns an *A* from George Pierce Baker in expository writing course (see pp. 291–92).

Charles Dana Gibson's drawings create the slim, beautiful "Gibson Girl."

Marconi's telegraph; Roentgen's X-rays.

Oscar Wilde's trial and imprisonment in England.

Thomas Hardy, *Jude the Obscure;* Oscar Wilde, *The Importance of Being Earnest;* Stephen Crane, *The Red Badge of Courage.*

1896

Gertrude, as coauthor with a graduate student, publishes "Normal Motor Automatism" in *The Harvard Psychological Review;* begins senior thesis project under William James.

Plessy v. Ferguson confirms constitutionality of "separate but equal" railway accommodations for African Americans and whites (Jim Crow laws hold until the 1950s).

Emily Dickinson (1830–1886), *Poems: Third Series;* Sarah Orne Jewett (1849–1909), *The Country of the Pointed Firs;* Anton Chekhov (1860–1904), *The Sea Gull;* Alfred Jarry (1873–1907), *Ubu Roi.*

1897

Gertrude begins medical school at Johns Hopkins (she will graduate from Radcliffe in 1898, after she passes an "entrance" Latin exam, with magna cum laude honors in philosophy). She and Leo set up housekeeping in Baltimore (Leo is a graduate student at Hopkins).

Klondike Gold Rush (Yukon); coal miners' strike.

Havelock Ellis (1859–1939), *Studies in the Psychology of Sex;* Bram Stoker (1847–1912), *Dracula;* William James, *The Will to Believe and Other Essays.*

1898

Stein's honors essay, "Cultivated Motor Automatism: A Study of Character in Relation to Attention," published in *The Harvard Psychological Review.*

Spanish-American War.

Charlotte Perkins Stetson (Gilman) (1860–1935), *Women and Economics;* Joseph Conrad (1857–1924), *The Nigger of the "Narcissus";* Oscar Wilde, *The Ballad of Reading Gaol.*

1899

Stein disagrees with the beliefs of some of her professors about women's medicine.

Henry James, *The Awkward Age;* Kate Chopin (1851–1904), *The Awakening;* Thorstein Veblen (1857–1929), *The Theory of the Leisure Class.*

1900

Stein's romance with May Bookstaver leads to her disinterest in school; Leo lives abroad.

Sigmund Freud (1856–1939), *The Interpretation of Dreams;* the death of Oscar Wilde.

1901–03

Stein does not graduate with her medical school class, and refuses to attend summer school. She studies the brain with Dr. Llewellys Barker, travels, and eventually moves to Paris to live with Leo at 27, rue de Fleurus. They purchase contemporary art.

Booker T. Washington (1856–1915), *Up from Slavery;* W. E. B. DuBois (1868–1963), *The Souls of Black Folk;* Theodore Dreiser (1871–1945), *Sister Carrie* (repressed until 1907); Henry James, *The Ambassadors.*

1904–05

With Michael and his wife Sally, Leo and Gertrude continue collecting Matisse, Picasso, Cézanne, and other works; their salons begin. Gertrude writes *Q.E.D., Fernhurst,* sections of *The Making of Americans,* and "The Good Anna" of *Three Lives.*

1906

As Picasso is painting her portrait, Gertrude writes "Melanctha" and sends the manuscript of *Three Lives* to publishers.

1907

Gertrude and newly arrived Alice B. Toklas become a couple; Toklas later moves in.

1909

Three Lives appears from a vanity press to good reviews.

1911
Gertrude finishes *The Making of Americans* and writes portraits of friends.

1912
Her portraits "Matisse" and "Picasso" appear in Alfred Steiglitz's *Camera Work.*

1913
With the Armory Show exhibit in the States, Gertrude becomes an icon of the European avant garde. Leo moves from the rue de Fleurus address.

1914
Gertrude's three-part poem collection, *Tender Buttons,* is published.

World War I begins; United States is neutral.

James Joyce (1882–1941), *Dubliners;* Robert Frost (1874–1963), *North of Boston.*

1915–1918
During the war, Stein and Toklas live for a time in Spain, but eventually work for the American Fund for French Wounded (Stein drives the Ford van, "Auntie") in several European countries.

1919–1921
Resuming their Paris life, Stein and Toklas find themselves central to expatriates visiting France: they meet Sherwood Anderson (1876–1941) and other American writers and artists.

U.S. ratifies the Eighteenth Amendment, prohibiting the use of alcohol (i.e., prohibition).

1922–24
Stein becomes a friend of Ernest and Hadley Hemingway; hundreds of expatriates attend the Stein-Toklas salons. *Geography and Plays* is published.

Mussolini heads the Fascist political party in Italy, becoming dictator.

Edith Sitwell (1887–1964), *Facade;* Virginia Woolf (1882–1941), *Jacob's Room;* e. e. cummings (1894–1962), *Enormous Room;* Jean Toomer (1894–1967), *Cane;* James Joyce, *Ulysses;* T. S. Eliot (1888–1965), *The Waste Land.*

1925
The Making of Americans is published in Paris after parts of it appeared in Ford Madox Ford's *transatlantic review.*

F. Scott Fitzgerald (1896–1940), *The Great Gatsby;* Ernest Hemingway (1899–1961), *In Our Time;* Willa Cather (1873–1947), *The Professor's House;* John Dos Passos (1896–1970), *Manhattan Transfer;* Theodore Dreiser, *An American Tragedy; New Yorker* founded; Alain Locke (1885–1954), *The New Negro.*

1926

Arranged by the Sitwells, Stein lectures in Cambridge and Oxford; the Woolfs publish her lecture.

Ernest Hemingway, *The Sun Also Rises;* Langston Hughes (1902–1967), *The Weary Blues.*

1927–29

Stein works to publish her writings; finally, in 1929, she and Toklas rent a summer home in Bilignin, where they subsequently spend months of each year.

Nella Larsen (1891–1964), *Quicksand;* D. H. Lawrence (1889–1930), *Lady Chatterley;* William Faulkner (1897–1962), *The Sound and the Fury;* Thomas Wolfe (1882–1941), *Look Homeward, Angel;* Ernest Hemingway, *A Farewell to Arms.*

In 1927, Nicola Sacco and Bartolomeo Vanzetti are executed; in October of 1929, the stock market crashes and the U.S. is plunged into the decade of the Great Depression.

1930–31

Toklas and Stein establish their own publishing firm, The Plain Edition. Stein's *Lucy Church Amiably* is their first book.

Sinclair Lewis is the first American to win the Nobel Prize for Literature.

1932–33

Stein writes *The Autobiography of Alice B. Toklas,* which becomes a Book of the Month selection and is excerpted in *The Atlantic.* For the first time, her art makes money.

Prohibition repealed.

William Faulkner, *Light in August;* James Farrell (1904–1979), *Young Lonigan;* Ernest Hemingway, *Winner Take Nothing;* John Dos Passos, *1919;* Fannie Hurst (1889–1968), *Imitation of Life.*

1934–35

With much anxiety, Stein and Toklas return to the U.S. for an eight-month speaking tour, based on the popular *The Autobiography* (and on Stein's and Virgil Thomson's opera, *Four Saints in Three Acts,* first performed in Hartford, Connecticut, with a black cast). Becomes friends with Thornton Wilder (1897–1975). *Portraits and Prayers* and *Lectures in America* published by Random House, her new publisher. Social Security Act passed by Congress; Works Projects Administration formed.

Adolf Hitler is president and chancellor of Germany; Jews there lose civil rights.

1936

Gertrude lectures again in England; she and Alice consider moving to the States but think they cannot afford to. Publishes *The Geographical History of America.*

William Faulkner, *Absalom, Absalom!*; Margaret Mitchell (1900–1949), *Gone with the Wind*; Carl Sandburg (1878–1967), *The People, Yes*; Mabel Dodge Luhan (1879–1962), *Movers and Shakers*; John Steinbeck (1902–1968), *In Dubious Battle.*

1937

Stein publishes *Everybody's Autobiography* with Random House.

1938

Stein and Toklas are forced to leave rue de Fleurus, and move to a rue Christine apartment. Toklas has sent many of Stein's manuscripts to Yale University for safekeeping. *Picasso* published.

Thornton Wilder, *Our Town.*

1939–41

Fearful, they close the Paris apartment and move to Bilignin for the duration of World War II. They live meagrely, and are sometimes in danger of being deported. *Paris France, What Are Masterpieces?* and *Ida* are published.

John Steinbeck, *The Grapes of Wrath*; Ernest Hemingway, *For Whom the Bell Tolls*; Richard Wright (1908–1960), *Native Son*; Raymond Chandler (1888–1959), *The Big Sleep*; Pietro Di Donato (1911–1992) *Christ in Concrete*; F. Scott Fitzgerald, *The Last Tycoon.*

1943

Forced to move from Bilignin to Culoz, Stein's home is sometimes occupied by Nazis.

1944

When the occupation ends, Stein flies with U.S. forces and writes and speaks about the American GIs. In December, she and Alice return to Paris.

1945

Stein and Toklas tour American Army bases in occupied Germany; Stein lectures in Brussels. *Wars I Have Seen,* Stein's third autobiography, is published.

1946

Stein dies of intestinal cancer; Toklas dies in Paris in 1967.

Stein's *Brewsie and Willie* (about U.S. GIs) and Carl Van Vechten (1932–1964), *Gertrude Stein: Selected Writings* are published.

A Note on the Text

This volume reprints the Grafton Press publication of Stein's *Three Lives* (1909). Copyright, 1909, renewed, 1936, by Gertrude Stein. All rights reserved under International and Pan-American Copyright Conventions.

Annotations have been included where words or terms may be unfamiliar to readers.

Wherever possible, the text of the documents in Part Two is that of the original edition, and unless otherwise indicated in the headnote for a document, the text has not been modified from the copy text. Students should expect to encounter some archaic or variant spelling and punctuation conventions.

Three Lives

Donc je suis un malheureux et ce
n'est ni ma faute ni celle de la vie.°
— Jules Laforgue

CONTENTS

° Thus I am unhappy and this is neither my fault nor that of life.

THE GOOD ANNA

Part I

The tradesmen of Bridgepoint learned to dread the sound of "Miss Mathilda", for with that name the good Anna always conquered.

The strictest of the one price stores found that they could give things for a little less, when the good Anna had fully said that "Miss Mathilda" could not pay so much and that she could buy it cheaper "by Lindheims."

Lindheims was Anna's favorite store, for there they had bargain days, when flour and sugar were sold for a quarter of a cent less for a pound, and there the heads of the departments were all her friends and always managed to give her the bargain prices, even on other days.

Anna led an arduous and troubled life.

Anna managed the whole little house for Miss Mathilda. It was a funny little house, one of a whole row of all the same kind that made a close pile like a row of dominoes that a child knocks over, for they were built along a street which at this point came down a steep hill. They were funny little houses, two stories high, with red brick fronts and long white steps.

This one little house was always very full with Miss Mathilda, an under servant, stray dogs and cats and Anna's voice that scolded, managed, grumbled all day long.

"Sallie! can't I leave you alone a minute but you must run to the door to see the butcher boy come down the street and there is Miss Mathilda calling for her shoes. Can I do everything while you go around always thinking about nothing at all? If I ain't after you every minute you would be forgetting all the time, and I take all this pains, and when you come to me you was as ragged as a buzzard and as dirty as a dog. Go and find Miss Mathilda her shoes where you put them this morning."

"Peter!", — her voice rose higher, — "Peter!", — Peter was the youngest and the favorite dog, — "Peter, if you don't leave Baby alone," — Baby was an old, blind terrier that Anna had loved for many years, — "Peter if you don't leave Baby alone, I take a rawhide to you, you bad dog."

The good Anna had high ideals for canine chastity and discipline. The three regular dogs, the three that always lived with Anna, Peter and old Baby, and the fluffy little Rags, who was always jumping up into the air just to show that he was happy, together with the transients, the

Most household labor in homes outside the South was drawn from the immigrant pool of German, Italian, and Irish, as in this photograph of the staff of a
Black River Falls (Wisconsin) household circa 1890. Photo by Charles Van
Schaick, reproduced with permission of the State Historical Society of Wisconsin, Visual Material Archives.

many stray ones that Anna always kept until she found them homes,
were all under strict orders never to be bad one with the other.

A sad disgrace did once happen in the family. A little transient terrier for whom Anna had found a home suddenly produced a crop of
pups. The new owners were certain that this Foxy had known no dog
since she was in their care. The good Anna held to it stoutly that her
Peter and her Rags were guiltless, and she made her statement with so
much heat that Foxy's owners were at last convinced that these results
were due to their neglect.

"You bad dog," Anna said to Peter that night, "you bad dog."

"Peter was the father of those pups," the good Anna explained to
Miss Mathilda, "and they look just like him too, and poor little Foxy,
they were so big that she could hardly have them, but Miss Mathilda, I
would never let those people know that Peter was so bad."

Periods of evil thinking came very regularly to Peter and to Rags
and to the visitors within their gates. At such times Anna would be

very busy and scold hard, and then too she always took great care to seclude the bad dogs from each other whenever she had to leave the house. Sometimes just to see how good it was that she had made them, Anna would leave the room a little while and leave them all together, and then she would suddenly come back. Back would slink all the wicked-minded dogs at the sound of her hand upon the knob, and then they would sit desolate in their corners like a lot of disappointed children whose stolen sugar has been taken from them.

Innocent blind old Baby was the only one who preserved the dignity becoming in a dog.

You see that Anna led an arduous and troubled life.

The good Anna was a small, spare, german woman, at this time about forty years of age. Her face was worn, her cheeks were thin, her mouth drawn and firm, and her light blue eyes were very bright. Sometimes they were full of lightning and sometimes full of humor, but they were always sharp and clear.

Her voice was a pleasant one, when she told the histories of bad Peter and of Baby and of little Rags. Her voice was a high and piercing one when she called to the teamsters[1] and to the other wicked men, what she wanted that should come to them, when she saw them beat a horse or kick a dog. She did not belong to any society that could stop them and she told them so most frankly, but her strained voice and her glittering eyes, and her queer piercing german english first made them afraid and then ashamed. They all knew too, that all the policemen on the beat were her friends. These always respected and obeyed Miss Annie, as they called her, and promptly attended to all of her complaints.

For five years Anna managed the little house for Miss Mathilda. In these five years there were four different under servants.

The one that came first was a pretty, cheerful irish girl. Anna took her with a doubting mind. Lizzie was an obedient, happy servant, and Anna began to have a little faith. This was not for long. The pretty, cheerful Lizzie disappeared one day without her notice and with all her baggage and returned no more.

This pretty, cheerful Lizzie was succeeded by a melancholy Molly.

Molly was born in America, of german parents. All her people had been long dead or gone away. Molly had always been alone. She was a tall, dark, sallow, thin-haired creature, and she was always troubled

---

[1] *teamsters:* Men who drove wagons or trucks for hauling goods, dependent on the strength of their animals — mules or horses.

with a cough, and she had a bad temper, and always said ugly dreadful swear words.

Anna found all this very hard to bear, but she kept Molly a long time out of kindness. The kitchen was constantly a battle-ground. Anna scolded and Molly swore strange oaths, and then Miss Mathilda would shut her door hard to show that she could hear it all.

At last Anna had to give it up. "Please Miss Mathilda won't you speak to Molly," Anna said, "I can't do a thing with her. I scold her, and she don't seem to hear and then she swears so that she scares me. She loves you Miss Mathilda, and you scold her please once."

"But Anna," cried poor Miss Mathilda, "I don't want to," and that large, cheerful, but faint hearted woman looked all aghast at such a prospect. "But you must, please Miss Mathilda!" Anna said.

Miss Mathilda never wanted to do any scolding. "But you must please Miss Mathilda," Anna said.

Miss Mathilda every day put off the scolding, hoping always that Anna would learn to manage Molly better. It never did get better and at last Miss Mathilda saw that the scolding simply had to be.

It was agreed between the good Anna and her Miss Mathilda that Anna should be away when Molly would be scolded. The next evening that it was Anna's evening out, Miss Mathilda faced her task and went down into the kitchen.

Molly was sitting in the little kitchen leaning her elbows on the table. She was a tall, thin, sallow girl, aged twenty-three, by nature slatternly and careless but trained by Anna into superficial neatness. Her drab striped cotton dress and gray black checked apron increased the length and sadness of her melancholy figure. "Oh, Lord!" groaned Miss Mathilda to herself as she approached her.

"Molly, I want to speak to you about your behaviour to Anna!", here Molly dropped her head still lower on her arms and began to cry.

"Oh! Oh!" groaned Miss Mathilda.

"It's all Miss Annie's fault, all of it," Molly said at last, in a trembling voice, "I do my best."

"I know Anna is often hard to please," began Miss Mathilda, with a twinge of mischief, and then she sobered herself to her task, "but you must remember, Molly, she means it for your good and she is really very kind to you."

"I don't want her kindness," Molly cried, "I wish you would tell me what to do, Miss Mathilda, and then I would be all right. I hate Miss Annie."

"This will never do Molly," Miss Mathilda said sternly, in her deepest, firmest tones, "Anna is the head of the kitchen and you must either obey her or leave."

"I don't want to leave you," whimpered melancholy Molly. "Well Molly then try and do better," answered Miss Mathilda, keeping a good stern front, and backing quickly from the kitchen. "Oh! Oh!" groaned Miss Mathilda, as she went back up the stairs.

Miss Mathilda's attempt to make peace between the constantly contending women in the kitchen had no real effect. They were very soon as bitter as before.

At last it was decided that Molly was to go away. Molly went away to work in a factory in the town, and she went to live with an old woman in the slums, a very bad old woman Anna said.

Anna was never easy in her mind about the fate of Molly. Sometimes she would see or hear of her. Molly was not well, her cough was worse, and the old woman really was a bad one.

After a year of this unwholesome life, Molly was completely broken down. Anna then again took her in charge. She brought her from her work and from the woman where she lived, and put her in a hospital to stay till she was well. She found a place for her as nursemaid to a little girl out in the country, and Molly was at last established and content.

Molly had had, at first, no regular successor. In a few months it was going to be the summer and Miss Mathilda would be gone away, and old Katie would do well to come in every day and help Anna with her work.

Old Katy was a heavy, ugly, short and rough old german woman, with a strange distorted german-english all her own. Anna was worn out now with her attempt to make the younger generation do all that it should and rough old Katy never answered back, and never wanted her own way. No scolding or abuse could make its mark on her uncouth and aged peasant hide. She said her "Yes, Miss Annie," when an answer had to come, and that was always all that she could say.

"Old Katy is just a rough old woman, Miss Mathilda," Anna said, "but I think I keep her here with me. She can work and she don't give me trouble like I had with Molly all the time."

Anna always had a humorous sense from this old Katy's twisted peasant english, from the roughness on her tongue of buzzing s's and from the queer ways of her brutish servile humor. Anna could not let old Katy serve at table — old Katy was too coarsely made from natural earth for that — and so Anna had all this to do herself and that she never liked, but even then this simple rough old creature was pleasanter to her than any of the upstart young.

Life went on very smoothly now in these few months before the summer came. Miss Mathilda every summer went away across the ocean to be gone for several months. When she went away this summer old Katy was so sorry, and on the day that Miss Mathilda went, old Katy cried hard for many hours. An earthy, uncouth, servile peasant creature old Katy surely was. She stood there on the white stone steps of the little red brick house, with her bony, square dull head with its thin, tanned, toughened skin and its sparse and kinky grizzled hair, and her strong, squat figure a little overmade on the right side, clothed in her blue striped cotton dress, all clean and always washed but rough and harsh to see — and she stayed there on the steps till Anna brought her in, blubbering, her apron to her face, and making queer guttural broken moans.

When Miss Mathilda early in the fall came to her house again old Katy was not there.

"I never thought old Katy would act so Miss Mathilda," Anna said, "when she was so sorry when you went away, and I gave her full wages all the summer, but they are all alike Miss Mathilda, there isn't one of them that's fit to trust. You know how Katy said she liked you, Miss Mathilda, and went on about it when you went away and then she was so good and worked all right until the middle of the summer, when I got sick, and then she went away and left me all alone and took a place out in the country, where they gave her some more money. She didn't say a word, Miss Mathilda, she just went off and left me there alone when I was sick after that awful hot summer that we had, and after all we done for her when she had no place to go, and all summer I gave her better things to eat than I had for myself. Miss Mathilda, there isn't one of them has any sense of what's the right way for a girl to do, not one of them."

Old Katy was never heard from any more.

No under servant was decided upon now for several months. Many came and many went, and none of them would do. At last Anna heard of Sallie.

Sallie was the oldest girl in a family of eleven and Sallie was just sixteen years old. From Sallie down they came always littler and littler in her family, and all of them were always out at work excepting only the few littlest of them all.

Sallie was a pretty blonde and smiling german girl, and stupid and a little silly. The littler they came in her family the brighter they all were.

The brightest of them all was a little girl of ten. She did a good day's work washing dishes for a man and wife in a saloon, and she earned a fair day's wage, and then there was one littler still. She only worked for half the day. She did the house work for a bachelor doctor. She did it all, all of the housework and received each week her eight cents for her wage. Anna was always indignant when she told that story.

"I think he ought to give her ten cents Miss Mathilda any way. Eight cents is so mean when she does all his work and she is such a bright little thing too, not stupid like our Sallie. Sallie would never learn to do a thing if I didn't scold her all the time, but Sallie is a good girl, and I take care and she will do all right."

Sallie was a good, obedient german child. She never answered Anna back, no more did Peter, old Baby and little Rags and so though always Anna's voice was sharply raised in strong rebuke and worn expostulation, they were a happy family all there together in the kitchen.

Anna was a mother now to Sallie, a good incessant german mother who watched and scolded hard to keep the girl from any evil step. Sallie's temptations and transgressions were much like those of naughty Peter and jolly little Rags, and Anna took the same way to keep all three from doing what was bad.

Sallie's chief badness besides forgetting all the time and never washing her hands clean to serve at table, was the butcher boy.

He was an unattractive youth enough, that butcher boy. Suspicion began to close in around Sallie that she spent the evenings when Anna was away, in company with this bad boy.

"Sallie is such a pretty girl, Miss Mathilda," Anna said, "and she is so dumb and silly, and she puts on that red waist,[2] and she crinkles up her hair with irons[3] so I have to laugh, and then I tell her if she only washed her hands clean it would be better than all that fixing all the time, but you can't do a thing with the young girls nowadays Miss Mathilda. Sallie is a good girl but I got to watch her all the time."

Suspicion closed in around Sallie more and more, that she spent Anna's evenings out with this boy sitting in the kitchen. One early morning Anna's voice was sharply raised.

"Sallie this ain't the same banana that I brought home yesterday,

---

[2] *red waist:* A blouse or garment that was worn above the waist. The bright color was considered a bit unseemly for women, especially in the days before cosmetics.

[3] *irons:* Curling irons were more primitive in the nineteenth century than they are today, and using them took more time.

for Miss Mathilda, for her breakfast, and you was out early in the street this morning, what was you doing there?"

"Nothing, Miss Annie, I just went out to see, that's all and that's the same banana, 'deed it is Miss Annie."

"Sallie, how can you say so and after all I do for you, and Miss Mathilda is so good to you. I never brought home no bananas yesterday with specks on it like that. I know better, it was that boy was here last night and ate it while I was away, and you was out to get another this morning. I don't want no lying Sallie."

Sallie was stout in her defence but then she gave it up and she said it was the boy who snatched it as he ran away at the sound of Anna's key opening the outside door. "But I will never let him in again, Miss Annie, 'deed I won't," said Sallie.

And now it was all peaceful for some weeks and then Sallie with fatuous simplicity began on certain evenings to resume her bright red waist, her bits of jewels and her crinkly hair.

One pleasant evening in the early spring, Miss Mathilda was standing on the steps beside the open door, feeling cheerful in the pleasant, gentle night. Anna came down the street, returning from her evening out. "Don't shut the door, please, Miss Mathilda," Anna said in a low voice, "I don't want Sallie to know I'm home."

Anna went softly through the house and reached the kitchen door. At the sound of her hand upon the knob there was a wild scramble and a bang, and then Sallie sitting there alone when Anna came into the room, but, alas, the butcher boy forgot his overcoat in his escape.

You see that Anna led an arduous and troubled life.

Anna had her troubles, too, with Miss Mathilda. "And I slave and slave to save the money and you go out and spend it all on foolishness," the good Anna would complain when her mistress, a large and careless woman, would come home with a bit of porcelain, a new etching and sometimes even an oil painting on her arm.

"But Anna," argued Miss Mathilda, "if you didn't save this money, don't you see I could not buy these things," and then Anna would soften and look pleased until she learned the price, and then wringing her hands, "Oh, Miss Mathilda, Miss Mathilda," she would cry, "and you gave all that money out for that, when you need a dress to go out in so bad." "Well, perhaps I will get one for myself next year, Anna," Miss Mathilda would cheerfully concede. "If we live till then Miss Mathilda, I see that you do," Anna would then answer darkly.

Anna had great pride in the knowledge and possessions of her cherished Miss Mathilda, but she did not like her careless way of wearing always her old clothes. "You can't go out to dinner in that dress, Miss Mathilda," she would say, standing firmly before the outside door, "You got to go and put on your new dress you always look so nice in." "But Anna, there isn't time." "Yes there is, I go up and help you fix it, please Miss Mathilda you can't go out to dinner in that dress and next year if we live till then, I make you get a new hat, too. It's a shame Miss Mathilda to go out like that."

The poor mistress sighed and had to yield. It suited her cheerful, lazy temper to be always without care but sometimes it was a burden to endure, for so often she had it all to do again unless she made a rapid dash out of the door before Anna had a chance to see.

Life was very easy always for this large and lazy Miss Mathilda, with the good Anna to watch and care for her and all her clothes and goods. But, alas, this world of ours is after all much what it should be and cheerful Miss Mathilda had her troubles too with Anna.

It was pleasant that everything for one was done, but annoying often that what one wanted most just then, one could not have when one had foolishly demanded and not suggested one's desire. And then Miss Mathilda loved to go out on joyous, country tramps when, stretching free and far with cheerful comrades, over rolling hills and cornfields, glorious in the setting sun, and dogwood white and shining underneath the moon and clear stars over head, and brilliant air and tingling blood, it was hard to have to think of Anna's anger at the late return, though Miss Mathilda had begged that there might be no hot supper cooked that night. And then when all the happy crew of Miss Mathilda and her friends, tired with fullness of good health and burning winds and glowing sunshine in the eyes, stiffened and justly worn and wholly ripe for pleasant food and gentle content, were all come together to the little house — it was hard for all that tired crew who loved the good things Anna made to eat, to come to the closed door and wonder there if it was Anna's evening in or out, and then the others must wait shivering on their tired feet, while Miss Mathilda softened Anna's heart, or if Anna was well out, boldly ordered youthful Sallie to feed all the hungry lot.

Such things were sometimes hard to bear and often grievously did Miss Mathilda feel herself a rebel with the cheerful Lizzies, the melancholy Mollies, the rough old Katies and the stupid Sallies.

Miss Mathilda had other troubles too, with the good Anna. Miss Mathilda had to save her Anna from the many friends, who in the kindly fashion of the poor, used up her savings and then gave her promises in place of payments.

The good Anna had many curious friends that she had found in the twenty years that she had lived in Bridgepoint, and Miss Mathilda would often have to save her from them all.

Part II
The Life of the Good Anna

Anna Federner, this good Anna, was of solid lower middle-class south german stock.

When she was seventeen years old she went to service in a bourgeois family, in the large city near her native town, but she did not stay there long. One day her mistress offered her maid — that was Anna — to a friend, to see her home. Anna felt herself to be a servant, not a maid, and so she promptly left the place.

Anna had always a firm old world sense of what was the right way for a girl to do.

No argument could bring her to sit an evening in the empty parlour, although the smell of paint when they were fixing up the kitchen made her very sick, and tired as she always was, she never would sit down during the long talks she held with Miss Mathilda. A girl was a girl and should act always like a girl, both as to giving all respect and as to what she had to eat.

A little time after she left this service, Anna and her mother made the voyage to America. They came second-class, but it was for them a long and dreary journey. The mother was already ill with consumption.

They landed in a pleasant town in the far South and there the mother slowly died.

Anna was now alone and she made her way to Bridgepoint where an older half brother was already settled. This brother was a heavy, lumbering, good natured german man, full of the infirmity that comes of excess of body.

He was a baker and married and fairly well to do.

Anna liked her brother well enough but was never in any way dependent on him.

When she arrived in Bridgepoint, she took service with Miss Mary Wadsmith.

Miss Mary Wadsmith was a large, fair, helpless woman, burdened with the care of two young children. They had been left her by her brother and his wife who had died within a few months of each other.

Anna soon had the household altogether in her charge.

Anna found her place with large, abundant women, for such were always lazy, careless or all helpless, and so the burden of their lives could fall on Anna, and give her just content. Anna's superiors must be always these large helpless women, or be men, for none others could give themselves to be made so comfortable and free.

Anna had no strong natural feeling to love children, as she had to love cats and dogs, and a large mistress. She never became deeply fond of Edgar and Jane Wadsmith. She naturally preferred the boy, for boys love always better to be done for and made comfortable and full of eating, while in the little girl she had to meet the feminine, the subtle opposition, showing so early always in a young girl's nature.

For the summer, the Wadsmiths had a pleasant house out in the country, and the winter months they spent in hotel apartments in the city.

Gradually it came to Anna to take the whole direction of their movements, to make all the decisions as to their journeyings to and fro, and for the arranging of the places where they were to live.

Anna had been with Miss Mary for three years, when little Jane began to raise her strength in opposition. Jane was a neat, pleasant little girl, pretty and sweet with a young girl's charm, and with two blonde braids carefully plaited down her back.

Miss Mary, like her Anna, had no strong natural feeling to love children, but she was fond of these two young ones of her blood, and yielded docilely to the stronger power in the really pleasing little girl. Anna always preferred the rougher handling of the boy, while Miss Mary found the gentle force and the sweet domination of the girl to please her better.

In a spring when all the preparations for the moving had been made, Miss Mary and Jane went together to the country home, and Anna, after finishing up the city matters was to follow them in a few days with Edgar, whose vacation had not yet begun.

Many times during the preparations for this summer, Jane had met Anna with sharp resistance, in opposition to her ways. It was simple for little Jane to give unpleasant orders, not from herself but from Miss Mary, large, docile, helpless Miss Mary Wadsmith who could never think out any orders to give Anna from herself.

Anna's eyes grew slowly sharper, harder, and her lower teeth thrust a little forward and pressing strongly up, framed always more slowly the "Yes, Miss Jane," to the quick, "Oh Anna! Miss Mary says she wants you to do it so!"

On the day of their migration, Miss Mary had been already put into the carriage. "Oh, Anna!" cried little Jane running back into the house, "Miss Mary says that you are to bring along the blue dressings out of her room and mine." Anna's body stiffened, "We never use them in the summer, Miss Jane," she said thickly. "Yes Anna, but Miss Mary thinks it would be nice, and she told me to tell you not to forget, good-by!" and the little girl skipped lightly down the steps into the carriage and they drove away.

Anna stood still on the steps, her eyes hard and sharp and shining, and her body and her face stiff with resentment. And then she went into the house, giving the door a shattering slam.

Anna was very hard to live with in those next three days. Even Baby, the new puppy, the pride of Anna's heart, a present from her friend the widow, Mrs. Lehntman — even this pretty little black and tan felt the heat of Anna's scorching flame. And Edgar, who had looked forward to these days, to be for him filled full of freedom and of things to eat — he could not rest a moment in Anna's bitter sight.

On the third day, Anna and Edgar went to the Wadsmith country home. The blue dressings out of the two rooms remained behind.

All the way, Edgar sat in front with the colored man and drove. It was an early spring day in the South. The fields and woods were heavy from the soaking rains. The horses dragged the carriage slowly over the long road, sticky with brown clay and rough with masses of stones thrown here and there to be broken and trodden into place by passing teams. Over and through the soaking earth was the feathery new spring growth of little flowers, of young leaves and of ferns. The tree tops were all bright with reds and yellows, with brilliant gleaming whites and gorgeous greens. All the lower air was full of the damp haze rising from heavy soaking water on the earth, mingled with a warm and pleasant smell from the blue smoke of the spring fires in all the open fields. And above all this was the clear, upper air, and the songs of birds and the joy of sunshine and of lengthening days.

The languor and the stir, the warmth and weight and the strong feel of life from the deep centres of the earth that comes always with the early, soaking spring, when it is not answered with an active fervent joy, gives always anger, irritation and unrest.

To Anna alone there in the carriage, drawing always nearer to the struggle with her mistress, the warmth, the slowness, the jolting over stones, the steaming from the horses, the cries of men and animals and birds, and the new life all round about were simply maddening. "Baby! if you don't lie still, I think I kill you. I can't stand it any more like this."

At this time Anna, about twenty-seven years of age, was not yet all thin and worn. The sharp bony edges and corners of her head and face were still rounded out with flesh, but already the temper and the humor showed sharply in her clean blue eyes, and the thinning was begun about the lower jaw, that was so often strained with the upward pressure of resolve.

To-day, alone there in the carriage, she was all stiff and yet all trembling with the sore effort of decision and revolt.

As the carriage turned into the Wadsmith gate, little Jane ran out to see. She just looked at Anna's face; she did not say a word about blue dressings.

Anna got down from the carriage with little Baby in her arms. She took out all the goods that she had brought and the carriage drove away. Anna left everything on the porch, and went in to where Miss Mary Wadsmith was sitting by the fire.

Miss Mary was sitting in a large armchair by the fire. All the nooks and crannies of the chair were filled full of her soft and spreading body. She was dressed in a black satin morning gown, the sleeves, great monster things, were heavy with the mass of her soft flesh. She sat there always, large, helpless, gentle. She had a fair, soft, regular, good-looking face, with pleasant, empty, grey-blue eyes, and heavy sleepy lids.

Behind Miss Mary was the little Jane, nervous and jerky with excitement as she saw Anna come into the room.

"Miss Mary," Anna began. She had stopped just within the door, her body and her face stiff with repression, her teeth closed hard and the white lights flashing sharply in the pale, clean blue of her eyes. Her bearing was full of the strange coquetry of anger and of fear, the stiffness, the bridling, the suggestive movement underneath the rigidness of forced control, all the queer ways the passions have to show themselves all one.

"Miss Mary," the words came slowly with thick utterance and with jerks, but always firm and strong. "Miss Mary, I can't stand it any more like this. When you tell me anything to do, I do it. I do everything I can and you know I work myself sick for you. The blue dressings in your room makes too much work to have for summer. Miss Jane don't know what work is. If you want to do things like that I go away."

Anna stopped still. Her words had not the strength of meaning they were meant to have, but the power in the mood of Anna's soul frightened and awed Miss Mary through and through.

Like in all large and helpless women, Miss Mary's heart beat weakly in the soft and helpless mass it had to govern. Little Jane's excitements had already tried her strength. Now she grew pale and fainted quite away.

"Miss Mary!" cried Anna running to her mistress and supporting all her helpless weight back in the chair. Little Jane, distracted, flew about as Anna ordered, bringing smelling salts and brandy and vinegar and water and chafing poor Miss Mary's wrists.

Miss Mary slowly opened her mild eyes. Anna sent the weeping little Jane out of the room. She herself managed to get Miss Mary quiet on the couch.

There was never a word more said about blue dressings.

Anna had conquered, and a few days later little Jane gave her a green parrot to make peace.

For six more years little Jane and Anna lived in the same house. They were careful and respectful to each other to the end.

Anna liked the parrot very well. She was fond of cats too and of horses, but best of all animals she loved the dog and best of all dogs, little Baby, the first gift from her friend, the widow Mrs. Lehntman.

The widow Mrs. Lehntman was the romance in Anna's life.

Anna met her first at the house of her half brother, the baker, who had known the late Mr. Lehntman, a small grocer, very well.

Mrs. Lehntman had been for many years a midwife. Since her husband's death she had herself and two young children to support.

Mrs. Lehntman was a good looking woman. She had a plump well rounded body, clear olive skin, bright dark eyes and crisp black curling hair. She was pleasant, magnetic, efficient and good. She was very attractive, very generous and very amiable.

She was a few years older than our good Anna, who was soon entirely subdued by her magnetic, sympathetic charm.

Mrs. Lehntman in her work loved best to deliver young girls who were in trouble. She would take these into her own house and care for them in secret, till they could guiltlessly go home or back to work, and then slowly pay her the money for their care. And so through this new friend Anna led a wider and more entertaining life, and often she used up her savings in helping Mrs. Lehntman through those times when she was giving very much more than she got.

It was through Mrs. Lehntman that Anna met Dr. Shonjen who employed her when at last it had to be that she must go away from her Miss Mary Wadsmith.

During the last years with her Miss Mary, Anna's health was very bad, as indeed it always was from that time on until the end of her strong life.

Anna was a medium sized, thin, hard working, worrying woman. She had always had bad headaches and now they came more often and more wearing.

Her face grew thin, more bony and more worn, her skin stained itself pale yellow, as it does with working sickly women, and the clear blue of her eyes went pale.

Her back troubled her a good deal, too. She was always tired at her work and her temper grew more difficult and fretful.

Miss Mary Wadsmith often tried to make Anna see a little to herself, and get a doctor, and the little Jane, now blossoming into a pretty, sweet young woman, did her best to make Anna do things for her good. Anna was stubborn always to Miss Jane, and fearful of interference in her ways. Miss Mary Wadsmith's mild advice she easily could always turn aside.

Mrs. Lehntman was the only one who had any power over Anna. She induced her to let Dr. Shonjen take her in his care.

No one but a Dr. Shonjen could have brought a good and german Anna first to stop her work and then submit herself to operation, but he knew so well how to deal with german and poor people. Cheery, jovial, hearty, full of jokes that made much fun and yet were full of simple common sense and reasoning courage, he could persuade even a good Anna to do things that were for her own good.

Edgar had now been for some years away from home, first at a school and then at work to prepare himself to be a civil engineer. Miss Mary and Jane promised to take a trip for all the time that Anna was away and so there would be no need for Anna's work, nor for a new girl to take Anna's place.

Anna's mind was thus a little set at rest. She gave herself to Mrs. Lehntman and the doctor to do what they thought best to make her well and strong.

Anna endured the operation very well, and was patient, almost docile, in the slow recovery of her working strength. But when she was once more at work for her Miss Mary Wadsmith, all the good effect of these several months of rest were soon worked and worried well away.

For all the rest of her strong working life Anna was never really well. She had bad headaches all the time and she was always thin and worn.

She worked away her appetite, her health and strength, and always for the sake of those who begged her not to work so hard. To her thinking, in her stubborn, faithful, german soul, this was the right way for a girl to do.

Anna's life with Miss Mary Wadsmith was now drawing to an end.

Miss Jane, now altogether a young lady, had come out into the world. Soon she would become engaged and then be married, and then perhaps Miss Mary Wadsmith would make her home with her.

In such a household Anna was certain that she would never take a place. Miss Jane was always careful and respectful and very good to Anna, but never could Anna be a girl in a household where Miss Jane would be the head. This much was very certain in her mind, and so these last two years with her Miss Mary were not as happy as before.

The change came very soon.

Miss Jane became engaged and in a few months was to marry a man from out of town, from Curden, an hour's railway ride from Bridgepoint.

Poor Miss Mary Wadsmith did not know the strong resolve Anna had made to live apart from her when this new household should be formed. Anna found it very hard to speak to her Miss Mary of this change.

The preparations for the wedding went on day and night.

Anna worked and sewed hard to make it all go well.

Miss Mary was much fluttered, but content and happy with Anna to make everything so easy for them all.

Anna worked so all the time to drown her sorrow and her conscience too, for somehow it was not right to leave Miss Mary so. But what else could she do? She could not live as her Miss Mary's girl, in a house where Miss Jane would be the head.

The wedding day grew always nearer. At last it came and passed.

The young people went on their wedding trip, and Anna and Miss Mary were left behind to pack up all the things.

Even yet poor Anna had not had the strength to tell Miss Mary her resolve, but now it had to be.

Anna every spare minute ran to her friend Mrs. Lehntman for comfort and advice. She begged her friend to be with her when she told the news to Miss Mary.

Perhaps if Mrs. Lehntman had not been in Bridgepoint, Anna would have tried to live in the new house. Mrs. Lehntman did not urge her to this thing nor even give her this advice, but feeling for Mrs. Lehntman as she did made even faithful Anna not quite so strong in her dependence on Miss Mary's need as she would otherwise have been.

Remember, Mrs. Lehntman was the romance in Anna's life.

All the packing was now done and in a few days Miss Mary was to go to the new house, where the young people were ready for her coming.

At last Anna had to speak.

Mrs. Lehntman agreed to go with her and help to make the matter clear to poor Miss Mary.

The two women came together to Miss Mary Wadsmith sitting placid by the fire in the empty living room. Miss Mary had seen Mrs. Lehntman many times before, and so her coming in with Anna raised no suspicion in her mind.

It was very hard for the two women to begin.

It must be very gently done, this telling to Miss Mary of the change. She must not be shocked by suddenness or with excitement.

Anna was all stiff, and inside all a quiver with shame, anxiety and grief. Even courageous Mrs. Lehntman, efficient, impulsive and complacent as she was and not deeply concerned in the event, felt awkward, abashed and almost guilty in that large, mild, helpless presence. And at her side to make her feel the power of it all, was the intense conviction of poor Anna, struggling to be unfeeling, self righteous and suppressed.

"Miss Mary" — with Anna when things had to come they came always sharp and short — "Miss Mary, Mrs. Lehntman has come here with me, so I can tell you about not staying with you there in Curden. Of course I go help you to get settled and then I think I come back and stay right here in Bridgepoint. You know my brother he is here and all his family, and I think it would be not right to go away from them so far, and you know you don't want me now so much Miss Mary when you are all together there in Curden."

Miss Mary Wadsmith was puzzled. She did not understand what Anna meant by what she said.

"Why Anna of course you can come to see your brother whenever you like to, and I will always pay your fare. I thought you understood all about that, and we will be very glad to have your nieces come to stay with you as often as they like. There will always be room enough in a big house like Mr. Goldthwaite's."

It was now for Mrs. Lehntman to begin her work.

"Miss Wadsmith does not understand just what you mean Anna," she began. "Miss Wadsmith, Anna feels how good and kind you are, and she talks about it all the time, and what you do for her in every way you can, and she is very grateful and never would want to go away from you, only she thinks it would be better now that Mrs. Goldthwaite has this big new house and will want to manage it in her own way, she thinks perhaps it would be better if Mrs. Goldthwaite had all new servants with her to begin with, and not a girl like Anna who knew her when she was little girl. That is what Anna feels about it now, and she asked me and I said to her that I thought it would be better for you all and you knew she liked you so much and that you were so good to her, and you would understand how she thought it would be better in the new house if she stayed on here in Bridgepoint, anyway for a little while until Mrs. Goldthwaite was used to her new house. Is'nt that it Anna that you wanted Miss Wadsmith to know?"

"Oh Anna," Miss Mary Wadsmith said it slowly and in a grieved tone of surprise that was very hard for the good Anna to endure, "Oh Anna, I didn't think that you would ever want to leave me after all these years."

"Miss Mary!" it came in one tense jerky burst, "Miss Mary it's only working under Miss Jane now would make me leave you so. I know how good you are and I work myself sick for you and for Mr. Edgar and for Miss Jane too, only Miss Jane she will want everything different from like the way we always did, and you know Miss Mary I can't have Miss Jane watching at me all the time, and every minute something new. Miss Mary, it would be very bad and Miss Jane don't really want me to come with you to the new house, I know that all the time. Please Miss Mary don't feel bad about it or think I ever want to go away from you if I could do things right for you the way they ought to be."

Poor Miss Mary. Struggling was not a thing for her to do. Anna would surely yield if she would struggle, but struggling was too much work and too much worry for peaceful Miss Mary to endure. If Anna would do so she must. Poor Miss Mary Wadsmith sighed, looked wistfully at Anna and then gave it up.

"You must do as you think best Anna," she said at last letting all of her soft self sink back into the chair. "I am very sorry and so I am sure will be Miss Jane when she hears what you have thought it best to do. It was very good of Mrs. Lehntman to come with you and I am sure she does it for your good. I suppose you want to go out a little now. Come back in an hour Anna and help me go to bed." Miss Mary closed her eyes and rested still and placid by the fire.

The two women went away.

This was the end of Anna's service with Miss Mary Wadsmith, and soon her new life taking care of Dr. Shonjen was begun.

Keeping house for a jovial bachelor doctor gave new elements of understanding to Anna's maiden german mind. Her habits were as firm fixed as before, but it always was with Anna that things that had been done once with her enjoyment and consent could always happen any time again, such as her getting up at any hour of the night to make a supper and cook hot chops and chicken fry for Dr. Shonjen and his bachelor friends.

Anna loved to work for men, for they could eat so much and with such joy. And when they were warm and full, they were content, and let her do whatever she thought best. Not that Anna's conscience ever slept, for neither with interference or without would she strain less to keep on saving every cent and working every hour of the day. But truly she loved it best when she could scold. Now it was not only other girls and the colored man, and dogs, and cats, and horses and her parrot, but her cheery master, jolly Dr. Shonjen, whom she could guide and constantly rebuke to his own good.

The doctor really loved her scolding as she loved his wickednesses and his merry joking ways.

These days were happy days with Anna.

Her freakish humor now first showed itself, her sense of fun in the queer ways that people had, that made her later find delight in brutish servile Katy, in Sally's silly ways and in the badness of Peter and of Rags. She loved to make sport with the skeletons the doctor had, to make them move and make strange noises till the negro boy shook in his shoes and his eyes rolled white in his agony of fear.

Then Anna would tell these histories to her doctor. Her worn, thin, lined, determined face would form for itself new and humorous creases, and her pale blue eyes would kindle with humour and with joy as her doctor burst into his hearty laugh. And the good Anna full of the coquetry of pleasing would bridle with her angular, thin, spinster body, straining her stories and herself to please.

These early days with jovial Dr. Shonjen were very happy days with the good Anna.

All of Anna's spare hours in these early days she spent with her friend, the widow Mrs. Lehntman. Mrs. Lehntman lived with her two children in a small house in the same part of the town as Dr. Shonjen. The older of these two children was a girl named Julia and was now about thirteen years of age. This Julia Lehntman was an unattractive girl enough, harsh featured, dull and stubborn as had been her heavy german father. Mrs. Lehntman did not trouble much with her, but gave her always all she wanted that she had, and let the girl do as she liked. This was not from indifference or dislike on the part of Mrs. Lehntman, it was just her usual way.

Her second child was a boy, two years younger than his sister, a bright, pleasant, cheery fellow, who too, did what he liked with his money and his time. All this was so with Mrs. Lehntman because she had so much in her head and in her house that clamoured for her concentration and her time.

This slackness and neglect in the running of the house, and the indifference in this mother for the training of her young was very hard for our good Anna to endure. Of course she did her best to scold, to save for Mrs. Lehntman, and to put things in their place the way they ought to be.

Even in the early days when Anna was first won by the glamour of Mrs. Lehntman's brilliancy and charm, she had been uneasy in Mrs. Lehntman's house with a need of putting things to rights. Now that the two children growing up were of more importance in the house, and now that long acquaintance had brushed the dazzle out of Anna's eyes, she began to struggle to make things go here as she thought was right.

She watched and scolded hard these days to make young Julia do the way she should. Not that Julia Lehntman was pleasant in the good Anna's sight, but it must never be that a young girl growing up should have no one to make her learn to do things right.

The boy was easier to scold, for scoldings never sank in very deep, and indeed he liked them very well for they brought with them new things to eat, and lively teasing, and good jokes.

Julia, the girl, grew very sullen with it all, and very often won her point, for after all Miss Annie was no relative of hers and had no business coming there and making trouble all the time. Appealing to the mother was no use. It was wonderful how Mrs. Lehntman could listen and not hear, could answer and yet not decide, could say and do what she was asked and yet leave things as they were before.

One day it got almost too bad for even Anna's friendship to bear out.

"Well, Julia, is your mamma out?" Anna asked, one Sunday summer afternoon, as she came into the Lehntman house.

Anna looked very well this day. She was always careful in her dress and sparing of new clothes. She made herself always fulfill her own ideal of how a girl should look when she took her Sundays out. Anna knew so well the kind of ugliness appropriate to each rank in life.

It was interesting to see how when she bought things for Miss Wadsmith and later for her cherished Miss Mathilda and always entirely from her own taste and often as cheaply as she bought things for her friends or for herself, that on the one hand she chose the things having the right air for a member of the upper class, and for the others always the things having the awkward ugliness that we call Dutch.[4] She knew the best thing in each kind, and she never in the course of her strong life compromised her sense of what was the right thing for a girl to wear.

On this bright summer Sunday afternoon she came to the Lehntmans', much dressed up in her new, brick red, silk waist trimmed with broad black beaded braid, a dark cloth skirt and a new stiff, shiny, black straw hat, trimmed with colored ribbons and a bird. She had on new gloves, and a feather boa[5] about her neck.

Her spare, thin, awkward body and her worn, pale yellow face though lit up now with the pleasant summer sun made a queer discord with the brightness of her clothes.

She came to the Lehntman house, where she had not been for several days, and opening the door that is always left unlatched in the houses of the lower middle class in the pleasant cities of the South, she found Julia in the family sitting-room alone.

"Well, Julia, where is your mamma?" Anna asked. "Ma is out but come in, Miss Annie, and look at our new brother." "What you talk so foolish for Julia," said Anna sitting down. "I ain't talkin' foolish, Miss Annie. Didn't you know mamma has just adopted a cute, nice little baby boy?" "You talk so crazy, Julia, you ought to know better than to say such things." Julia turned sullen. "All right Miss Annie, you don't need to believe what I say, but the little baby is in the kitchen and ma will tell you herself when she comes in."

It sounded most fantastic, but Julia had an air of truth and Mrs. Lehntman was capable of doing stranger things. Anna was disturbed. "What you mean Julia," she said. "I don't mean nothin' Miss Annie, you don't believe the baby is in there, well you can go and see it for yourself."

Anna went into the kitchen. A baby was there all right enough, and a lusty little boy he seemed. He was very tight asleep in a basket that stood in the corner by the open door.

---

[4] *Dutch:* Slang for poor taste, derived from the class-based ideas about what immigrants of German descent might choose to wear, given their poverty.

[5] *feather boa:* A long wrap that covered the neck and upper chest, made of feathers and intended to be wrapped and thrown casually across one shoulder.

"You mean your mamma is just letting him stay here a little while," Anna said to Julia who had followed her into the kitchen to see Miss Annie get real mad. "No that ain't it Miss Annie. The mother was that girl, Lily that came from Bishop's place out in the country, and she don't want no children, and ma liked the little boy so much, she said she'd keep him here and adopt him for her own child."

Anna, for once, was fairly dumb with astonishment and rage. The front door slammed.

"There's ma now," cried Julia in an uneasy triumph, for she was not quite certain in her mind which side of the question she was on, "There's ma now, and you can ask her for yourself if I ain't told you true."

Mrs. Lehntman came into the kitchen where they were. She was bland, impersonal and pleasant, as it was her wont to be. Still to-day, through this her usual manner that gave her such success in her practice as a midwife, there shone an uneasy consciousness of guilt, for like all who had to do with the good Anna, Mrs. Lehntman dreaded her firm character, her vigorous judgments and the bitter fervour of her tongue.

It had been plain to see in the six years these women were together, how Anna gradually had come to lead. Not really lead, of course, for Mrs. Lehntman never could be led, she was so very devious in her ways; but Anna had come to have direction whenever she could learn what Mrs. Lehntman meant to do before the deed was done. Now it was hard to tell which would win out. Mrs. Lehntman had her unhearing mind and her happy way of giving a pleasant well diffused attention, and then she had it on her side that, after all, this thing was already done.

Anna was, as usual, determined for the right. She was stiff and pale with her anger and her fear, and nervous, and all a tremble as was her usual way when a bitter fight was near.

Mrs. Lehntman was easy and pleasant as she came into the room. Anna was stiff and silent and very white.

"We haven't seen you for a long time, Anna," Mrs. Lehntman cordially began. "I was just gettin' worried thinking you was sick. My! but it's a hot day to-day. Come into the sittin'-room, Anna, and Julia will make us some ice tea."

Anna followed Mrs. Lehntman into the other room in a stiff silence, and when there she did not, as invited, take a chair.

As always with Anna when a thing had to come it came very short and sharp. She found it hard to breathe just now, and every word came with a jerk.

"Mrs. Lehntman, it ain't true what Julia said about your taking that Lily's boy to keep. I told Julia when she told me she was crazy to talk so."

Anna's real excitements stopped her breath, and made her words come sharp and with a jerk. Mrs. Lehntman's feelings spread her breath, and made her words come slow, but more pleasant and more easy even than before.

"Why Anna," she began, "don't you see Lily couldn't keep her boy for she is working at the Bishops' now, and he is such a cute dear little chap, and you know how fond I am of little fellers, and I thought it would be nice for Julia and for Willie to have a little brother. You know Julia always loves to play with babies, and I have to be away so much, and Willie he is running in the streets every minute all the time, and you see a baby would be sort of nice company for Julia, and you know you are always saying Anna, Julia should not be on the streets so much and the baby will be so good to keep her in."

Anna was every minute paler with indignation and with heat.

"Mrs. Lehntman, I don't see what business it is for you to take another baby for your own, when you can't do what's right by Julia and Willie you got here already. There's Julia, nobody tells her a thing when I ain't here, and who is going to tell her now how to do things for that baby? She ain't got no sense what's the right way to do with children, and you out all the time, and you ain't got no time for your own neither, and now you want to be takin' up with strangers. I know you was careless, Mrs. Lehntman, but I didn't think that you could do this so. No, Mrs. Lehntman, it ain't your duty to take up with no others, when you got two children of your own, that got to get along just any way they can, and you know you ain't got any too much money all the time, and you are all so careless here and spend it all the time, and Julia and Willie growin' big. It ain't right, Mrs. Lehntman, to do so."

This was as bad as it could be. Anna had never spoken her mind so to her friend before. Now it was too harsh for Mrs. Lehntman to allow herself to really hear. If she really took the meaning in these words she could never ask Anna to come into her house again, and she liked Anna very well, and was used to depend on her savings and her strength. And then too Mrs. Lehntman could not really take in harsh ideas. She was too well diffused to catch the feel of any sharp firm edge.

Now she managed to understand all this in a way that made it easy for her to say, "Why, Anna, I think you feel too bad about seeing what the children are doing every minute in the day. Julia and Willie are real good, and they play with all the nicest children in the square. If you had some, all your own, Anna, you'd see it don't do no harm to let them do a little as they like, and Julia likes this baby so, and sweet dear little boy, it would be so kind of bad to send him to a 'sylum[6] now, you know it would Anna, when you like children so yourself, and are so good to my Willie all the time. No indeed Anna, it's easy enough to say I should send this poor, cute little boy to a 'sylum when I could keep him here so nice, but you know Anna, you wouldn't like to do it yourself, now you really know you wouldn't, Anna, though you talk to me so hard. — My, it's hot to-day, what you doin' with that ice tea in there Julia, when Miss Annie is waiting all this time for her drink?"

Julia brought in the ice tea. She was so excited with the talk she had been hearing from the kitchen, that she slopped it on the plate out of the glasses a good deal. But she was safe, for Anna felt this trouble so deep down that she did not even see those awkward, bony hands, adorned to-day with a new ring, those stupid, foolish hands that always did things the wrong way.

"Here Miss Annie," Julia said, "Here, Miss Annie, is your glass of tea, I know you like it good and strong."

"No, Julia, I don't want no ice tea here. Your mamma ain't able to afford now using her money upon ice tea for her friends. It ain't right she should now any more. I go out now to see Mrs. Drehten. She does all she can, and she is sick now working so hard taking care of her own children. I go there now. Good by Mrs. Lehntman, I hope you don't get no bad luck doin' what it ain't right for you to do."

"My, Miss Annie is real mad now," Julia said, as the house shook, as the good Anna shut the outside door with a concentrated shattering slam.

It was some months now that Anna had been intimate with Mrs. Drehten.

Mrs. Drehten had had a tumor and had come to Dr. Shonjen to be treated. During the course of her visits there, she and Anna had learned to like each other very well. There was no fever in this friendship, it was just the interchange of two hard working, worrying women, the one large and motherly, with the pleasant, patient, soft, worn, tolerant face, that comes with a german husband to obey, and seven solid girls and boys to bear and rear, and the other was our good

---

[6] *'sylum:* Asylum — orphanage or mental institution, where poor people were marginally cared for by the state or federal government, or by private charities.

Anna with her spinster body, her firm jaw, her humorous, light, clean eyes and her lined, worn, thin, pale yellow face.

Mrs. Drehten lived a patient, homely, hard-working life. Her husband an honest, decent man enough, was a brewer, and somewhat given to over drinking, and so he was often surly and stingy and unpleasant. The family of seven children was made up of four stalwart, cheery, filial sons, and three hard working obedient simple daughters.

It was a family life the good Anna very much approved and also she was much liked by them all. With a german woman's feeling for the masterhood in men, she was docile to the surly father and rarely rubbed him the wrong way. To the large, worn, patient, sickly mother she was a sympathetic listener, wise in council and most efficient in her help. The young ones too, liked her very well. The sons teased her all the time and roared with boisterous pleasure when she gave them back sharp hits. The girls were all so good that her scoldings here were only in the shape of good advice, sweetened with new trimmings for their hats, and ribbons, and sometimes on their birthdays, bits of jewels.

It was here that Anna came for comfort after her grievous stroke at her friend the widow, Mrs. Lehntman. Not that Anna would tell Mrs. Drehten of this trouble. She could never lay bare the wound that came to her through this idealised affection. Her affair with Mrs. Lehntman was too sacred and too grievous ever to be told. But here in this large household, in busy movement and variety in strife, she could silence the uneasiness and pain of her own wound.

The Drehtens lived out in the country in one of the wooden, ugly houses that lie in groups outside of our large cities.

The father and the sons all had their work here making beer, and the mother and her girls scoured and sewed and cooked.

On Sundays they were all washed very clean, and smelling of kitchen soap. The sons, in their Sunday clothes, loafed around the house or in the village, and on special days went on picnics with their girls. The daughters in their awkward, colored finery went to church most of the day and then walking with their friends.

They always came together for their supper, where Anna always was most welcome, the jolly Sunday evening supper that german people love. Here Anna and the boys gave it to each other in sharp hits and hearty boisterous laughter, the girls made things for them to eat, and waited on them all, the mother loved all her children all the time, and the father joined in with his occasional unpleasant word that made a bitter feeling but which they had all learned to pass as if it were not said.

It was to the comfort of this house that Anna came that Sunday summer afternoon, after she had left Mrs. Lehntman and her careless ways.

The Drehten house was open all about. No one was there but Mrs. Drehten resting in her rocking chair, out in the pleasant, scented, summer air.

Anna had had a hot walk from the cars.[7]

She went into the kitchen for a cooling drink, and then came out and sat down on the steps near Mrs. Drehten.

Anna's anger had changed. A sadness had come to her. Now with the patient, friendly, gentle mother talk of Mrs. Drehten, this sadness changed to resignation and to rest.

As the evening came on the young ones dropped in one by one. Soon the merry Sunday evening supper was begun.

It had not been all comfort for our Anna, these months of knowing Mrs. Drehten. It had made trouble for her with the family of her half brother, the fat baker.

Her half brother, the fat baker, was a queer kind of a man. He was a huge, unwieldy creature, all puffed out all over, and no longer able to walk much, with his enormous body and the big, swollen, bursted veins in his great legs. He did not try to walk much now. He sat around his place, leaning on his great thick stick, and watching his workmen at their work.

On holidays, and sometimes of a Sunday, he went out in his bakery wagon. He went then to each customer he had and gave them each a large, sweet, raisined loaf of caky bread. At every house with many groans and gasps he would descend his heavy weight out of the wagon, his good featured, black haired, flat, good natured face shining with oily perspiration, with pride in labor and with generous kindness. Up each stoop he hobbled with the help of his big stick, and into the nearest chair in the kitchen or in the parlour, as the fashion of the house demanded, and there he sat and puffed, and then presented to the mistress or the cook the raisined german loaf his boy supplied him.

Anna had never been a customer of his. She had always lived in another part of the town, but he never left her out in these bakery progresses of his, and always with his own hand he gave her her festive loaf.

Anna liked her half brother well enough. She never knew him really well, for he rarely talked at all and least of all to women, but he

---

[7] *the cars:* Street car or trolley travel was the chief method of conveyance in turn of the century urban America.

seemed to her, honest, and good and kind, and he never tried to interfere in Anna's ways. And then Anna liked the loaves of raisined bread, for in the summer she and the second girl could live on them, and not be buying bread with the household money all the time.

But things were not so simple with our Anna, with the other members of her half brother's house.

Her half brother's family was made up of himself, his wife, and their two daughters.

Anna never liked her brother's wife.

The youngest of the two daughters was named after her aunt Anna.

Anna never liked her half brother's wife. This woman had been very good to Anna, never interfering in her ways, always glad to see her and to make her visits pleasant, but she had not found favour in our good Anna's sight.

Anna had too, no real affection for her nieces. She never scolded them or tried to guide them for their good. Anna never criticised or interfered in the running of her half brother's house.

Mrs. Federner was a good looking, prosperous woman, a little harsh and cold within her soul perhaps, but trying always to be pleasant, good and kind. Her daughters were well trained, quiet, obedient, well dressed girls, and yet our good Anna loved them not, nor their mother, nor any of their ways.

It was in this house that Anna had first met her friend, the widow, Mrs. Lehntman.

The Federners had never seemed to feel it wrong in Anna, her devotion to this friend and her care of her and of her children. Mrs. Lehntman and Anna and her feelings were all somehow too big for their attack. But Mrs. Federner had the mind and tongue that blacken things. Not really to blacken black, of course, but just to roughen and to rub on a little smut. She could somehow make even the face of the Almighty seem pimply and a little coarse, and so she always did this with her friends, though not with the intent to interfere.

This was really true with Mrs. Lehntman that Mrs. Federner did not mean to interfere, but Anna's friendship with the Drehtens was a very different matter.

Why should Mrs. Drehten, that poor common working wife of a man who worked for others in a brewery and who always drank too much, and was not like a thrifty, decent german man, why should that Mrs. Drehten and her ugly, awkward daughters be getting presents from her husband's sister all the time, and her husband always so good to Anna, and one of the girls having her name too, and those Drehtens all strangers to her and never going to come to any good? It was not right for Anna to do so.

Mrs. Federner knew better than to say such things straight out to her husband's fiery, stubborn sister, but she lost no chance to let Anna feel and see what they all thought.

It was easy to blacken all the Drehtens, their poverty, the husband's drinking, the four big sons carrying on and always lazy, the awkward, ugly daughters dressing up with Anna's help and trying to look so fine, and the poor, weak, hard-working sickly mother, so easy to degrade with large dosings of contemptuous pity.

Anna could not do much with these attacks for Mrs. Federner always ended with, "And you so good to them Anna all the time. I don't see how they could get along at all if you didn't help them all the time, but you are so good Anna, and got such a feeling heart, just like your brother, that you give anything away you got to anybody that will ask you for it, and that's shameless enough to take it when they ain't no relatives of yours. Poor Mrs. Drehten, she is a good woman. Poor thing it must be awful hard for her to have to take things from strangers all the time, and her husband spending it on drink. I was saying to Mrs. Lehntman, Anna, only yesterday, how I never was so sorry for any one as Mrs. Drehten, and how good it was for you to help them all the time."

All this meant a gold watch and chain to her god daughter for her birthday, the next month, and a new silk umbrella for the elder sister. Poor Anna, and she did not love them very much, these relatives of hers, and they were the only kin she had.

Mrs. Lehntman never joined in, in these attacks. Mrs. Lehntman was diffuse and careless in her ways, but she never worked such things for her own ends, and she was too sure of Anna to be jealous of her other friends.

All this time Anna was leading her happy life with Dr. Shonjen. She had every day her busy time. She cooked and saved and sewed and scrubbed and scolded. And every night she had her happy time, in seeing her Doctor like the fine things she bought so cheap and cooked so good for him to eat. And then he would listen and laugh so loud, as she told him stories of what had happened on that day.

The Doctor, too, liked it better all the time and several times in these five years he had of his own motion raised her wages.

Anna was content with what she had and grateful for all her doctor did for her.

So Anna's serving and her giving life went on, each with its varied pleasures and its pains.

The adopting of the little boy did not put an end to Anna's friendship for the widow Mrs. Lehntman. Neither the good Anna nor the careless Mrs. Lehntman would give each other up excepting for the gravest cause. Mrs. Lehntman was the only romance Anna ever knew. A certain magnetic brilliancy in person and in manner made Mrs. Lehntman a woman other women loved. Then, too, she was generous and good and honest, though she was so careless always in her ways. And then she trusted Anna and liked her better than any of her other friends, and Anna always felt this very much.

No, Anna could not give up Mrs. Lehntman, and soon she was busier than before making Julia do things right for little Johnny.

And now new schemes were working strong in Mrs. Lehntman's head, and Anna must listen to her plans and help her make them work.

Mrs. Lehntman always loved best in her work to deliver young girls who were in trouble. She would keep these in her house until they could go to their homes or to their work, and slowly pay her back the money for their care.

Anna had always helped her friend to do this thing, for like all the good women of the decent poor, she felt it hard that girls should not be helped, not girls that were really bad of course, these she condemned and hated in her heart and with her tongue, but honest, decent, good, hard working, foolish girls who were in trouble.

For such as these Anna always liked to give her money and her strength.

Now Mrs. Lehntman thought that it would pay to take a big house for herself to take in girls and to do everything in a big way.

Anna did not like this plan.

Anna was never daring in her ways. Save and you will have the money you have saved, was all that she could know.

Not that the good Anna had it so.

She saved and saved and always saved, and then here and there, to this friend and to that, to one in her trouble and to the other in her joy, in sickness, death, and weddings, or to make young people happy, it always went, the hard earned money she had saved.

Anna could not clearly see how Mrs. Lehntman could make a big house pay. In the small house where she had these girls, it did not pay, and in a big house there was so much more that she would spend.

Such things were hard for the good Anna to very clearly see. One day she came into the Lehntman house. "Anna," Mrs. Lehntman said, "you know that nice big house on the next corner that we saw to rent. I took it for a year just yesterday. I paid a little down you know so I could have it sure all right and now you fix it up just like you want. I let you do just what you like with it."

Anna knew that it was now too late. However, "But Mrs. Lehntman you said you would not take another house, you said so just last week. Oh, Mrs. Lehntman I didn't think that you would do this so!"

Anna knew so well it was too late.

"I know, Anna, but it was such a good house, just right you know and some one else was there to see, and you know you said it suited very well, and if I didn't take it the others said they would, and I wanted to ask you only there wasn't time, and really Anna, I don't need much help, it will go so well I know. I just need a little to begin and to fix up with and that's all Anna that I need, and I know it will go awful well. You wait Anna and you'll see, and I let you fix it up just like you want, and you will make it look so nice, you got such sense in all these things. It will be a good place. You see Anna if I ain't right in what I say."

Of course Anna gave the money for this thing though she could not believe that it was best. No, it was very bad. Mrs. Lehntman could never make it pay and it would cost so much to keep. But what could our poor Anna do? Remember Mrs. Lehntman was the only romance Anna knew.

Anna's strength in her control of what was done in Mrs. Lehntman's house, was not now what it had been before that Lily's little Johnny came. That thing had been for Anna a defeat. There had been no fighting to a finish but Mrs. Lehntman had very surely won.

Mrs. Lehntman needed Anna just as much as Anna needed Mrs. Lehntman, but Mrs. Lehntman was more ready to risk Anna's loss, and so the good Anna grew always weaker in her power to control.

In friendship, power always has its downward curve. One's strength to manage rises always higher until there comes a time one does not win, and though one may not really lose, still from the time that victory is not sure, one's power slowly ceases to be strong. It is only in a close tie such as marriage, that influence can mount and grow always stronger with the years and never meet with a decline. It can only happen so when there is no way to escape.

Friendship goes by favour. There is always danger of a break or of a stronger power coming in between. Influence can only be a steady march when one can surely never break away.

Anna wanted Mrs. Lehntman very much and Mrs. Lehntman needed Anna, but there were always other ways to do and if Anna had once given up she might do so again, so why should Mrs. Lehntman have real fear?

No, while the good Anna did not come to open fight she had been stronger. Now Mrs. Lehntman could always hold out longer. She knew too, that Anna had a feeling heart. Anna could never stop doing all she could for any one that really needed help. Poor Anna had no power to say no.

And then, too, Mrs. Lehntman was the only romance Anna ever knew. Romance is the ideal in one's life and it is very lonely living with it lost.

So the good Anna gave all her savings for this place, although she knew that this was not the right way for her friend to do.

For some time now they were all very busy fixing up the house. It swallowed all Anna's savings fixing up this house, for when Anna once began to make it nice, she could not leave it be until it was as good as for the purpose it should be.

Somehow it was Anna now that really took the interest in the house. Mrs. Lehntman, now the thing was done seemed very lifeless, without interest in the house, uneasy in her mind and restless in her ways, and more diffuse even than before in her attention. She was good and kind to all the people in her house, and let them do whatever they thought best.

Anna did not fail to see that Mrs. Lehntman had something on her mind that was all new. What was it that disturbed Mrs. Lehntman so? She kept on saying it was all in Anna's head. She had no trouble now at all. Everybody was so good and it was all so nice in the new house. But surely there was something here that was all wrong.

Anna heard a good deal of all this from her half brother's wife, the hard speaking Mrs. Federner.

Through the fog of dust and work and furnishing in the new house, and through the disturbed mind of Mrs. Lehntman, and with the dark hints of Mrs. Federner, there loomed up to Anna's sight a man, a new doctor that Mrs. Lehntman knew.

Anna had never met the man but she heard of him very often now. Not from her friend, the widow Mrs. Lehntman. Anna knew that Mrs. Lehntman made of him a mystery that Anna had not the strength just then to vigorously break down.

Mrs. Federner gave always dark suggestions and unpleasant hints. Even good Mrs. Drehten talked of it.

Mrs. Lehntman never spoke of the new doctor more than she could help. This was most mysterious and unpleasant and very hard for our good Anna to endure.

Anna's troubles came all of them at once.

Here in Mrs. Lehntman's house loomed up dismal and forbidding, a mysterious, perhaps an evil man. In Dr. Shonjen's house were beginning signs of interest in the doctor in a woman.

This, too, Mrs. Federner often told to the poor Anna. The doctor surely would be married soon, he liked so much now to go to Mr. Weingartner's house where there was a daughter who loved Doctor, everybody knew.

In these days the living room in her half brother's house was Anna's torture chamber. And worst of all there was so much reason for her half sister's words. The Doctor certainly did look like marriage and Mrs. Lehntman acted very queer.

Poor Anna. Dark were these days and much she had to suffer.

The Doctor's trouble came to a head the first. It was true Doctor was engaged and to be married soon. He told Anna so himself.

What was the good Anna now to do? Dr. Shonjen wanted her of course to stay. Anna was so sad with all these troubles. She knew here in the Doctor's house it would be bad when he was married, but she had not the strength now to be firm and go away. She said at last that she would try and stay.

Doctor got married now very soon. Anna made the house all beautiful and clean and she really hoped that she might stay. But this was not for long.

Mrs. Shonjen was a proud, unpleasant woman. She wanted constant service and attention and never even a thank you to a servant. Soon all Doctor's old people went away. Anna went to Doctor and explained. She told him what all the servants thought of his new wife. Anna bade him a sad farewell and went away.

Anna was now most uncertain what to do. She could go to Curden to her Miss Mary Wadsmith who always wrote how much she needed Anna, but Anna still dreaded Miss Jane's interfering ways. Then too, she could not yet go away from Bridgepoint and from Mrs. Lehntman, unpleasant as it always was now over there.

Through one of Doctor's friends Anna heard of Miss Mathilda. Anna was very doubtful about working for a Miss Mathilda. She did not think it would be good working for a woman any more. She had found it very good with Miss Mary but she did not think that many women would be so.

Most women were interfering in their ways.

Anna heard that Miss Mathilda was a great big woman, not so big perhaps as her Miss Mary, still she was big, and the good Anna liked them better so. She did not like them thin and small and active and always looking in and always prying.

Anna could not make up her mind what was the best thing now for her to do. She could sew and this way make a living, but she did not like such business very well.

Mrs. Lehntman urged the place with Miss Mathilda. She was sure Anna would find it better so. The good Anna did not know.

"Well Anna," Mrs. Lehntman said, "I tell you what we do. I go with you to that woman that tells fortunes, perhaps she tell us something that will show us what is the best way for you now to do."

It was very bad to go to a woman who tells fortunes. Anna was of strong South German Catholic religion and the german priests in the churches always said that it was very bad to do things so. But what else now could the good Anna do? She was so mixed and bothered in her mind, and troubled with this life that was all wrong, though she did try so hard to do the best she knew. "All right, Mrs. Lehntman," Anna said at last, "I think I go there now with you."

This woman who told fortunes was a medium. She had a house in the lower quarter of the town. Mrs. Lehntman and the good Anna went to her.

The medium opened the door for them herself. She was a loose made, dusty, dowdy woman with a persuading, conscious and embracing manner and very greasy hair.

The woman let them come into the house.

The street door opened straight into the parlor, as is the way in the small houses of the south. The parlor had a thick and flowered carpet on the floor. The room was full of dirty things all made by hand. Some hung upon the wall, some were on the seats and over backs of chairs and some on tables and on those what-nots that poor people love. And everywhere were little things that break. Many of these little things were broken and the place was stuffy and not clean.

No medium uses her parlor for her work. It is always in her eating room that she has her trances.

The eating room in all these houses is the living room in winter. It has a round table in the centre covered with a decorated woolen cloth, that has soaked in the grease of many dinners, for though it should be always taken off, it is easier to spread the cloth upon it than change it for the blanket deadener that one owns. The upholstered chairs are dark and worn, and dirty. The carpet has grown dingy with the food that's fallen from the table, the dirt that's scraped from off the shoes, and the dust that settles with the ages. The sombre greenish colored paper on the walls has been smoked a dismal dirty grey, and all pervading is the smell of soup made out of onions and fat chunks of meat.

The medium brought Mrs. Lehntman and our Anna into this eating room, after she had found out what it was they wanted. They all three sat around the table and then the medium went into her trance.

The medium first closed her eyes and then they opened very wide and lifeless. She took a number of deep breaths, choked several times and swallowed very hard. She waved her hand back every now and then, and she began to speak in a monotonous slow, even tone.

"I see — I see — don't crowd so on me, — I see — I see — too many forms — don't crowd so on me — I see — I see — you are thinking of something — you don't know whether you want to do it now. I see — I see — don't crowd so on me — I see — I see — you are not sure, — I see — I see — a house with trees around it, — it is dark — it is evening — I see — I see — you go in the house — I see — I see you come out — it will be all right — you go and do it — do what you are not certain about — it will come out all right — it is best and you should do it now."

She stopped, she made deep gulps, her eyes rolled back into her head, she swallowed hard and then she was her former dingy and bland self again.

"Did you get what you wanted that the spirit should tell you?" the woman asked. Mrs. Lehntman answered yes, it was just what her friend had wanted so bad to know. Anna was uneasy in this house with superstition, with fear of her good priest, and with disgust at all the dirt and grease, but she was most content for now she knew what it was best for her to do.

Anna paid the woman for her work and then they came away.

"There Anna didn't I tell you how it would all be? You see the spirit says so too. You must take the place with Miss Mathilda, that is what I told you was the best thing for you to do. We go out and see her where she lives to-night. Ain't you glad, Anna, that I took you to this place, so you know now what you will do?"

Mrs. Lehntman and Anna went that evening to see Miss Mathilda. Miss Mathilda was staying with a friend who lived in a house that did have trees about. Miss Mathilda was not there herself to talk with Anna.

If it had not been that it was evening, and so dark, and that this house had trees all round about, and that Anna found herself going in and coming out just as the woman that day said that she would do, had it not all been just as the medium said, the good Anna would never have taken the place with Miss Mathilda.

Anna did not see Miss Mathilda and she did not like the friend who acted in her place.

This friend was a dark, sweet, gentle little mother woman, very easy to be pleased in her own work and very good to servants, but she felt that acting for her young friend, the careless Miss Mathilda, she must be very careful to examine well and see that all was right and that Anna would surely do the best she knew. She asked Anna all about her ways and her intentions and how much she would spend, and how often she went out and whether she could wash and cook and sew.

The good Anna set her teeth fast to endure and would hardly answer anything at all. Mrs. Lehntman made it all go fairly well.

The good Anna was all worked up with her resentment, and Miss Mathilda's friend did not think that she would do.

However, Miss Mathilda was willing to begin and as for Anna, she knew that the medium said it must be so. Mrs. Lehntman, too, was sure, and said she knew that this was the best thing for Anna now to do. So Anna sent word at last to Miss Mathilda, that if she wanted her, she would try if it would do.

So Anna began a new life taking care of Miss Mathilda.

Anna fixed up the little red brick house where Miss Mathilda was going to live and made it very pleasant, clean and nice. She brought over her dog, Baby, and her parrot. She hired Lizzie for a second girl to be with her and soon they were all content. All except the parrot, for Miss Mathilda did not like its scream. Baby was all right but not the parrot. But then Anna never really loved the parrot, and so she gave it to the Drehten girls to keep.

Before Anna could really rest content with Miss Mathilda, she had to tell her good german priest what it was that she had done, and how very bad it was that she had been and how she would never do so again.

Anna really did believe with all her might. It was her fortune never to live with people who had any faith, but then that never worried Anna. She prayed for them always as she should, and she was very sure that they were good. The doctor loved to tease her with his doubts and Miss Mathilda liked to do so too, but with the tolerant spirit of her church, Anna never thought that such things were bad for them to do.

Anna found it hard to always know just why it was that things went wrong. Sometimes her glasses broke and then she knew that she had not done her duty by the church, just in the way that she should do.

Sometimes she was so hard at work that she would not go to mass. Something always happened then. Anna's temper grew irritable and

The Baltimore row house where Gertrude and Leo Stein lived during their years as students at Johns Hopkins (Gertrude in the medical program; Leo finishing his undergraduate degree — after years at University of California, Berkeley, and Harvard — and beginning a postbaccalaureate program). While here, they were cared for by Lena Lebender, a housekeeper of German origin. In "The Good Anna," Miss Mathilda is usually read as Gertrude. Wagner-Martin photo.

her ways uncertain and distraught. Everybody suffered and then her glasses broke. That was always very bad because they cost so much to fix. Still in a way it always ended Anna's troubles, because she knew then that all this was because she had been bad. As long as she could scold it might be just the bad ways of all the thoughtless careless world, but when her glasses broke that made it clear. That meant that it was she herself who had been bad.

No, it was no use for Anna not to do the way she should, for things always then went wrong and finally cost money to make whole, and this was the hardest thing for the good Anna to endure.

Anna almost always did her duty. She made confession and her mission whenever it was right. Of course she did not tell the father when she deceived people for their good, or when she wanted them to give something for a little less.

When Anna told such histories to her doctor and later to her cherished Miss Mathilda, her eyes were always full of humor and enjoyment as she explained that she had said it so, and now she would not have to tell the father for she had not really made a sin.

But going to a fortune teller Anna knew was really bad. That had to be told to the father just as it was and penance had then to be done.

Anna did this and now her new life was well begun, making Miss Mathilda and the rest do just the way they should.

Yes, taking care of Miss Mathilda were the happiest days of all the good Anna's strong hard working life.

With Miss Mathilda Anna did it all. The clothes, the house, the hats, what she should wear and when and what was always best for her to do. There was nothing Miss Mathilda would not let Anna manage, and only be too glad if she would do.

Anna scolded and cooked and sewed and saved so well, that Miss Mathilda had so much to spend, that it kept Anna still busier scolding all the time about the things she bought, that made so much work for Anna and the other girl to do. But for all the scolding, Anna was proud almost to bursting of her cherished Miss Mathilda with all her knowledge and her great possessions, and the good Anna was always telling of it all to everybody that she knew.

Yes these were the happiest days of all her life with Anna, even though with her friends there were great sorrows. But these sorrows did not hurt the good Anna now, as they had done in the years that went before.

Miss Mathilda was not a romance in the good Anna's life, but Anna gave her so much strong affection that it almost filled her life as full.

It was well for the good Anna that her life with Miss Mathilda was so happy, for now in these days, Mrs. Lehntman went altogether bad. The doctor she had learned to know, was too certainly an evil as well as a mysterious man, and he had power over the widow and midwife, Mrs. Lehntman.

Anna never saw Mrs. Lehntman at all now any more.

Mrs. Lehntman had borrowed some more money and had given Anna a note then for it all, and after that Anna never saw her any more. Anna now stopped altogether going to the Lehntmans'. Julia, the tall, gawky, good, blonde, stupid daughter, came often to see Anna, but she could tell little of her mother.

It certainly did look very much as if Mrs. Lehntman had now gone altogether bad. This was a great grief to the good Anna, but not so great a grief as it would have been had not Miss Mathilda meant so much to her now.

Mrs. Lehntman went from bad to worse. The doctor, the mysterious and evil man, got into trouble doing things that were not right to do.

Mrs. Lehntman was mixed up in this affair.

It was just as bad as it could be, but they managed, both the doctor and Mrs. Lehntman, finally to come out safe.

Everybody was so sorry about Mrs. Lehntman. She had been really a good woman before she met this doctor, and even now she certainly had not been really bad.

For several years now Anna never even saw her friend.

But Anna always found new people to befriend, people who, in the kindly fashion of the poor, used up her savings and then gave promises in place of payments. Anna never really thought that these people would be good, but when they did not do the way they should, and when they did not pay her back the money she had loaned, and never seemed the better for her care, then Anna would grow bitter with the world.

No, none of them had any sense of what was the right way for them to do. So Anna would repeat in her despair.

The poor are generous with their things. They give always what they have, but with them to give or to receive brings with it no feeling that they owe the giver for the gift.

Even a thrifty german Anna was ready to give all that she had saved, and so not be sure that she would have enough to take care of herself if she fell sick, or for old age, when she could not work. Save and you will have the money you have saved was true only for the day of saving, even for a thrifty german Anna. There was no certain way to have it for old age, for the taking care of what is saved can never be relied on, for it must always be in strangers' hands in a bank or in investments by a friend.

And so when any day one might need life and help from others of the working poor, there was no way a woman who had a little saved could say them no.

So the good Anna gave her all to friends and strangers, to children, dogs and cats, to anything that asked or seemed to need her care.

It was in this way that Anna came to help the barber and his wife who lived around the corner, and who somehow could never make ends meet. They worked hard, were thrifty, had no vices, but the barber was one of them who never can make money. Whoever owed him money did not pay. Whenever he had a chance at a good job he fell sick and could not take it. It was never his own fault that he had trouble, but he never seemed to make things come out right.

His wife was a blonde, thin, pale, german little woman, who bore her children very hard, and worked too soon, and then till she was sick. She too, always had things that went wrong.

They both needed constant help and patience, and the good Anna gave both to them all the time.

Another woman who needed help from the good Anna, was one who was in trouble from being good to others.

This woman's husband's brother, who was very good, worked in a shop where there was a Bohemian,[8] who was getting sick with a consumption. This man got so much worse he could not do his work, but he was not so sick that he could stay in a hospital. So this woman had him living there with her. He was not a nice man, nor was he thankful for all the woman did for him. He was cross to her two children and made a great mess always in her house. The doctor said he must have many things to eat, and the woman and the brother of the husband got them for him.

There was no friendship, no affection, no liking even for the man this woman cared for, no claim of common country or of kin, but in the kindly fashion of the poor this woman gave her all and made her house a nasty place, and for a man who was not even grateful for the gift.

---

[8] *Bohemian:* Literally, a man from Bohemia, a less-prestigious European point of origin than even Germany; its meaning was often derogatory. Country of origin was often used to characterize people in this highly nationalistic period.

Then, of course, the woman herself got into trouble. Her husband's brother was now married. Her husband lost his job. She did not have the money for the rent. It was the good Anna's savings that were handy.

So it went on. Sometimes a little girl, sometimes a big one was in trouble and Anna heard of them and helped them to find places.

Stray dogs and cats Anna always kept until she found them homes. She was always careful to learn whether these people would be good to animals.

Out of the whole collection of stray creatures, it was the young Peter and the jolly little Rags, Anna could not find it in her heart to part with. These became part of the household of the good Anna's Miss Mathilda.

Peter was a very useless creature, a foolish, silly, cherished, coward male. It was wild to see him rush up and down in the back yard, barking and bouncing at the wall, when there was some dog out beyond, but when the very littlest one there was got inside of the fence and only looked at Peter, Peter would retire to his Anna and blot himself out between her skirts.

When Peter was left downstairs alone, he howled. "I am all alone," he wailed, and then the good Anna would have to come and fetch him up. Once when Anna stayed a few nights in a house not far away, she had to carry Peter all the way, for Peter was afraid when he found himself on the street outside his house. Peter was a good sized creature and he sat there and he howled, and the good Anna carried him all the way in her own arms. He was a coward was this Peter, but he had kindly, gentle eyes and a pretty collie head, and his fur was very thick and white and nice when he was washed. And then Peter never strayed away, and he looked out of his nice eyes and he liked it when you rubbed him down, and he forgot you when you went away, and he barked whenever there was any noise.

When he was a little pup he had one night been put into the yard and that was all of his origin she knew. The good Anna loved him well and spoiled him as a good german mother always does her son.

Little Rags was very different in his nature. He was a lively creature made out of ends of things, all fluffy and dust color, and he was always bounding up into the air and darting all about over and then under silly Peter and often straight into solemn fat, blind, sleepy Baby, and then in a wild rush after some stray cat.

Rags was a pleasant, jolly little fellow. The good Anna liked him very well, but never with her strength as she loved her good looking coward, foolish young man, Peter.

Baby was the dog of her past life and she held Anna with old ties of past affection. Peter was the spoiled, good looking young man, of her middle age, and Rags was always something of a toy. She liked him but he never struck in very deep. Rags had strayed in somehow one day and then when no home for him was quickly found, he had just stayed right there.

It was a very happy family there all together in the kitchen, the good Anna and Sally and old Baby and young Peter and the jolly little Rags.

The parrot had passed out of Anna's life. She had really never loved the parrot and now she hardly thought to ask for him, even when she visited the Drehtens.

Mrs. Drehten was the friend Anna always went to, for her Sundays. She did not get advice from Mrs. Drehten as she used to from the widow, Mrs. Lehntman, for Mrs. Drehten was a mild, worn, unaggressive nature that never cared to influence or to lead. But they could mourn together for the world these two worn, working german women, for its sadness and its wicked ways of doing. Mrs. Drehten knew so well what one could suffer.

Things did not go well in these days with the Drehtens. The children were all good, but the father with his temper and his spending kept everything from being what it should.

Poor Mrs. Drehten still had trouble with her tumor. She could hardly do any work now any more. Mrs. Drehten was a large, worn, patient german woman, with a soft face, lined, yellow brown in color and the look that comes from a german husband to obey, and many solid girls and boys to bear and rear, and from being always on one's feet and never having any troubles cured.

Mrs. Drehten was always getting worse, and now the doctor thought it would be best to take the tumor out.

It was no longer Dr. Shonjen who treated Mrs. Drehten. They all went now to a good old german doctor they all knew.

"You see, Miss Mathilda," Anna said, "All the old german patients don't go no more now to Doctor. I stayed with him just so long as I could stand it, but now he is moved away up town too far for poor people, and his wife, she holds her head up so and always is spending so much money just for show, and so he can't take right care of us poor people any more. Poor man, he has got always to be thinking about making money now. I am awful sorry about Doctor, Miss Mathilda, but he neglected Mrs. Drehten shameful when she had her trouble, so now I never see him any more. Doctor Herman is a good, plain, german doctor and he would never do things so, and Miss

Mathilda, Mrs. Drehten is coming in to-morrow to see you before she goes to the hospital for her operation. She could not go comfortable till she had seen you first to see what you would say."

All Anna's friends reverenced the good Anna's cherished Miss Mathilda. How could they not do so and still remain friends with the good Anna? Miss Mathilda rarely really saw them but they were always sending flowers and words of admiration through her Anna. Every now and then Anna would bring one of them to Miss Mathilda for advice.

It is wonderful how poor people love to take advice from people who are friendly and above them, from people who read in books and who are good.

Miss Mathilda saw Mrs. Drehten and told her she was glad that she was going to the hospital for operation for that surely would be best, and so good Mrs. Drehten's mind was set at rest.

Mrs. Drehten's tumor came out very well. Mrs. Drehten was afterwards never really well, but she could do her work a little better, and be on her feet and yet not get so tired.

And so Anna's life went on, taking care of Miss Mathilda and all her clothes and goods, and being good to every one that asked or seemed to need her help.

Now, slowly, Anna began to make it up with Mrs. Lehntman. They could never be as they had been before. Mrs. Lehntman could never be again the romance in the good Anna's life, but they could be friends again, and Anna could help all the Lehntmans in their need. This slowly came about.

Mrs. Lehntman had now left the evil and mysterious man who had been the cause of all her trouble. She had given up, too, the new big house that she had taken. Since her trouble her practice had been very quiet. Still she managed to do fairly well. She began to talk of paying the good Anna. This, however, had not gotten very far.

Anna saw Mrs. Lehntman a good deal now. Mrs. Lehntman's crisp, black, curly hair had gotten streaked with gray. Her dark, full, good looking face had lost its firm outline, gone flabby and a little worn. She had grown stouter and her clothes did not look very nice. She was as bland as ever in her ways, and as diffuse as always in her attention, but through it all there was uneasiness and fear and uncertainty lest some danger might be near.

She never said a word of her past life to the good Anna, but it was very plain to see that her experience had not left her easy, nor yet altogether free.

It had been hard for this good woman, for Mrs. Lehntman was really a good woman, it had been a very hard thing for this german woman to do what everybody knew and thought was wrong. Mrs. Lehntman was strong and she had courage, but it had been very hard to bear. Even the good Anna did not speak to her with freedom. There always remained a mystery and a depression in Mrs. Lehntman's affair.

And now the blonde, foolish, awkward daughter, Julia was in trouble. During the years the mother gave her no attention, Julia kept company with a young fellow who was a clerk somewhere in a store down in the city. He was a decent, dull young fellow, who did not make much money and could never save it for he had an old mother he supported. He and Julia had been keeping company for several years and now it was needful that they should be married. But then how could they marry? He did not make enough to start them and to keep on supporting his old mother too. Julia was not used to working much and she said, and she was stubborn, that she would not live with Charley's dirty, cross, old mother. Mrs. Lehntman had no money. She was just beginning to get on her feet. It was of course, the good Anna's savings that were handy.

However it paid Anna to bring about this marriage, paid her in scoldings and in managing the dull, long, awkward Julia, and her good, patient, stupid Charley. Anna loved to buy things cheap, and fix up a new place.

Julia and Charley were soon married and things went pretty well with them. Anna did not approve their slack, expensive ways of doing.

"No Miss Mathilda," she would say, "The young people nowadays have no sense for saving and putting money by so they will have something to use when they need it. There's Julia and her Charley. I went in there the other day, Miss Mathilda, and they had a new table with a marble top and on it they had a grand new plush album. 'Where you get that album?' I asked Julia. 'Oh, Charley he gave it to me for my birthday,' she said, and I asked her if it was paid for and she said not yet but it would be soon. Now I ask you what business have they Miss Mathilda, when they ain't paid for anything they got already, what business have they to be buying new things for her birthday. Julia she don't do no work, she just sits around and thinks how she can spend the money, and Charley he never puts one cent by. I never see anything like the people nowadays Miss Mathilda, they don't seem to have any sense of being careful about money. Julia and Charley when they have any children they won't have nothing to bring them up with right. I said that to Julia, Miss Mathilda, when she showed me those silly things that Charley bought her, and she just said in her silly, giggling way, perhaps they won't have any children. I told her she ought to be ashamed of talking so, but I don't know, Miss Mathilda, the young people nowadays have no sense at all of what's the right way for them to do, and perhaps its better if they don't have any children, and then Miss Mathilda you know there is Mrs. Lehntman. You know she regular adopted little Johnny just so she could pay out some more money just as if she didn't have trouble enough taking care of her own children. No Miss Mathilda, I never see how people can do things so. People don't seem to have no sense of right or wrong or anything these days Miss Mathilda, they are just careless and thinking always of themselves and how they can always have a happy time. No, Miss Mathilda I don't see how people can go on and do things so."

The good Anna could not understand the careless and bad ways of all the world and always she grew bitter with it all. No, not one of them had any sense of what was the right way for them to do.

Anna's past life was now drawing to an end. Her old blind dog, Baby, was sick and like to die. Baby had been the first gift from her friend the widow, Mrs. Lehntman in the old days when Anna had been with Miss Mary Wadsmith, and when these two women had first come together.

Through all the years of change, Baby had stayed with the good Anna, growing old and fat and blind and lazy. Baby had been active and a ratter when she was young, but that was so long ago it was forgotten, and for many years now Baby had wanted only her warm basket and her dinner.

Anna in her active life found need of others, of Peter and the funny little Rags, but always Baby was the eldest and held her with the ties of old affection. Anna was harsh when the young ones tried to keep poor Baby out and use her basket. Baby had been blind now for some years as dogs get, when they are no longer active. She got weak and fat and breathless and she could not even stand long any more. Anna had always to see that she got her dinner and that the young active ones did not deprive her.

Baby did not die with a real sickness. She just got older and more blind and coughed and then more quiet, and then slowly one bright summer's day she died.

There is nothing more dreary than old age in animals. Somehow it is all wrong that they should have grey hair and withered skin, and blind old eyes, and decayed and useless teeth. An old man or an old woman almost always has some tie that seems to bind them to the younger, realer life. They have children or the remembrance of old duties, but a dog that's old and so cut off from all its world of struggle, is like a dreary, deathless Struldbrug,[9] the dreary dragger on of death through life.

And so one day old Baby died. It was dreary, more than sad, for the good Anna. She did not want the poor old beast to linger with its weary age, and blind old eyes and dismal shaking cough, but this death left Anna very empty. She had the foolish young man Peter, and the jolly little Rags for comfort, but Baby had been the only one that could remember.

The good Anna wanted a real graveyard for her Baby, but this could not be in a Christian country, and so Anna all alone took her old friend done up in decent wrappings and put her into the ground in some quiet place that Anna knew of.

The good Anna did not weep for poor old Baby. Nay, she had not time even to feel lonely, for with the good Anna it was sorrow upon sorrow. She was now no longer to keep house for Miss Mathilda.

When Anna had first come to Miss Mathilda she had known that it might only be for a few years, for Miss Mathilda was given to much wandering and often changed her home, and found new places where she went to live. The good Anna did not then think much about this, for when she first went to Miss Mathilda she had not thought that she would like it and so she had not worried about staying. Then in those happy years that they had been together, Anna had made herself forget it. This last year when she knew that it was coming she had tried hard to think it would not happen.

"We won't talk about it now Miss Mathilda, perhaps we all be dead by then," she would say when Miss Mathilda tried to talk it over. Or, "If we live till then Miss Mathilda, perhaps you will be staying on right here."

No, the good Anna could not talk as if this thing were real, it was too weary to be once more left with strangers.

Both the good Anna and her cherished Miss Mathilda tried hard to think that this would not really happen. Anna made missions and all kinds of things to keep her Miss Mathilda and Miss Mathilda thought out all the ways to see if the good Anna could not go with her, but neither the missions nor the plans had much success. Miss Mathilda

[9] *Struldbrug:* One of the groups in Jonathan Swift's *Gulliver's Travels* characterized by sluggishness, awkwardly large size, and longevity.

would go, and she was going far away to a new country where Anna could not live, for she would be too lonesome.

There was nothing that these two could do but part. Perhaps we all be dead by then, the good Anna would repeat, but even that did not really happen. If we all live till then Miss Mathilda, came out truer. They all did live till then, all except poor old blind Baby, and they simply had to part.

Poor Anna and poor Miss Mathilda. They could not look at each other that last day. Anna could not keep herself busy working. She just went in and out and sometimes scolded.

Anna could not make up her mind what she should do now for her future. She said that she would for a while keep this little red brick house that they had lived in. Perhaps she might just take in a few boarders. She did not know, she would write about it later and tell it all to Miss Mathilda.

The dreary day dragged out and then all was ready and Miss Mathilda left to take her train. Anna stood strained and pale and dry eyed on the white stone steps of the little red brick house that they had lived in. The last thing Miss Mathilda heard was the good Anna bidding foolish Peter say good bye and be sure to remember Miss Mathilda.

Part III
The Death of the Good Anna

Every one who had known of Miss Mathilda wanted the good Anna now to take a place with them, for they all knew how well Anna could take care of people and all their clothes and goods. Anna too could always go to Curden to Miss Mary Wadsmith, but none of all these ways seemed very good to Anna.

It was not now any longer that she wanted to stay near Mrs. Lehntman. There was no one now that made anything important, but Anna was certain that she did not want to take a place where she would be under some new people. No one could ever be for Anna as had been her cherished Miss Mathilda. No one could ever again so freely let her do it all. It would be better Anna thought in her strong strained weary body, it would be better just to keep on there in the little red brick house that was all furnished, and make a living taking in some boarders. Miss Mathilda had let her have the things, so it would not cost any money to begin. She could perhaps manage to live on so. She could do all the work and do everything as she thought best, and she was too weary with the changes to do more than she just had to, to keep living. So she stayed on in the house where they had lived, and she found some men, she would not take in women, who took her rooms and who were her boarders.

Things soon with Anna began to be less dreary. She was very popular with her few boarders. They loved her scolding and the good things she made for them to eat. They made good jokes and laughed loud and always did whatever Anna wanted, and soon the good Anna got so that she liked it very well. Not that she did not always long for Miss Mathilda. She hoped and waited and was very certain that sometime, in one year or in another Miss Mathilda would come back, and then of course would want her, and then she could take all good care of her again.

Anna kept all Miss Mathilda's things in the best order. The boarders were well scolded if they ever made a scratch on Miss Mathilda's table.

Some of the boarders were hearty good south german fellows and Anna always made them go to mass. One boarder was a lusty german student who was studying in Bridgepoint to be a doctor. He was Anna's special favourite and she scolded him as she used to her old doctor so that he always would be good. Then, too, this cheery fellow always sang when he was washing, and that was what Miss Mathilda always used to do. Anna's heart grew warm again with this young fellow who seemed to bring back to her everything she needed.

And so Anna's life in these days was not all unhappy. She worked and scolded, she had her stray dogs and cats and people, who all asked and seemed to need her care, and she had hearty german fellows who loved her scoldings and ate so much of the good things that she knew so well the way to make.

No, the good Anna's life in these days was not all unhappy. She did not see her old friends much, she was too busy, but once in a great while she took a Sunday afternoon and went to see good Mrs. Drehten.

The only trouble was that Anna hardly made a living. She charged so little for her board and gave her people good things to eat, that she could only just make both ends meet. The good german priest to whom she always told her troubles tried to make her have the boarders pay a little higher, and Miss Mathilda always in her letters urged her to this thing, but the good Anna somehow could not do it. Her boarders were nice men but she knew they did not have much money, and then she could not raise on those who had been with her and she could not ask the new ones to pay higher, when those who were already there were paying just what they had paid before. So Anna let it go just as she had begun it. She worked and worked all day and thought all night how she could save, and with all the work she just managed to keep living. She could not make enough to lay any money by.

Anna got so little money that she had all the work to do herself. She could not pay even the little Sally enough to keep her with her.

Not having little Sally nor having any one else working with her, made it very hard for Anna ever to go out, for she never thought that it was right to leave a house all empty. Once in a great while of a Sunday, Sally who was now working in a factory would come and stay in the house for the good Anna, who would then go out and spend the afternoon with Mrs. Drehten.

No, Anna did not see her old friends much any more. She went sometimes to see her half brother and his wife and her nieces, and they always came to her on her birthdays to give presents, and her half brother never left her out of his festive raisined bread giving progresses. But these relatives of hers had never meant very much to the good Anna. Anna always did her duty by them all, and she liked her half brother very well and the loaves of raisined bread that he supplied her were most welcome now, and Anna always gave her god daughter and her sister handsome presents, but no one in this family had ever made a way inside to Anna's feelings.

Mrs. Lehntman she saw very rarely. It is hard to build up new on an old friendship when in that friendship there has been bitter disillusion. They did their best, both these women, to be friends, but they were never able to again touch one another nearly. There were too many things between them that they could not speak of, things that had never been explained nor yet forgiven. The good Anna still did her best for foolish Julia and still every now and then saw Mrs. Lehntman, but this family had now lost all its real hold on Anna.

Mrs. Drehten was now the best friend that Anna knew. Here there was never any more than the mingling of their sorrows. They talked over all the time the best way for Mrs. Drehten now to do; poor Mrs. Drehten who with her chief trouble, her bad husband, had really now no way that she could do. She just had to work and to be patient and to love her children and be very quiet. She always had a soothing mother influence on the good Anna who with her irritable, strained, worn-out body would come and sit by Mrs. Drehten and talk all her troubles over.

Of all the friends that the good Anna had had in these twenty years in Bridgepoint, the good father and patient Mrs. Drehten were the only ones that were now near to Anna and with whom she could talk her troubles over.

Anna worked, and thought, and saved, and scolded, and took care of all the boarders, and of Peter and of Rags, and all the others. There was never any end to Anna's effort and she grew always more tired, more pale yellow, and in her face more thin and worn and worried. Sometimes she went farther in not being well, and then she went to see Dr. Herman who had operated on good Mrs. Drehten.

The things that Anna really needed were to rest sometimes and eat more so that she could get stronger, but these were the last things that Anna could bring herself to do. Anna could never take a rest. She must work hard through the summer as well as through the winter, else she could never make both ends meet. The doctor gave her medicines to make her stronger but these did not seem to do much good.

Anna grew always more tired, her headaches came oftener and harder, and she was now almost always feeling very sick. She could not sleep much in the night. The dogs with their noises disturbed her and everything in her body seemed to pain her.

The doctor and the good father tried often to make her give herself more care. Mrs. Drehten told her that she surely would not get well unless for a little while she would stop working. Anna would then promise to take care, to rest in bed a little longer and to eat more so that she would get stronger, but really how could Anna eat when she always did the cooking and was so tired of it all, before it was half ready for the table?

Anna's only friendship now was with good Mrs. Drehten who was too gentle and too patient to make a stubborn faithful german Anna ever do the way she should, in the things that were for her own good.

Anna grew worse all through this second winter. When the summer came the doctor said that she simply could not live on so. He said she must go to his hospital and there he would operate upon her. She would then be well and strong and able to work hard all next winter.

Anna for some time would not listen. She could not do this so, for she had her house all furnished and she simply could not let it go. At last a woman came and said she would take care of Anna's boarders and then Anna said that she was prepared to go.

Anna went to the hospital for her operation. Mrs. Drehten was herself not well but she came into the city, so that some friend would be with the good Anna. Together, then, they went to this place where the doctor had done so well by Mrs. Drehten.

In a few days they had Anna ready. Then they did the operation, and then the good Anna with her strong, strained, worn-out body died.

Mrs. Drehten sent word of her death to Miss Mathilda.

"Dear Miss Mathilda," wrote Mrs. Drehten, "Miss Annie died in the hospital yesterday after a hard operation. She was talking about you and Doctor and Miss Mary Wadsmith all the time. She said she hoped you would take Peter and the little Rags to keep when you came back to America to live. I will keep them for you here Miss Mathilda. Miss Annie died easy, Miss Mathilda, and sent you her love."

FINIS

MELANCTHA

Each One as She May

Rose Johnson made it very hard to bring her baby to birth.

Melanctha Herbert who was Rose Johnson's friend, did everything that any woman could. She tended Rose, and she was patient, submissive, soothing, and untiring, while the sullen, childish, cowardly, black Rosie grumbled and fussed and howled and made herself to be an abomination and like a simple beast.

The child though it was healthy after it was born, did not live long. Rose Johnson was careless and negligent and selfish, and when Melanctha had to leave for a few days, the baby died. Rose Johnson had liked the baby well enough and perhaps she just forgot it for awhile, anyway the child was dead and Rose and Sam her husband were very sorry but then these things came so often in the negro world in Bridgepoint, that they neither of them thought about it very long.

Rose Johnson and Melanctha Herbert had been friends now for some years. Rose had lately married Sam Johnson a decent honest kindly fellow, a deck hand on a coasting steamer.

Melanctha Herbert had not yet been really married.

Rose Johnson was a real black, tall, well built, sullen, stupid, childlike, good looking negress. She laughed when she was happy and grumbled and was sullen with everything that troubled.

Rose Johnson was a real black negress but she had been brought up quite like their own child by white folks.

Rose laughed when she was happy but she had not the wide, abandoned laughter[1] that makes the warm broad glow of negro sunshine. Rose was never joyous with the earth-born, boundless joy of negroes. Hers was just ordinary, any sort of woman laughter.

Rose Johnson was careless and was lazy, but she had been brought up by white folks and she needed decent comfort. Her white training had only made for habits, not for nature. Rose had the simple, promiscuous unmorality of the black people.

Rose Johnson and Melanctha Herbert like many of the twos with women were a curious pair to be such friends.

---

[1] *wide, abandoned laughter:* In this section, Stein uses a number of racial stereotypes that have little credibility, and to a modern reader, are embarrassing. See also her use of the sexual stereotype that follows, the phrase "simple, promiscuous unmorality."

This black woman suggests the kind of African American woman Melanctha, as Stein described her, might have been. Detail from an untitled photograph, St. Paul, Minnesota, circa 1910. Reproduced with permission of CORBIS/ Minnesota Historical Society.

Melanctha Herbert was a graceful, pale yellow, intelligent, attractive negress. She had not been raised like Rose by white folks but then she had been half made[2] with real white blood. She and Rose Johnson were both of the better sort of negroes, there, in Bridgepoint.

"No, I ain't no common nigger," said Rose Johnson, "for I was raised by white folks, and Melanctha she is so bright and learned so much in school, she ain't no common nigger either, though she ain't got no husband to be married to like I am to Sam Johnson."

Why did the subtle, intelligent, attractive, half white girl Melanctha Herbert love and do for and demean herself in service to this coarse, decent, sullen, ordinary, black childish Rose, and why was this unmoral, promiscuous, shiftless Rose married, and that's not so common either, to a good man of the negroes, while Melanctha with her white blood and attraction and her desire for a right position had not yet been really married.

Sometimes the thought of how all her world was made, filled the complex, desiring Melanctha with despair. She wondered, often, how she could go on living when she was so blue.[3]

Melanctha told Rose one day how a woman whom she knew had killed herself because she was so blue. Melanctha said, sometimes, she thought this was the best thing for her herself to do.

Rose Johnson did not see it the least bit that way.

"I don't see Melanctha why you should talk like you would kill yourself just because you're blue. I'd never kill myself Melanctha just 'cause I was blue. I'd maybe kill somebody else Melanctha 'cause I was blue, but I'd never kill myself. If I ever killed myself Melanctha it'd be by accident, and if I ever killed myself by accident Melanctha, I'd be awful sorry."

Rose Johnson and Melanctha Herbert had first met, one night, at church. Rose Johnson did not care much for religion. She had not enough emotion to be really roused by a revival. Melanctha Herbert had not come yet to know how to use religion. She was still too complex with desire. However, the two of them in negro fashion went very often to the negro church, along with all their friends, and they slowly came to know each other very well.

---

[2] *had been half made:* Melanctha's mother was a white woman. As above, regretfully, Stein interprets Melanctha's mother's "yellow" skin tone as signifying white parentage; she equates "whiteness" with positive characteristics.

[3] *blue:* Melancholy, sad, sometimes for no explicable reason.

Rose Johnson had been raised not as a servant but quite like their own child by white folks. Her mother who had died when Rose was still a baby, had been a trusted servant in the family. Rose was a cute, attractive, good looking little black girl and these people had no children of their own and so they kept Rose in their house.

As Rose grew older she drifted from her white folks back to the colored people, and she gradually no longer lived in the old house. Then it happened that these people went away to some other town to live, and somehow Rose stayed behind in Bridgepoint. Her white folks left a little money to take care of Rose, and this money she got every little while.

Rose now in the easy fashion of the poor lived with one woman in her house, and then for no reason went and lived with some other woman in her house. All this time, too, Rose kept company, and was engaged, first to this colored man and then to that, and always she made sure she was engaged, for Rose had strong the sense of proper conduct.

"No, I ain't no common nigger just to go around with any man, nor you Melanctha shouldn't neither," she said one day when she was telling the complex and less sure Melanctha what was the right way for her to do. "No Melanctha, I ain't no common nigger to do so, for I was raised by white folks. You know very well Melanctha that I'se always been engaged to them."

And so Rose lived on, always comfortable and rather decent and very lazy and very well content.

After she had lived some time this way, Rose thought it would be nice and very good in her position to get regularly really married. She had lately met Sam Johnson somewhere, and she liked him and she knew he was a good man, and then he had a place where he worked every day and got good wages. Sam Johnson liked Rose very well and he was quite ready to be married. One day they had a grand real wedding and were married. Then with Melanctha Herbert's help to do the sewing and the nicer work, they furnished comfortably a little red brick house. Sam then went back to his work as deck hand on a coasting steamer, and Rose stayed home in her house and sat and bragged to all her friends how nice it was to be married really to a husband.

Life went on very smoothly with them all the year. Rose was lazy but not dirty and Sam was careful but not fussy, and then there was Melanctha to come in every day and help to keep things neat.

When Rose's baby was coming to be born, Rose came to stay in the house where Melanctha Herbert lived just then, with a big good natured colored woman who did washing.

Rose went there to stay, so that she might have the doctor from the hospital near by to help her have the baby, and then, too, Melanctha could attend to her while she was sick.

Here the baby was born, and here it died, and then Rose went back to her house again with Sam.

Melanctha Herbert had not made her life all simple like Rose Johnson. Melanctha had not found it easy with herself to make her wants and what she had, agree.

Melanctha Herbert was always losing what she had in wanting all the things she saw. Melanctha was always being left when she was not leaving others.

Melanctha Herbert always loved too hard and much too often. She was always full with mystery and subtle movements and denials and vague distrusts and complicated disillusions. Then Melanctha would be sudden and impulsive and unbounded in some faith, and then she would suffer and be strong in her repression.

Melanctha Herbert was always seeking rest and quiet, and always she could only find new ways to be in trouble.

Melanctha wondered often how it was she did not kill herself when she was so blue. Often she thought this would be really the best way for her to do.

Melanctha Herbert had been raised to be religious, by her mother. Melanctha had not liked her mother very well. This mother, 'Mis' Herbert, as her neighbors called her, had been a sweet appearing and dignified and pleasant, pale yellow, colored woman. 'Mis' Herbert had always been a little wandering⁴ and mysterious and uncertain in her ways.

Melanctha was pale yellow and mysterious and a little pleasant like her mother, but the real power in Melanctha's nature came through her robust and unpleasant and unendurable black father.

Melanctha's father only used to come to where Melanctha and her mother lived, once in a while.

It was many years now that Melanctha had not heard or seen or known of anything her father did.

Melanctha Herbert almost always hated her black father, but she loved very well the power in herself that came through him. And so her feeling was really closer to her black coarse father, than her feeling had ever been toward her pale yellow, sweet-appearing mother. The things she had in her of her mother never made her feel respect.

---

⁴ *wandering:* Used in almost a coded way, this word appears throughout the story, often applied to Melanctha in her adolescence. It suggests an impropriety, whether a sexual or a physical one.

Melanctha Herbert had not loved herself in childhood. All of her youth was bitter to remember.

Melanctha had not loved her father and her mother and they had found it very troublesome to have her.

Melanctha's mother and her father had been regularly married. Melanctha's father was a big black virile negro. He only came once in a while to where Melanctha and her mother lived, but always that pleasant, sweet-appearing, pale yellow woman, mysterious and uncertain and wandering in her ways, was close in sympathy and thinking to her big black virile husband.

James Herbert was a common, decent enough, colored workman, brutal and rough to his one daughter, but then she was a most disturbing child to manage.

The young Melanctha did not love her father and her mother, and she had a break neck courage, and a tongue that could be very nasty. Then, too, Melanctha went to school and was very quick in all the learning, and she knew very well how to use this knowledge to annoy her parents who knew nothing.

Melanctha Herbert had always had a break neck courage. Melanctha always loved to be with horses; she loved to do wild things, to ride the horses and to break and tame them.

Melanctha, when she was a little girl, had had a good chance to live with horses. Near where Melanctha and her mother lived was the stable of the Bishops, a rich family who always had fine horses.

John, the Bishops' coachman, liked Melanctha very well and he always let her do anything she wanted with the horses. John was a decent, vigorous mulatto with a prosperous house and wife and children. Melanctha Herbert was older than any of his children. She was now a well grown girl of twelve and just beginning as a woman.

James Herbert, Melanctha's father, knew this John, the Bishops' coachman very well.

One day James Herbert came to where his wife and daughter lived, and he was furious.

"Where's that Melanctha girl of yours," he said fiercely, "if she is to the Bishops' stables again, with that man John, I swear I kill her. Why don't you see to that girl better you, you're her mother."

James Herbert was a powerful, loose built, hard handed, black, angry negro. Herbert never was a joyous negro. Even when he drank with other men, and he did that very often, he was never really joyous. In the days when he had been most young and free and open, he had never had the wide abandoned laughter that gives the broad glow to negro sunshine.

His daughter, Melanctha Herbert, later always made a hard forced laughter. She was only strong and sweet and in her nature when she was really deep in trouble, when she was fighting so with all she really had, that she did not use her laughter. This was always true of poor Melanctha who was so certain that she hated trouble. Melanctha Herbert was always seeking peace and quiet, and she could always only find new ways to get excited.

James Herbert was often a very angry negro. He was fierce and serious, and he was very certain that he often had good reason to be angry with Melanctha, who knew so well how to be nasty, and to use her learning with a father who knew nothing.

James Herbert often drank with John, the Bishops' coachman. John in his good nature sometimes tried to soften Herbert's feeling toward Melanctha. Not that Melanctha ever complained to John of her home life or her father. It was never Melanctha's way, even in the midst of her worst trouble to complain to any one of what happened to her, but nevertheless somehow every one who knew Melanctha always knew how much she suffered. It was only while one really loved Melanctha that one understood how to forgive her, that she never once complained nor looked unhappy, and was always handsome and in spirits, and yet one always knew how much she suffered.

The father, James Herbert, never told his troubles either, and he was so fierce and serious that no one ever thought of asking.

'Mis' Herbert as her neighbors called her was never heard even to speak of her husband or her daughter. She was always pleasant, sweet-appearing, mysterious and uncertain, and a little wandering in her ways.

The Herberts were a silent family with their troubles, but somehow every one who knew them always knew everything that happened.

The morning of one day when in the evening Herbert and the coachman John were to meet to drink together, Melanctha had to come to the stable joyous and in the best of humors. Her good friend John on this morning felt very firmly how good and sweet she was and how very much she suffered.

John was a very decent colored coachman. When he thought about Melanctha it was as if she were the eldest of his children. Really he felt very strongly the power in her of a woman. John's wife always liked Melanctha and she always did all she could to make things pleasant. And Melanctha all her life loved and respected kind and good and considerate people. Melanctha always loved and wanted peace and gentleness and goodness and all her life for herself poor Melanctha could only find new ways to be in trouble.

This evening after John and Herbert had drunk awhile together, the good John began to tell the father what a fine girl he had for a daughter. Perhaps the good John had been drinking a good deal of liquor, perhaps there was a gleam of something softer than the feeling of a friendly elder in the way John then spoke of Melanctha. There had been a good deal of drinking and John certainly that very morning had felt strongly Melanctha's power as a woman. James Herbert was always a fierce, suspicious, serious negro, and drinking never made him feel more open. He looked very black and evil as he sat and listened while John grew more and more admiring as he talked half to himself, half to the father, of the virtues and the sweetness of Melanctha.

Suddenly between them there came a moment filled full with strong black curses, and then sharp razors flashed in the black hands, that held them flung backward in the negro fashion, and then for some minutes there was fierce slashing.

John was a decent, pleasant, good natured, light brown negro, but he knew how to use a razor to do bloody slashing.

When the two men were pulled apart by the other negroes who were in the room drinking, John had not been much wounded but James Herbert had gotten one good strong cut that went from his right shoulder down across the front of his whole body. Razor fighting does not wound very deeply, but it makes a cut that looks most nasty, for it is so very bloody.

Herbert was held by the other negroes until he was cleaned and plastered, and then he was put to bed to sleep off his drink and fighting.

The next day he came to where his wife and daughter lived and he was furious.

"Where's that Melanctha, of yours?" he said to his wife, when he saw her. "If she is to the Bishops' stables now with that yellow John, I swear I kill her. A nice way she is going for a decent daughter. Why don't you see to that girl better you, ain't you her mother!"

Melanctha Herbert had always been old in all her ways and she knew very early how to use her power as a woman, and yet Melanctha with all her inborn intense wisdom was really very ignorant of evil. Melanctha had not yet come to understand what they meant, the things she so often heard around her, and which were just beginning to stir strongly in her.

Now when her father began fiercely to assail her, she did not really know what it was that he was so furious to force from her. In every way that he could think of in his anger, he tried to make her say a thing she did not really know. She held out and never answered anything he asked her, for Melanctha had a breakneck courage and she just then badly hated her black father.

When the excitement was all over, Melanctha began to know her power, the power she had so often felt stirring within her and which she now knew she could use to make her stronger.

James Herbert did not win this fight with his daughter. After awhile he forgot it as he soon forgot John and the cut of his sharp razor.

Melanctha almost forgot to hate her father, in her strong interest in the power she now knew she had within her.

Melanctha did not care much now, any longer, to see John or his wife or even the fine horses. This life was too quiet and accustomed and no longer stirred her to any interest or excitement.

Melanctha now really was beginning as a woman. She was ready, and she began to search in the streets and in dark corners to discover men and to learn their natures and their various ways of working.

In the next years Melanctha learned many ways that lead to wisdom. She learned the ways, and dimly in the distance she saw wisdom. These years of learning led very straight to trouble for Melanctha, though in these years Melanctha never did or meant anything that was really wrong.

Girls who are brought up with care and watching can always find moments to escape into the world, where they may learn the ways that lead to wisdom. For a girl raised like Melanctha Herbert, such escape was always very simple. Often she was alone, sometimes she was with a fellow seeker, and she strayed and stood, sometimes by railroad yards, sometimes on the docks or around new buildings where many men were working. Then when the darkness covered everything all over, she would begin to learn to know this man or that. She would advance, they would respond, and then she would withdraw a little, dimly, and always she did not know what it was that really held her. Sometimes she would almost go over, and then the strength in her of not really knowing, would stop the average man in his endeavor. It was a strange experience of ignorance and power and desire. Melanctha did not know what it was that she so badly wanted. She was afraid, and yet she did not understand that here she really was a coward.

Boys had never meant much to Melanctha. They had always been too young to content her. Melanctha had a strong respect for any kind of successful power. It was this that always kept Melanctha nearer, in her feeling toward her virile and unendurable black father, than she ever was in her feeling for her pale yellow, sweet-appearing mother. The things she had in her of her mother, never made her feel respect.

In these young days, it was only men that for Melanctha held anything there was of knowledge and power. It was not from men however that Melanctha learned to really understand this power.

From the time that Melanctha was twelve until she was sixteen she
wandered, always seeking but never more than very dimly seeing wisdom. All this time Melanctha went on with her school learning; she
went to school rather longer than do most of the colored children.

Melanctha's wanderings after wisdom she always had to do in
secret and by snatches, for her mother was then still living and 'Mis'
Herbert always did some watching, and Melanctha with all her hard
courage dreaded that there should be much telling to her father, who
came now quite often to where Melanctha lived with her mother.

In these days Melanctha talked and stood and walked with many
kinds of men, but she did not learn to know any of them very deeply.
They all supposed her to have world knowledge and experience. They,
believing that she knew all, told her nothing, and thinking that she
was deciding with them, asked for nothing, and so though Melanctha
wandered widely, she was really very safe with all the wandering.

It was a very wonderful experience this safety of Melanctha in these
days of her attempted learning. Melanctha herself did not feel the
wonder, she only knew that for her it all had no real value.

Melanctha all her life was very keen in her sense for real experience.
She knew she was not getting what she so badly wanted, but with all
her break neck courage Melanctha here was a coward, and so she
could not learn to really understand.

Melanctha liked to wander, and to stand by the railroad yard, and
watch the men and the engines and the switches and everything that
was busy there, working. Railroad yards are a ceaseless fascination.
They satisfy every kind of nature. For the lazy man whose blood flows
very slowly, it is a steady soothing world of motion which supplies him
with the sense of a strong moving power. He need not work and yet
he has it very deeply; he has it even better than the man who works
in it or owns it. Then for natures that like to feel emotion without the
trouble of having any suffering, it is very nice to get the swelling in the
throat, and the fullness, and the heart beats, and all the flutter of
excitement that comes as one watches the people come and go, and
hear the engine pound and give a long drawn whistle. For a child
watching through a hole in the fence above the yard, it is a wonder
world of mystery and movement. The child loves all the noise, and
then it loves the silence of the wind that comes before the full rush of
the pounding train, that bursts out from the tunnel where it lost itself

and all its noise in darkness, and the child loves all the smoke, that sometimes comes in rings, and always puffs with fire and blue color. For Melanctha the yard was full of the excitement of many men, and perhaps a free and whirling future. Melanctha came here very often and watched the men and all the things that were so busy working. The men always had time for, "Hullo sis, do you want to sit on my engine," and, "Hullo, that's a pretty lookin' yaller girl, do you want to come and see him cookin." All the colored porters liked Melanctha. They often told her exciting things that had happened; how in the West they went through big tunnels where there was no air to breathe, and then out and winding around edges of great canyons on thin high spindling trestles, and sometimes cars, and sometimes whole trains fell from the narrow bridges, and always up from the dark places death and all kinds of queer devils looked up and laughed in their faces. And then they would tell how sometimes when the train went pounding down steep slippery mountains, great rocks would racket and roll down around them, and sometimes would smash in the car and kill men; and as the porters told these stories their round, black, shining faces would grow solemn, and their color would go grey beneath the greasy black, and their eyes would roll white in the fear and wonder of the things they could scare themselves by telling.

There was one, big, serious, melancholy, light brown porter who often told Melanctha stories, for he liked the way she had of listening with intelligence and sympathetic feeling, when he told how the white men in the far South tried to kill him because he made one of them who was drunk and called him a damned nigger, and who refused to pay money for his chair to a nigger, get off the train between stations. And then this porter had to give up going to that part of the Southern country, for all the white men swore that if he ever came there again they would surely kill him.

Melanctha liked this serious, melancholy light brown negro very well, and all her life Melanctha wanted and respected gentleness and goodness, and this man always gave her good advice and serious kindness, and Melanctha felt such things very deeply, but she could never let them help her or affect her to change the ways that always made her keep herself in trouble.

Melanctha spent many of the last hours of the daylight with the porters and with other men who worked hard, but when darkness came it was always different. Then Melanctha would find herself with the, for her, gentlemanly classes. A clerk, or a young express agent would begin to know her, and they would stand, or perhaps, walk a little while together.

Melanctha always made herself escape but often it was with an effort. She did not know what it was that she so badly wanted, but with all her courage Melanctha here was a coward, and so she could not learn to understand.

Melanctha and some man would stand in the evening and would talk together. Sometimes Melanctha would be with another girl and then it was much easier to stay or to escape, for then they could make way for themselves together, and by throwing words and laughter to each other, could keep a man from getting too strong in his attention.

But when Melanctha was alone, and she was so, very often, she would sometimes come very near to making a long step on the road that leads to wisdom. Some man would learn a good deal about her in the talk, never altogether truly, for Melanctha all her life did not know how to tell a story wholly. She always, and yet not with intention, managed to leave out big pieces which make a story very different, for when it came to what had happened and what she had said and what it was that she had really done, Melanctha never could remember right. The man would sometimes come a little nearer, would detain her, would hold her arm or make his jokes a little clearer, and then Melanctha would always make herself escape. The man thinking that she really had world wisdom would not make his meaning clear, and believing that she was deciding with him he never went so fast that he could stop her when at last she made herself escape.

And so Melanctha wandered on the edge of wisdom. "Say, Sis, why don't you when you come here stay a little longer?" they would all ask her, and they would hold her for an answer, and she would laugh, and sometimes she did stay longer, but always just in time she made herself escape.

Melanctha Herbert wanted very much to know and yet she feared the knowledge. As she grew older she often stayed a good deal longer, and sometimes it was almost a balanced struggle, but she always made herself escape.

Next to the railroad yard it was the shipping docks that Melanctha loved best when she wandered. Often she was alone, sometimes she was with some better kind of black girl, and she would stand a long time and watch the men working at unloading, and see the steamers do their coaling, and she would listen with full feeling to the yowling of the free swinging negroes, as they ran, with their powerful loose jointed bodies and their childish savage yelling, pushing, carrying, pulling great loads from the ships to the warehouses. The men would call out, "Say, Sis, look out or we'll come and catch yer," or "Hi, there, you yaller girl, come here and we'll take you sailin'." And then, too, Melanctha would learn to know some of the serious foreign sailors who told her all sorts of wonders, and a cook would sometimes take her and her friends over a ship and show where he made his messes and where the men slept, and where the shops were, and how everything was made by themselves, right there, on ship board.

Melanctha loved to see these dark and smelly places. She always loved to watch and talk and listen with men who worked hard. But it was never from these rougher people that Melanctha tried to learn the ways that lead to wisdom. In the daylight she always liked to talk with rough men and to listen to their lives and about their work and their various ways of doing, but when the darkness covered everything all over, Melanctha would meet, and stand, and talk with a clerk or a young shipping agent who had seen her watching, and so it was that she would try to learn to understand.

And then Melanctha was fond of watching men work on new buildings. She loved to see them hoisting, digging, sawing and stone cutting. Here, too, in the daylight, she always learned to know the common workmen. "Heh, Sis, look out or that rock will fall on you and smash you all up into little pieces. Do you think you would make a nice jelly?" And then they would all laugh and feel that their jokes were very funny. And "Say, you pretty yaller girl, would it scare you bad to stand up here on top where I be? See if you've got grit and come up here where I can hold you. All you got to do is to sit still on that there rock that they're just hoistin', and then when you get here I'll hold you tight, don't you be scared Sis."

Sometimes Melanctha would do some of these things that had much danger, and always with such men, she showed her power and her break neck courage. Once she slipped and fell from a high place. A workman caught her and so she was not killed, but her left arm was badly broken.

All the men crowded around her. They admired her boldness in doing and in bearing pain when her arm was broken. They all went along with her with great respect to the doctor, and then they took her home in triumph and all of them were bragging about her not squealing.

James Herbert was home where his wife lived, that day. He was furious when he saw the workmen and Melanctha. He drove the men away with curses so that they were all very nearly fighting, and he would not let a doctor come in to attend Melanctha. "Why don't you see to that girl better, you, you're her mother."

James Herbert did not fight things out now any more with his daughter. He feared her tongue, and her school learning, and the way she had of saying things that were very nasty to a brutal black man who knew nothing. And Melanctha just then hated him very badly in her suffering.

And so this was the way Melanctha lived the four years of her beginning as a woman. And many things happened to Melanctha, but she knew very well that none of them had led her on to the right way, that certain way that was to lead her to world wisdom.

Melanctha Herbert was sixteen when she first met Jane Harden. Jane was a negress, but she was so white that hardly any one could guess it. Jane had had a good deal of education. She had been two years at a colored college. She had had to leave because of her bad conduct. She taught Melanctha many things. She taught her how to go the ways that lead to wisdom.

Jane Harden was at this time twenty-three years old and she had had much experience. She was very much attracted by Melanctha, and Melanctha was very proud that this Jane would let her know her.

Jane Harden was not afraid to understand. Melanctha who had strong the sense for real experience, knew that here was a woman who had learned to understand.

Jane Harden had many bad habits. She drank a great deal, and she wandered widely. She was safe though now, when she wanted to be safe, in this wandering.

Melanctha Herbert soon always wandered with her. Melanctha tried the drinking and some of the other habits, but she did not find that she cared very much to do them. But every day she grew stronger in her desire to really understand.

It was now no longer, even in the daylight, the rougher men that these two learned to know in their wanderings, and for Melanctha the better classes were now a little higher. It was no longer express agents and clerks that she learned to know, but men in business, commercial travelers, and even men above these, and Jane and she would talk and walk and laugh and escape from them all very often. It was still the same, the knowing of them and the always just escaping, only now for Melanctha somehow it was different, for though it was always the same thing that happened it had a different flavor, for now Melanctha was with a woman who had wisdom, and dimly she began to see what it was that she should understand.

It was not from the men that Melanctha learned her wisdom. It was always Jane Harden herself who was making Melanctha begin to understand.

Jane was a roughened woman. She had power and she liked to use it, she had much white blood and that made her see clear, she liked drinking and that made her reckless. Her white blood was strong in her and she had grit and endurance and a vital courage. She was always game, however much she was in trouble. She liked Melanctha Herbert for the things that she had like her, and then Melanctha was young, and she had sweetness, and a way of listening with intelligence and sympathetic interest, to the stories that Jane Harden often told out of her experience.

Jane grew always fonder of Melanctha. Soon they began to wander, more to be together than to see men and learn their various ways of working. Then they began not to wander, and Melanctha would spend long hours with Jane in her room, sitting at her feet and listening to her stories, and feeling her strength and the power of her affection, and slowly she began to see clear before her one certain way that would be sure to lead to wisdom.

Before the end came, the end of the two years in which Melanctha spent all her time when she was not at school or in her home, with Jane Harden, before these two years were finished, Melanctha had come to see very clear, and she had come to be very certain, what it is that gives the world its wisdom.

Jane Harden always had a little money and she had a room in the lower part of the town. Jane had once taught in a colored school. She had had to leave that too on account of her bad conduct. It was her drinking that always made all the trouble for her, for that can never be really covered over.

Jane's drinking was always growing worse upon her. Melanctha had tried to do the drinking but it had no real attraction for her.

In the first year, between Jane Harden and Melanctha Herbert, Jane had been much the stronger. Jane loved Melanctha and she found her always intelligent and brave and sweet and docile, and Jane meant to, and before the year was over she had taught Melanctha what it is that gives many people in the world their wisdom.

Jane had many ways in which to do this teaching. She told Melanctha many things. She loved Melanctha hard and made Melanctha feel it very deeply. She would be with other people and with men and with Melanctha, and she would make Melanctha understand what everybody wanted, and what one did with power when one had it.

Melanctha sat at Jane's feet for many hours in these days and felt Jane's wisdom. She learned to love Jane and to have this feeling very deeply. She learned a little in these days to know joy, and she was taught too how very keenly she could suffer. It was very different this suffering from that Melanctha sometimes had from her mother and from her very unendurable black father. Then she was fighting and she could be strong and valiant in her suffering, but here with Jane Harden she was longing and she bent and pleaded with her suffering.

It was a very tumultuous, very mingled year, this time for Melanctha, but she certainly did begin to really understand.

In every way she got it from Jane Harden. There was nothing good or bad in doing, feeling, thinking or in talking, that Jane spared her. Sometimes the lesson came almost too strong for Melanctha, but somehow she always managed to endure it and so slowly, but always with increasing strength and feeling, Melanctha began to really understand.

Then slowly, between them, it began to be all different. Slowly now between them, it was Melanctha Herbert, who was stronger. Slowly now they began to drift apart from one another.

Melanctha Herbert never really lost her sense that it was Jane Harden who had taught her, but Jane did many things that Melanctha now no longer needed. And then, too, Melanctha never could remember right when it came to what she had done and what had happened. Melanctha now sometimes quarreled with Jane, and they no longer went about together, and sometimes Melanctha really forgot how much she owed to Jane Harden's teaching.

Melanctha began now to feel that she had always had world wisdom. She really knew of course, that it was Jane who had taught her, but all that began to be covered over by the trouble between them, that was now always getting stronger.

Jane Harden was a roughened woman. Once she had been very strong, but now she was weakened in all her kinds of strength by her drinking. Melanctha had tried the drinking but it had had no real attraction for her.

Jane's strong and roughened nature and her drinking made it always harder for her to forgive Melanctha, that now Melanctha did not really need her any longer. Now it was Melanctha who was stronger and it was Jane who was dependent on her.

Melanctha was now come to be about eighteen years old. She was a graceful, pale yellow, good looking, intelligent, attractive negress, a little mysterious sometimes in her ways, and always good and pleasant, and always ready to do things for people.

Melanctha from now on saw very little of Jane Harden. Jane did not like that very well and sometimes she abused Melanctha, but her drinking soon covered everything all over.

It was not in Melanctha's nature to really lose her sense for Jane Harden. Melanctha all her life was ready to help Jane out in any of her trouble, and later, when Jane really went to pieces, Melanctha always did all that she could to help her.

But Melanctha Herbert was ready now herself to do teaching. Melanctha could do anything now that she wanted. Melanctha knew now what everybody wanted.

Melanctha had learned how she might stay a little longer; she had learned that she must decide when she wanted really to stay longer, and she had learned how when she wanted to, she could escape.

And so Melanctha began once more to wander. It was all now for her very different. It was never rougher men now that she talked to, and she did not care much now to know white men of the, for her, very better classes. It was now something realler that Melanctha wanted, something that would move her very deeply, something that would fill her fully with the wisdom that was planted now within her, and that she wanted badly, should really wholly fill her.

Melanctha these days wandered very widely. She was always alone now when she wandered. Melanctha did not need help now to know, or to stay longer, or when she wanted, to escape.

Melanctha tried a great many men, in these days before she was really suited. It was almost a year that she wandered and then she met with a young mulatto. He was a doctor who had just begun to practice. He would most likely do well in the future, but it was not this that concerned Melanctha. She found him good and strong and gentle and very intellectual, and all her life Melanctha liked and wanted good and considerate people, and then too he did not at first believe in Melanctha. He held off and did not know what it was that Melanctha wanted. Melanctha came to want him very badly. They began to know each other better. Things began to be very strong between them. Melanctha wanted him so badly that now she never wandered. She just gave herself to this experience.

Melanctha Herbert was now, all alone, in Bridgepoint. She lived now with this colored woman and now with that one, and she sewed, and sometimes she taught a little in a colored school as substitute for some teacher. Melanctha had now no home nor any regular employment. Life was just commencing for Melanctha. She had youth and had learned wisdom, and she was graceful and pale yellow and very pleasant, and always ready to do things for people, and she was mysterious in her ways and that only made belief in her more fervent.

During the year before she met Jefferson Campbell, Melanctha had tried many kinds of men but they had none of them interested Melanctha very deeply. She met them, she was much with them, she left them, she would think perhaps this next time it would be more exciting, and always she found that for her it all had no real meaning. She could now do everything she wanted, she knew now everything that everybody wanted, and yet it all had no excitement for her. With these men, she knew she could learn nothing. She wanted some one that could teach her very deeply and now at last she was sure that she had found him, yes she really had it, before she had thought to look if in this man she would find it.

During this year 'Mis' Herbert as her neighbors called her, Melanctha's pale yellow mother was very sick, and in this year she died.

Melanctha's father during these last years did not come very often to the house where his wife lived and Melanctha. Melanctha was not sure that her father was now any longer here in Bridgepoint. It was Melanctha who was very good now to her mother. It was always Melanctha's way to be good to any one in trouble.

Melanctha took good care of her mother. She did everything that any woman could, she tended and soothed and helped her pale yellow mother, and she worked hard in every way to take care of her, and make her dying easy. But Melanctha did not in these days like her mother any better, and her mother never cared much for this daughter who was always a hard child to manage, and who had a tongue that always could be very nasty.

Melanctha did everything that any woman could, and at last her mother died, and Melanctha had her buried. Melanctha's father was not heard from, and Melanctha in all her life after, never saw or heard or knew of anything that her father did.

It was the young doctor, Jefferson Campbell, who helped Melanctha toward the end, to take care of her sick mother. Jefferson Campbell had often before seen Melanctha Herbert, but he had never liked her very well, and he had never believed that she was any good. He had heard something about how she wandered. He knew a little too of

Jane Harden, and he was sure that this Melanctha Herbert, who was her friend and who wandered, would never come to any good. Dr. Jefferson Campbell was a serious, earnest, good young joyous doctor. He liked to take care of everybody and he loved his own colored people. He always found life very easy did Jeff Campbell, and everybody liked to have him with them. He was so good and sympathetic, and he was so earnest and so joyous. He sang when he was happy, and he laughed, and his was the free abandoned laughter that gives the warm broad glow to negro sunshine.

Jeff Campbell had never yet in his life had real trouble. Jefferson's father was a good, kind, serious, religious man. He was a very steady, very intelligent, and very dignified, light brown, grey haired negro. He was a butler and he had worked for the Campbell family many years, and his father and his mother before him had been in the service of this family as free people.

Jefferson Campbell's father and his mother had of course been regularly married. Jefferson's mother was a sweet, little, pale brown, gentle woman who reverenced and obeyed her good husband, and who worshipped and admired and loved hard her good, earnest, cheery, hard working doctor boy who was her only child.

Jeff Campbell had been raised religious by his people but religion had never interested Jeff very much. Jefferson was very good. He loved his people and he never hurt them, and he always did everything they wanted and that he could to please them, but he really loved best science and experimenting and to learn things, and he early wanted to be a doctor, and he was always very interested in the life of the colored people.

The Campbell family had been very good to him and had helped him on with his ambition. Jefferson studied hard, he went to a colored college, and then he learnt to be a doctor.

It was now two or three years, that he had started in to practice. Everybody liked Jeff Campbell, he was so strong and kindly and cheerful and understanding, and he laughed so with pure joy, and he always liked to help all his own colored people.

Dr. Jeff knew all about Jane Harden. He had taken care of her in some of her bad trouble. He knew about Melanctha too, though until her mother was taken sick he had never met her. Then he was called in to help Melanctha to take care of her sick mother. Dr. Campbell did not like Melanctha's ways and he did not think that she would ever come to any good.

Dr. Campbell had taken care of Jane Harden in some of her bad trouble. Jane sometimes had abused Melanctha to him. What right had that Melanctha Herbert who owed everything to her, Jane Harden, what right had a girl like that to go away to other men and leave her, but Melanctha Herbert never had any sense of how to act to anybody. Melanctha had a good mind, Jane never denied her that, but she never used it to do anything decent with it. But what could you expect when Melanctha had such a brute of a black nigger father, and Melanctha was always abusing her father and yet she was just like him, and really she admired him so much and he never had any sense of what he owed to anybody, and Melanctha was just like him and she was proud of it too, and it made Jane so tired to hear Melanctha talk all the time as if she wasn't. Jane Harden hated people who had good minds and didn't use them, and Melanctha always had that weakness, and wanting to keep in with people, and never really saying that she wanted to be like her father, and it was so silly of Melanctha to abuse her father, when she was so much like him and she really liked it. No, Jane Harden had no use for Melanctha. Oh yes, Melanctha always came around to be good to her. Melanctha was always sure to do that. She never really went away and left one. She didn't use her mind enough to do things straight out like that. Melanctha Herbert had a good mind, Jane never denied that to her, but she never wanted to see or hear about Melanctha Herbert any more, and she wished Melanctha wouldn't come in any more to see her. She didn't hate her, but she didn't want to hear about her father and all that talk Melanctha always made, and that just meant nothing to her. Jane Harden was very tired of all that now. She didn't have any use now any more for Melanctha, and if Dr. Campbell saw her he better tell her Jane didn't want to see her, and she could take her talk to somebody else, who was ready to believe her. And then Jane Harden would drop away and forget Melanctha and all her life before, and then she would begin to drink and so she would cover everything all over.

Jeff Campbell heard all this very often, but it did not interest him very deeply. He felt no desire to know more of this Melanctha. He heard her, once, talking to another girl outside of the house, when he was paying a visit to Jane Harden. He did not see much in the talk that he heard her do. He did not see much in the things Jane Harden said when she abused Melanctha to him. He was more interested in Jane herself than in anything he heard about Melanctha. He knew Jane Harden had a good mind, and she had had power, and she could really have done things, and now this drinking covered everything all over.

Jeff Campbell was always very sorry when he had to see it. Jane Harden was a roughened woman, and yet Jeff found a great many strong good things in her, that still made him like her. Jeff Campbell did everything he could for Jane Harden. He did not care much to hear about Melanctha. He had no feeling, much, about her. He did not find that he took any interest in her. Jane Harden was so much a stronger woman, and Jane really had had a good mind, and she had used it to do things with it, before this drinking business had taken such a hold upon her.

Dr. Campbell was helping Melanctha Herbert to take care of her sick mother. He saw Melanctha now for long times and very often, and they sometimes talked a good deal together, but Melanctha never said anything to him about Jane Harden. She never talked to him about anything that was not just general matters, or about medicine, or to tell him funny stories. She asked him many questions and always listened very well to all he told her, and she always remembered everything she heard him say about doctoring, and she always remembered everything that she had learned from all the others.

Jeff Campbell never found that all this talk interested him very deeply. He did not find that he liked Melanctha when he saw her so much, any better. He never found that he thought much about Melanctha. He never found that he believed much in her having a good mind, like Jane Harden. He found he liked Jane Harden always better, and that he wished very much that she had never begun that bad drinking.

Melanctha Herbert's mother was now always getting sicker. Melanctha really did everything that any woman could. Melanctha's mother never liked her daughter any better. She never said much, did 'Mis' Herbert, but anybody could see that she did not think much of this daughter.

Dr. Campbell now often had to stay a long time to take care of 'Mis' Herbert. One day 'Mis' Herbert was much sicker and Dr. Campbell thought that this night, she would surely die. He came back late to the house, as he had said he would, to sit up and watch 'Mis' Herbert, and to help Melanctha, if she should need anybody to be with her. Melanctha Herbert and Jeff Campbell sat up all that night together. 'Mis' Herbert did not die. The next day she was a little better.

This house where Melanctha had always lived with her mother was a little red brick, two story house. They had not much furniture to fill it and some of the windows were broken and not mended. Melanctha did not have much money to use now on the house, but with a colored woman, who was their neighbor and good natured and who had always helped them, Melanctha managed to take care of her mother and to keep the house fairly clean and neat.

Melanctha's mother was in bed in a room upstairs, and the steps from below led right up into it. There were just two rooms on this upstairs floor. Melanctha and Dr. Campbell sat down on the steps, that night they watched together, so that they could hear and see Melanctha's mother and yet the light would be shaded, and they could sit and read, if they wanted to, and talk low some, and yet not disturb 'Mis' Herbert.

Dr. Campbell was always very fond of reading. Dr. Campbell had not brought a book with him that night. He had just forgotten it. He had meant to put something in his pocket to read, so that he could amuse himself, while he was sitting there and watching. When he was through with taking care of 'Mis' Herbert, he came and sat down on the steps just above where Melanctha was sitting. He spoke about how he had forgotten to bring his book with him. Melanctha said there were some old papers in the house, perhaps Dr. Campbell could find something in them that would help pass the time for a while for him. All right, Dr. Campbell said, that would be better than just sitting there with nothing. Dr. Campbell began to read through the old papers that Melanctha gave him. When anything amused him in them, he read it out to Melanctha. Melanctha was now pretty silent, with him. Dr. Campbell began to feel a little, about how she responded to him. Dr. Campbell began to see a little that perhaps Melanctha had a good mind. Dr. Campbell was not sure yet that she had a good mind, but he began to think a little that perhaps she might have one.

Jefferson Campbell always liked to talk to everybody about the things he worked at and about his thinking about what he could do for the colored people. Melanctha Herbert never thought about these things the way that he did. Melanctha had never said much to Dr. Campbell about what she thought about them. Melanctha did not feel the same as he did about being good and regular in life, and not having excitements all the time, which was the way that Jefferson Campbell wanted that everybody should be, so that everybody would be wise and yet be happy. Melanctha always had strong the sense for real experience. Melanctha Herbert did not think much of this way of coming to real wisdom.

Dr. Campbell soon got through with his reading, in the old newspapers, and then somehow he began to talk along about the things he was always thinking. Dr. Campbell said he wanted to work so that he could understand what troubled people, and not to just have excitements, and he believed you ought to love your father and your mother and to be regular in all your life, and not to be always wanting new things and excitements, and to always know where you were, and what you wanted, and to always tell everything just as you meant it. That's the only kind of life he knew or believed in, Jeff Campbell repeated. "No I ain't got any use for all the time being in excitements and wanting to have all kinds of experience all the time. I got plenty of experience just living regular and quiet and with my family, and doing my work, and taking care of people, and trying to understand it. I don't believe much in this running around business and I don't want to see the colored people do it. I am a colored man and I ain't sorry, and I want to see the colored people like what is good and what I want them to have, and that's to live regular and work hard and understand things, and that's enough to keep any decent man excited." Jeff Campbell spoke now with some anger. Not to Melanctha, he did not think of her at all when he was talking. It was the life he wanted that he spoke to, and the way he wanted things to be with the colored people.

But Melanctha Herbert had listened to him say all this. She knew he meant it, but it did not mean much to her, and she was sure some day he would find out, that it was not all, of real wisdom. Melanctha knew very well what it was to have real wisdom. "But how about Jane Harden?" said Melanctha to Jeff Campbell, "seems to me Dr. Campbell you find her to have something in her, and you go there very often, and you talk to her much more than you do to the nice girls that stay at home with their people, the kind you say you are really wanting. It don't seem to me Dr. Campbell, that what you say and what you do seem to have much to do with each other. And about your being so good Dr. Campbell," went on Melanctha, "You don't care about going to church much yourself, and yet you always are saying you believe so much in things like that, for people. It seems to me, Dr. Campbell you want to have a good time just like all us others, and then you just keep on saying that it's right to be good and you ought not to have excitements, and yet you really don't want to do it Dr. Campbell, no more than me or Jane Harden. No, Dr. Campbell, it certainly does seem to me you don't know very well yourself, what you mean, when you are talking."

Jefferson had been talking right along, the way he always did when he got started, and now Melanctha's answer only made him talk a little harder. He laughed a little, too, but very low, so as not to disturb 'Mis' Herbert who was sleeping very nicely, and he looked brightly at Melanctha to enjoy her, and then he settled himself down to answer.

"Yes," he began, "it certainly does sound a little like I didn't know very well what I do mean, when you put it like that to me, Miss Melanctha, but that's just because you don't understand enough about what I meant, by what I was just saying to you. I don't say, never, I don't want to know all kinds of people, Miss Melanctha, and I don't say there ain't many kinds of people, and I don't say ever, that I don't find some like Jane Harden very good to know and talk to, but it's the strong things I like in Jane Harden, not all her excitements. I don't admire the bad things she does, Miss Melanctha, but Jane Harden is a strong woman and I always respect that in her. No I know you don't believe what I say, Miss Melanctha, but I mean it, and it's all just because you don't understand it when I say it. And as for religion, that just ain't my way of being good, Miss Melanctha, but it's a good way for many people to be good and regular in their way of living, and if they believe it, it helps them to be good, and if they're honest in it, I like to see them have it. No, what I don't like, Miss Melanctha, is this what I see so much with the colored people, their always wanting new things just to get excited."

Jefferson Campbell here stopped himself in this talking. Melanctha Herbert did not make any answer. They both sat there very quiet.

Jeff Campbell then began again on the old papers. He sat there on the steps just above where Melanctha was sitting, and he went on with his reading, and his head went moving up and down, and sometimes he was reading, and sometimes he was thinking about all the things he wanted to be doing, and then he would rub the back of his dark hand over his mouth, and in between he would be frowning with his thinking, and sometimes he would be rubbing his head hard to help his thinking. And Melanctha just sat still and watched the lamp burning, and sometimes he turned it down a little, when the wind caught it and it would begin to get to smoking.

And so Jeff Campbell and Melanctha Herbert sat there on the steps, very quiet, a long time, and they didn't seem to think much, that they were together. They sat there so, for about an hour, and then it came to Jefferson very slowly and as a strong feeling that he was sitting there on the steps, alone, with Melanctha. He did not know if Melanctha Herbert was feeling very much about their being there alone together. Jefferson began to wonder about it a little. Slowly he felt that surely they must both have this feeling. It was so important that he knew that she must have it. They both sat there, very quiet, a long time. At last Jefferson began to talk about how the lamp was smelling. Jefferson began to explain what it is that makes a lamp get to smelling. Melanctha let him talk. She did not answer, and then he stopped in his talking. Soon Melanctha began to sit up straighter and then she started in to question.

"About what you was just saying Dr. Campbell about living regular and all that, I certainly don't understand what you meant by what you was just saying. You ain't a bit like good people Dr. Campbell, like the goodpeople you are always saying are just like you. I know good people Dr. Campbell, and you ain't a bit like men who are good and got religion. You are just as free and easy as any man can be Dr. Campbell, and you always like to be with Jane Harden, and she is a pretty bad one and you don't look down on her and you never tell her she is a bad one. I know you like her just like a friend Dr. Campbell, and so I certainly don't understand just what it is you mean by all that you was just saying to me. I know you mean honest Dr. Campbell, and I am always trying to believe you, but I can't say as I see just what you mean when you say you want to be good and real pious, because I am very certain Dr. Campbell that you ain't that kind of a man at all, and you ain't never ashamed to be with queer folks Dr. Campbell, and you seem to be thinking what you are doing is just like what you are always saying, and Dr. Campbell, I certainly don't just see what you mean by what you say."

Dr. Campbell almost laughed loud enough to wake 'Mis' Herbert. He did enjoy the way Melanctha said these things to him. He began to feel very strongly about it that perhaps Melanctha really had a good mind. He was very free now in his laughing, but not so as to make Melanctha angry. He was very friendly with her in his laughing, and then he made his face get serious, and he rubbed his head to help him in his thinking.

"I know Miss Melanctha" he began, "It ain't very easy for you to understand what I was meaning by what I was just saying to you, and perhaps some of the good people I like so wouldn't think very much, any more than you do, Miss Melanctha, about the ways I have to be good. But that's no matter Miss Melanctha. What I mean Miss Melanctha by what I was just saying to you is, that I don't, no, never, believe in doing things just to get excited. You see Miss Melanctha I mean the way so many of the colored people do it. Instead of just working hard and caring about their working and living regular with their families and saving up all their money, so they will have some to bring up their children better, instead of living regular and doing like that and getting all their new ways from just decent living, the colored people just keep running round and perhaps drinking and doing everything bad they can ever think of, and not just because they like all those bad things that they are always doing, but only just because they want to get excited. No Miss Melanctha, you see I am a colored man myself and I ain't sorry, and I want to see the colored people being good and careful and always honest and living always just as regular as can be, and I am sure Miss Melanctha, that that way everybody can have a good time, and be happy and keep right and be busy, and not always have to be doing bad things for new ways to get excited. Yes Miss Melanctha, I certainly do like everything to be good, and quiet, and I certainly do think that is the best way for all us colored people. No, Miss Melanctha too, I don't mean this except only just the way I say it. I ain't got any other meaning Miss Melanctha, and it's that what I mean when I am saying about being really good. It ain't Miss Melanctha to be pious and not liking every kind of people, and I don't say ever Miss Melanctha that when other kind of people come regular into your life you shouldn't want to know them always. What I mean Miss Melanctha by what I am always saying is, you shouldn't try to know everybody just to run around and get excited. It's that kind of way of doing that I hate so always Miss Melanctha, and that is so bad for all us colored people. I don't know as you understand now any better what I mean by what I was just saying to you. But you certainly do know now Miss Melanctha, that I always mean it what I say when I am talking."

"Yes I certainly do understand you when you talk so Dr. Campbell. I certainly do understand now what you mean by what you was always saying to me. I certainly do understand Dr. Campbell that you mean you don't believe it's right to love anybody." "Why sure no, yes I do Miss Melanctha, I certainly do believe strong in loving, and in being good to everybody, and trying to understand what they all need, to help them." "Oh I know all about that way of doing Dr. Campbell, but that certainly ain't the kind of love I mean when I am talking. I mean real, strong, hot love Dr. Campbell, that makes you do anything for somebody that loves you." "I don't know much about that kind of love yet Miss Melanctha. You see it's this way with me always Miss Melanctha. I am always so busy with my thinking about my work I am doing and so I don't have time for just fooling, and then too, you see

Miss Melanctha, I really certainly don't ever like to get excited, and
that kind of loving hard does seem always to mean just getting all the
time excited. That certainly is what I always think from what I see
of them that have it bad Miss Melanctha, and that certainly would
never suit a man like me. You see Miss Melanctha I am a very quiet
kind of fellow, and I believe in a quiet life for all the colored people.
No Miss Melanctha I certainly never have mixed myself up in that
kind of trouble."

"Yes I certainly do see that very clear Dr. Campbell," said Melanctha, "I see that's certainly what it is always made me not know right
about you and that's certainly what it is that makes you really mean
what you was always saying. You certainly are just too scared Dr.
Campbell to really feel things way down in you. All you are always
wanting Dr. Campbell, is just to talk about being good, and to play
with people just to have a good time, and yet always to certainly keep
yourself out of trouble. It don't seem to me Dr. Campbell that I admire
that way to do things very much. It certainly ain't really to me being
very good. It certainly ain't any more to me Dr. Campbell, but that you
certainly are awful scared about really feeling things way down in you,
and that's certainly the only way Dr. Campbell I can see that you can
mean, by what it is that you are always saying to me."

"I don't know about that Miss Melanctha, I certainly don't think I
can't feel things very deep in me, though I do say I certainly do like to
have things nice and quiet, but I don't see harm in keeping out of danger Miss Melanctha, when a man knows he certainly don't want to get
killed in it, and I don't know anything that's more awful dangerous
Miss Melanctha than being strong in love with somebody. I don't
mind sickness or real trouble Miss Melanctha, and I don't want to be
talking about what I can do in real trouble, but you know something
about that Miss Melanctha, but I certainly don't see much in mixing
up just to get excited, in that awful kind of danger. No Miss Melanctha I certainly do only know just two kinds of ways of loving. One
kind of loving seems to me, is like one has a good quiet feeling in a
family when one does his work, and is always living good and being
regular, and then the other way of loving is just like having it like any
animal that's low in the streets together, and that don't seem to me
very good Miss Melanctha, though I don't say ever that it's not all
right when anybody likes it, and that's all the kinds of love I know
Miss Melanctha, and I certainly don't care very much to get mixed up
in that kind of a way just to be in trouble."

Jefferson stopped and Melanctha thought a little.

"That certainly does explain to me Dr. Campbell what I been thinking about you this long time. I certainly did wonder how you could be so live, and knowing everything, and everybody, and talking so big always about everything, and everybody always liking you so much, and you always looking as if you was thinking, and yet you really was never knowing about anybody and certainly not being really very understanding. It certainly is all Dr. Campbell because you is so afraid you will be losing being good so easy, and it certainly do seem to me Dr. Campbell that it certainly don't amount to very much that kind of goodness."

"Perhaps you are right Miss Melanctha," Jefferson answered. "I don't say never, perhaps you ain't right Miss Melanctha. Perhaps I ought to know more about such ways Miss Melanctha. Perhaps it would help me some, taking care of the colored people, Miss Melanctha. I don't say, no, never, but perhaps I could learn a whole lot about women the right way, if I had a real good teacher."

'Mis' Herbert just then stirred a little in her sleep. Melanctha went up the steps to the bed to attend her. Dr. Campbell got up too and went to help her. 'Mis' Herbert woke up and was a little better. Now it was morning and Dr. Campbell gave his directions to Melanctha, and then left her.

Melanctha Herbert all her life long, loved and wanted good, kind and considerate people. Jefferson Campbell was all the things that Melanctha had ever wanted. Jefferson was a strong, well built, good looking, cheery, intelligent and good mulatto. And then at first he had not cared to know Melanctha, and when he did begin to know her he had not liked her very well, and he had not thought that she would ever come to any good. And then Jefferson Campbell was so very gentle. Jefferson never did some things like other men, things that now were beginning to be ugly, for Melanctha. And then too Jefferson Campbell did not seem to know very well what it was that Melanctha really wanted, and all this was making Melanctha feel his power with her always getting stronger.

Dr. Campbell came in every day to see 'Mis' Herbert. 'Mis' Herbert, after that night they watched together, did get a little better, but 'Mis' Herbert was really very sick, and soon it was pretty sure that she would have to die. Melanctha certainly did everything, all the time, that any woman could. Jefferson never thought much better of Melanctha while she did it. It was not her being good, he wanted to find in her. He knew very well Jane Harden was right, when she said Melanctha was always being good to everybody but that that did not make Melanctha any better for her. Then too, 'Mis' Herbert never liked Melanctha any better, even on the last day of her living, and so Jefferson really never thought much of Melanctha's always being good to her mother.

Jefferson and Melanctha now saw each other, very often. They now always liked to be with each other, and they always now had a good time when they talked to one another. They, mostly in their talking to each other, still just talked about outside things and what they were thinking. Except just in little moments, and not those very often, they never said anything about their feeling. Sometimes Melanctha would tease Jefferson a little just to show she had not forgotten, but mostly she listened to his talking, for Jefferson still always liked to talk along about the things he believed in. Melanctha was liking Jefferson Campbell better every day, and Jefferson was beginning to know that Melanctha certainly had a good mind, and he was beginning to feel a little her real sweetness. Not in her being good to 'Mis' Herbert, that never seemed to Jefferson to mean much in her, but there was a strong kind of sweetness in Melanctha's nature that Jefferson began now to feel when he was with her.

'Mis' Herbert was now always getting sicker. One night again Dr. Campbell felt very certain that before it was morning she would surely die. Dr. Campbell said he would come back to help Melanctha watch her, and to do anything he could to make 'Mis' Herbert's dying more easy for her. Dr. Campbell came back that evening, after he was through with his other patients, and then he made 'Mis' Herbert easy, and then he came and sat down on the steps just above where Melanctha was sitting with the lamp, and looking very tired. Dr. Campbell was pretty tired too, and they both sat there very quiet.

"You look awful tired to-night, Dr. Campbell," Melanctha said at last, with her voice low and very gentle, "Don't you want to go lie down and sleep a little? You're always being much too good to everybody, Dr. Campbell. I like to have you stay here watching to-night with me, but it don't seem right you ought to stay here when you got so much always to do for everybody. You are certainly very kind to come back, Dr. Campbell, but I can certainly get along to-night without you. I can get help next door sure if I need it. You just go 'long home to bed, Dr. Campbell. You certainly do look as if you need it."

Jefferson was silent for some time, and always he was looking very gently at Melanctha.

"I certainly never did think, Miss Melanctha, I would find you to be so sweet and thinking, with me." "Dr. Campbell" said Melanctha, still more gentle, "I certainly never did think that you would ever feel it good to like me. I certainly never did think you would want to see for yourself if I had sweet ways in me."

They both sat there very tired, very gentle, very quiet, a long time. At last Melanctha in a low, even tone began to talk to Jefferson Campbell.

"You are certainly a very good man, Dr. Campbell, I certainly do feel that more every day I see you. Dr. Campbell, I sure do want to be friends with a good man like you, now I know you. You certainly, Dr. Campbell, never do things like other men, that's always ugly for me. Tell me true, Dr. Campbell, how you feel about being always friends with me. I certainly do know, Dr. Campbell, you are a good man, and if you say you will be friends with me, you certainly never will go back on me, the way so many kinds of them do to every girl they ever get to like them. Tell me for true, Dr. Campbell, will you be friends with me."

"Why, Miss Melanctha," said Campbell slowly, "why you see I just can't say that right out that way to you. Why sure you know Miss Melanctha, I will be very glad if it comes by and by that we are always friends together, but you see, Miss Melanctha, I certainly am a very slow-minded quiet kind of fellow though I do say quick things all the time to everybody, and when I certainly do want to mean it what I am saying to you, I can't say things like that right out to everybody till I know really more for certain all about you, and how I like you, and what I really mean to do better for you. You certainly do see what I mean, Miss Melanctha." "I certainly do admire you for talking honest to me, Jeff Campbell," said Melanctha. "Oh, I am always honest, Miss Melanctha. It's easy enough for me always to be honest, Miss Melanctha. All I got to do is always just to say right out what I am thinking. I certainly never have got any real reason for not saying it right out like that to anybody."

They sat together, very silent. "I certainly do wonder, Miss Melanctha," at last began Jeff Campbell, "I certainly do wonder, if we know very right, you and me, what each other is really thinking. I certainly do wonder, Miss Melanctha, if we know at all really what each other means by what we are always saying." "That certainly do mean, by what you say, that you think I am a bad one, Jeff Campbell," flashed out Melanctha. "Why no, Miss Melanctha, why sure I don't mean any thing like that at all, by what I am saying to you. You know well I do, Miss Melanctha, I think better of you every day I see you, and I like to talk with you all the time now, Miss Melanctha, and I certainly do think we both like it very well when we are together, and it seems to me always more, you are very good and sweet always to everybody. It only is, I am really so slow-minded in my ways, Miss Melanctha, for all I talk so quick to everybody, and I don't like to say to you what I don't know for very sure, and I certainly don't know for sure I know just all what you mean by what you are always saying to me. And you see, Miss Melanctha, that's what makes me say what I was just saying to you when you asked me."

"I certainly do thank you again for being honest to me, Dr. Campbell," said Melanctha. "I guess I leave you now, Dr. Campbell. I think I go in the other room and rest a little. I leave you here, so perhaps if I ain't here you will maybe sleep and rest yourself a little. Good night now, Dr. Campbell, I call you if I need you later to help me, Dr. Campbell, I hope you rest well, Dr. Campbell."

Jeff Campbell, when Melanctha left him, sat there and he was very quiet and just wondered. He did not know very well just what Melanctha meant by what she was always saying to him. He did not know very well how much he really knew about Melanctha Herbert. He wondered if he should go on being so much all the time with her. He began to think about what he should do now with her. Jefferson Campbell was a man who liked everybody and many people liked very much to be with him. Women liked him, he was so strong, and good, and understanding, and innocent, and firm, and gentle. Sometimes they seemed to want very much he should be with them. When they got so, they always had made Campbell very tired. Sometimes he would play a little with them, but he never had had any strong feeling for them. Now with Melanctha Herbert everything seemed different. Jefferson was not sure that he knew here just what he wanted. He was not sure he knew just what it was that Melanctha wanted. He knew if it was only play, with Melanctha, that he did not want to do it. But he remembered always how she had told him he never knew how to feel things very deeply. He remembered how she told him he was afraid to let himself ever know real feeling, and then too, most of all to him, she had told him he was not very understanding. That always troubled Jefferson very keenly, he wanted very badly to be really understanding. If Jefferson only knew better just what Melanctha meant by what she said. Jefferson always had thought he knew something about women. Now he found that really he knew nothing. He did not know the least bit about Melanctha. He did not know what it was right that he should do about it. He wondered if it was just a little play that they were doing. If it was a play he did not want to go on playing, but if it was really that he was not very understanding, and that with Melanctha Herbert he could learn to really understand, then he was very certain he did not want to be a coward. It was very hard for him to know what he wanted. He thought and thought, and always he did not seem to know any better what he wanted. At last he gave up this thinking. He felt sure it was only play with Melanctha. "No, I certainly won't go on fooling with her any more this way," he said at last out loud to himself, when he was through with this thinking. "I certainly will stop fooling, and begin to go on with my thinking about my work and what's the matter with people like 'Mis' Herbert," and Jefferson took out his book from his pocket, and drew near to the lamp, and began with some hard scientific reading.

Jefferson sat there for about an hour reading, and he had really forgotten all about his trouble with Melanctha's meaning. Then 'Mis' Herbert had some trouble with her breathing. She woke up and was gasping. Dr. Campbell went to her and gave her something that would help her. Melanctha came out from the other room and did things as he told her. They together made 'Mis' Herbert more comfortable and easy, and soon she was again in her deep sleep.

Dr. Campbell went back to the steps where he had been sitting. Melanctha came and stood a little while beside him, and then she sat down and watched him reading. By and by they began with their talking. Jeff Campbell began to feel that perhaps it was all different. Perhaps it was not just play, with Melanctha. Anyway he liked it very well that she was with him. He began to tell her about the book he was just reading.

Melanctha was very intelligent always in her questions. Jefferson knew now very well that she had a good mind. They were having a very good time, talking there together. And then they began again to get quiet.

"It certainly was very good in you to come back and talk to me Miss Melanctha," Jefferson said at last to her, for now he was almost certain, it was no game she was playing. Melanctha really was a good woman, and she had a good mind, and she had a real, strong sweetness, and she could surely really teach him. "Oh I always like to talk to you Dr. Campbell" said Melanctha, "And then you was only just honest to me, and I always like it when a man is really honest to me." Then they were again very silent, sitting there together, with the lamp between them, that was always smoking. Melanctha began to lean a little more toward Dr. Campbell, where he was sitting, and then she took his hand between her two and pressed it hard, but she said nothing to him. She let it go then and leaned a little nearer to him. Jefferson moved a little but did not do anything in answer. At last, "Well," said Melanctha sharply to him. "I was just thinking" began Dr. Campbell slowly, "I was just wondering," he was beginning to get ready to go on with his talking. "Don't you ever stop with your thinking long enough ever to have any feeling Jeff Campbell," said Melanctha a little sadly. "I don't know," said Jeff Campbell slowly, "I don't know Miss Melanctha much about that. No, I don't stop thinking much Miss Melanctha and if I can't ever feel without stopping thinking, I certainly am very much afraid Miss Melanctha that I never will do much with that kind of feeling. Sure you ain't worried Miss Melanctha, about my really not feeling very much all the time. I certainly do think I feel some, Miss Melanctha, even though I always do it without ever knowing how to stop with my thinking." "I am certainly afraid I don't think much of your kind of feeling Dr. Campbell." "Why I think you certainly are wrong Miss Melanctha I certainly do think I feel as much for you Miss Melanctha, as you ever feel about me, sure I do. I don't think you know me right when you talk like that to me. Tell me just straight out how much do you care about me, Miss Melanctha." "Care about you Jeff Campbell," said Melanctha slowly. "I certainly do care for you Jeff Campbell less than you are always thinking and much more than you are ever knowing."

Jeff Campbell paused on this, and he was silent with the power of Melanctha's meaning. They sat there together very silent, a long time. "Well Jeff Campbell," said Melanctha. "Oh," said Dr. Campbell and he moved himself a little, and then they were very silent a long time. "Haven't you got nothing to say to me Jeff Campbell?" said Melanctha. "Why yes, what was it we were just saying about to one another. You see Miss Melanctha I am a very quiet, slow minded kind of fellow, and I am never sure I know just exactly what you mean by all that you are always saying to me. But I do like you very much Miss Melanctha and I am very sure you got very good things in you all the time. You sure do believe what I am saying to you Miss Melanctha." "Yes I believe it when you say it to me, Jeff Campbell," said Melanctha, and then she was silent and there was much sadness in it. "I guess I go in and lie down again Dr. Campbell," said Melanctha. "Don't go leave me Miss Melanctha," said Jeff Campbell quickly. "Why not, what you want of me Jeff Campbell?" said Melanctha. "Why," said Jeff Campbell slowly, "I just want to go on talking with you. I certainly do like talking about all kinds of things with you. You certainly know that all right, Miss Melanctha." "I guess I go lie down again and leave you

here with your thinking," said Melanctha gently. "I certainly am very
tired to night Dr. Campbell. Good night I hope you rest well Dr.
Campbell." Melanctha stooped over him, where he was sitting, to say
this good night, and then, very quick and sudden, she kissed him and
then, very quick again, she went away and left him.

Dr. Campbell sat there very quiet, with only a little thinking and
sometimes a beginning feeling, and he was alone until it began to be
morning, and then he went, and Melanctha helped him, and he made
'Mis' Herbert more easy in her dying. 'Mis' Herbert lingered on till
about ten o'clock the next morning, and then slowly and without
much pain she died away. Jeff Campbell staid till the last moment,
with Melanctha, to make her mother's dying easy for her. When it was
over he sent in the colored woman from next door to help Melanctha
fix things, and then he went away to take care of his other patients. He
came back very soon to Melanctha. He helped her to have a funeral
for her mother. Melanctha then went to live with the good natured
woman, who had been her neighbor. Melanctha still saw Jeff Campbell very often. Things began to be very strong between them.

Melanctha now never wandered, unless she was with Jeff Campbell. Sometimes she and he wandered a good deal together. Jeff Campbell had not got over his way of talking to her all the time about all the
things he was always thinking. Melanctha never talked much, now,
when they were together. Sometimes Jeff Campbell teased her about
her not talking to him. "I certainly did think Melanctha you was a
great talker from the way Jane Harden and everybody said things to
me, and from the way I heard you talk so much when I first met you.
Tell me true Melanctha, why don't you talk more now to me, perhaps
it is I talk so much I don't give you any chance to say things to me, or
perhaps it is you hear me talk so much you don't think so much now
of a whole lot of talking. Tell me honest Melanctha, why don't you
talk more to me." "You know very well Jeff Campbell," said Melanctha "You certainly do know very well Jeff, you don't think really
much, of my talking. You think a whole lot more about everything
than I do Jeff, and you don't care much what I got to say about it. You
know that's true what I am saying Jeff, if you want to be real honest,
the way you always are when I like you so much." Jeff laughed and
looked fondly at her. "I don't say ever I know, you ain't right, when
you say things like that to me, Melanctha. You see you always like to
be talking just what you think everybody wants to be hearing from
you, and when you are like that, Melanctha, honest, I certainly don't
care very much to hear you, but sometimes you say something that is

what you are really thinking, and then I like a whole lot to hear you talking." Melanctha smiled, with her strong sweetness, on him, and she felt her power very deeply. "I certainly never do talk very much when I like anybody really, Jeff. You see, Jeff, it ain't much use to talk about what a woman is really feeling in her. You see all that, Jeff, better, by and by, when you get to really feeling. You won't be so ready then always with your talking. You see, Jeff, if it don't come true what I am saying." "I don't ever say you ain't always right, Melanctha," said Jeff Campbell. "Perhaps what I call my thinking ain't really so very understanding. I don't say, no never now any more, you ain't right, Melanctha, when you really say things to me. Perhaps I see it all to be very different when I come to really see what you mean by what you are always saying to me." "You is very sweet and good to me always, Jeff Campbell," said Melanctha. " 'Deed I certainly am not good to you, Melanctha. Don't I bother you all the time with my talking, but I really do like you a whole lot, Melanctha." "And I like you, Jeff Campbell, and you certainly are mother, and father, and brother, and sister, and child and everything, always to me. I can't say much about how good you been to me, Jeff Campbell, I never knew any man who was good and didn't do things ugly, before I met you to take care of me, Jeff Campbell. Good-by, Jeff, come see me to-morrow, when you get through with your working." "Sure Melanctha, you know that already," said Jeff Campbell, and then he went away and left her.

These months had been an uncertain time for Jeff Campbell. He never knew how much he really knew about Melanctha. He saw her now for long times and very often. He was beginning always more and more to like her. But he did not seem to himself to know very much about her. He was beginning to feel he could almost trust the goodness in her. But then, always, really, he was not very sure about her. Melanctha always had ways that made him feel uncertain with her, and yet he was so near, in his feeling for her. He now never thought about all this in real words any more. He was always letting it fight itself out in him. He was now never taking any part in this fighting that was always going on inside him.

Jeff always loved now to be with Melanctha and yet he always hated to go to her. Somehow he was always afraid when he was to go to her, and yet he had made himself very certain that here he would not be a coward. He never felt any of this being afraid, when he was with her. Then they always were very true, and near to one another. But always when he was going to her, Jeff would like anything that could happen that would keep him a little longer from her.

It was a very uncertain time, all these months, for Jeff Campbell. He did not know very well what it was that he really wanted. He was very certain that he did not know very well what it was that Melanctha wanted. Jeff Campbell had always all his life loved to be with people, and he had loved all his life always to be thinking, but he was still only a great boy, was Jeff Campbell, and he had never before had any of this funny kind of feeling. Now, this evening, when he was free to go and see Melanctha, he talked to anybody he could find who would detain him, and so it was very late when at last he came to the house where Melanctha was waiting to receive him.

Jeff came in to where Melanctha was waiting for him, and he took off his hat and heavy coat, and then drew up a chair and sat down by the fire. It was very cold that night, and Jeff sat there, and rubbed his hands and tried to warm them. He had only said "How do you do" to Melanctha, he had not yet begun to talk to her. Melanctha sat there, by the fire, very quiet. The heat gave a pretty pink glow to her pale yellow and attractive face. Melanctha sat in a low chair, her hands, with their long, fluttering fingers, always ready to show her strong feeling, were lying quiet in her lap. Melanctha was very tired with her waiting for Jeff Campbell. She sat there very quiet and just watching. Jeff was a robust, dark, healthy, cheery negro. His hands were firm and kindly and unimpassioned. He touched women always with his big hands, like a brother. He always had a warm broad glow, like southern sunshine. He never had anything mysterious in him. He was open, he was pleasant, he was cheery, and always he wanted, as Melanctha once had wanted, always now he too wanted really to understand.

Jeff sat there this evening in his chair and was silent a long time, warming himself with the pleasant fire. He did not look at Melanctha who was watching. He sat there and just looked into the fire. At first his dark, open face was smiling, and he was rubbing the back of his black-brown hand over his mouth to help him in his smiling. Then he was thinking, and he frowned and rubbed his head hard, to help him in his thinking. Then he smiled again, but now his smiling was not very pleasant. His smile was now wavering on the edge of scorning. His smile changed more and more, and then he had a look as if he were deeply down, all disgusted. Now his face was darker, and he was bitter in his smiling, and he began, without looking from the fire, to talk to Melanctha, who was now very tense with her watching.

"Melanctha Herbert", began Jeff Campbell, "I certainly after all this time I know you, I certainly do know little, real about you. You see, Melanctha, it's like this way with me"; Jeff was frowning, with his thinking and looking hard into the fire, "You see it's just this way, with me now, Melanctha. Sometimes you seem like one kind of a girl to me, and sometimes you are like a girl that is all different to me, and the two kinds of girls is certainly very different to each other, and I can't see any way they seem to have much to do, to be together in you. They certainly don't seem to be made much like as if they could have anything really to do with each other. Sometimes you are a girl to me I certainly never would be trusting, and you got a laugh then so hard, it just rattles, and you got ways so bad, I can't believe you mean them hardly, and yet all that I just been saying is certainly you one way I often see you, and it's what your mother and Jane Harden always found you, and it's what makes me hate so, to come near you. And then certainly sometimes, Melanctha, you certainly is all a different creature, and sometimes then there comes out in you what is certainly a thing, like a real beauty. I certainly, Melanctha, never can tell just how it is that it comes so lovely. Seem to me when it comes it's got a real sweetness, that is more wonderful than a pure flower, and a gentleness, that is more tender than the sunshine, and a kindness, that makes one feel like summer, and then a way to know, that makes everything all over, and all that, and it does certainly seem to be real for the little while it's lasting, for the little while that I can surely see it, and it gives me to feel like I certainly had got real religion. And then when I got rich with such a feeling, comes all that other girl, and then that seems more likely that that is really you what's honest, and then I certainly do get awful afraid to come to you, and I certainly never do feel I could be very trusting with you. And then I certainly don't know anything at all about you, and I certainly don't know which is a real Melanctha Herbert, and I certainly don't feel no longer, I ever want to talk to you. Tell me honest, Melanctha, which is the way that is you really, when you are alone, and real, and all honest. Tell me, Melanctha, for I certainly do want to know it."

Melanctha did not make him any answer, and Jeff, without looking at her, after a little while, went on with his talking. "And then, Melanctha, sometimes you certainly do seem sort of cruel, and not to care about people being hurt or in trouble, something so hard about you it makes me sometimes real nervous, sometimes somehow like you always, like your being, with 'Mis' Herbert. You sure did do everything that any woman could, Melanctha, I certainly never did see anybody do things any better, and yet, I don't know how to say just what I mean, Melanctha, but there was something awful hard about your feeling, so different from the way I'm always used to see good people feeling, and so it was the way Jane Harden and 'Mis' Herbert talked when they felt strong to talk about you, and yet, Melanctha, somehow I feel so really near to you, and you certainly have got an awful wonderful, strong kind of sweetness. I certainly would like to know for sure, Melanctha, whether I got really anything to be afraid for. I certainly did think once, Melanctha, I knew something about all kinds of women. I certainly know now really, how I don't know anything sure at all about you, Melanctha, though I been with you so long, and so many times for whole hours with you, and I like so awful much to be with you, and I can always say anything I am thinking to you. I certainly do awful wish, Melanctha, I really was more understanding. I certainly do that same, Melanctha."

Jeff stopped now and looked harder than before into the fire. His face changed from his thinking back into that look that was so like as if he was all through and through him, disgusted with what he had been thinking. He sat there a long time, very quiet, and then slowly, somehow, it came strongly to him that Melanctha Herbert, there beside him, was trembling and feeling it all to be very bitter. "Why, Melanctha," cried Jeff Campbell, and he got up and put his arm around her like a brother. "I stood it just so long as I could bear it, Jeff," sobbed Melanctha, and then she gave herself away, to her misery, "I was awful ready, Jeff, to let you say anything you liked that gave you any pleasure. You could say all about me what you wanted, Jeff, and I would try to stand it, so as you would be sure to be liking it, Jeff, but you was too cruel to me. When you do that kind of seeing how much you can make a woman suffer, you ought to give her a little rest, once sometimes, Jeff. They can't any of us stand it so for always, Jeff. I certainly did stand it just as long as I could, so you would like it, but I, — oh Jeff, you went on too long to-night Jeff. I couldn't stand it not a minute longer the way you was doing of it, Jeff. When you want to be seeing how the way a woman is really made of, Jeff, you shouldn't never be so cruel, never to be thinking how much she can stand, the strong way you always do it, Jeff." "Why, Melanctha," cried Jeff Campbell, in his horror, and then he was very tender to her, and like a good, strong, gentle brother in his soothing of her, "Why Melanctha dear, I certainly don't now see what it is you mean by what you was just saying to me. Why Melanctha, you poor little girl, you certainly never did believe I ever knew I was giving you real suffering. Why, Melanctha, how could you ever like me if you thought I ever could be so like a red Indian?" "I didn't just know, Jeff," and Melanctha nestled to him, "I certainly never did know just what it was you wanted to be doing with me, but I certainly wanted you should do anything you liked, you wanted, to make me more understanding for you. I tried awful hard to stand it, Jeff, so as you could do anything you wanted with me." "Good Lord and Jesus Christ, Melanctha!" cried Jeff Campbell. "I certainly never can know anything about you real, Melanctha, you poor little girl," and Jeff drew her closer to him, "But I certainly do admire and trust you a whole lot now, Melanctha. I certainly do, for I certainly never did think I was hurting you at all, Melanctha, by the things I always been saying to you. Melanctha, you poor little, sweet, trembling baby now, be good, Melanctha. I certainly can't ever tell you how awful sorry I am to hurt you so, Melanctha. I do anything I can to show you how I never did mean to hurt you, Melanctha." "I know, I know," murmured Melanctha, clinging to him. "I know you are a good man, Jeff. I always know that, no matter how much you can hurt me." "I sure don't see how you can think so, Melanctha, if you certainly did think I was trying so hard just to hurt you." "Hush, you are only a great big boy, Jeff Campbell, and you don't know nothing yet about real hurting," said Melanctha, smiling up through her crying, at him. "You see, Jeff, I never knew anybody I could know real well and yet keep on always respecting, till I came to know you real well, Jeff." "I sure don't understand that very well, Melanctha. I ain't a bit better than just lots of others of the colored people. You certainly have been unlucky with the kind you met before me, that's all, Melanctha. I certainly ain't very good, Melanctha." "Hush, Jeff, you don't know nothing at all about what you are," said Melanctha. "Perhaps you are right, Melanctha. I don't say ever any more, you ain't right, when you say things to me, Melanctha," and Jefferson sighed, and then he smiled, and then they were quiet a long time together, and then after some more kindness, it was late, and then Jeff left her.

Jeff Campbell, all these months, had never told his good mother anything about Melanctha Herbert. Somehow he always kept his seeing her so much now, to himself. Melanctha too had never had any of her other friends meet him. They always acted together, these two, as if their being so much together was a secret, but really there was no one who would have made it any harder for them. Jeff Campbell did not really know how it had happened that they were so secret. He did not know if it was what Melanctha wanted. Jeff had never spoken to her at all about it. It just seemed as if it were well understood between them that nobody should know that they were so much together. It was as if it were agreed between them, that they should be alone by themselves always, and so they would work out together what they meant by what they were always saying to each other.

Jefferson often spoke to Melanctha about his good mother. He never said anything about whether Melanctha would want to meet her. Jefferson never quite understood why all this had happened so, in secret. He never really knew what it was that Melanctha really wanted. In all these ways he just, by his nature, did, what he sort of felt Melanctha wanted. And so they continued to be alone and much together, and now it had come to be the spring time, and now they had all out-doors to wander.

They had many days now when they were very happy. Jeff every day found that he really liked Melanctha better. Now surely he was beginning to have real, deep feeling in him. And still he loved to talk himself out to Melanctha, and he loved to tell her how good it all was to him, and how he always loved to be with her, and to tell her always all about it. One day, now Jeff arranged, that Sunday they would go out and have a happy, long day in the bright fields, and they would be all day just alone together. The day before, Jeff was called in to see Jane Harden.

Jane Harden was very sick almost all day and Jeff Campbell did everything he could to make her better. After a while Jane became more easy and then she began to talk to Jeff about Melanctha. Jane did not know how much Jeff was now seeing of Melanctha. Jane these days never saw Melanctha. Jane began to talk of the time when she first knew Melanctha. Jane began to tell how in these days Melanctha had very little understanding. She was young then and she had a good mind. Jane Harden never would say Melanctha never had a good mind, but in those days Melanctha certainly had not been very understanding. Jane began to explain to Jeff Campbell how in every way, she Jane, had taught Melanctha. Jane then began to explain how eager Melanctha always had been for all that kind of learning. Jane Harden began to tell how they had wandered. Jane began to tell how Melanctha once had loved her, Jane Harden. Jane began to tell Jeff of all the bad ways Melanctha had used with her. Jane began to tell all she knew of the way Melanctha had gone on, after she had left her. Jane began to tell all about the different men, white ones and blacks, Melanctha never was particular about things like that, Jane Harden said in passing, not that Melanctha was a bad one, and she had a good mind, Jane Harden never would say that she hadn't, but Melanctha always liked to use all the understanding ways that Jane had taught her, and so she wanted to know everything, always, that they knew how to teach her.

Jane was beginning to make Jeff Campbell see much clearer. Jane Harden did not know what it was that she was really doing with all this talking. Jane did not know what Jeff was feeling. Jane was always honest when she was talking, and now it just happened she had started talking about her old times with Melanctha Herbert. Jeff understood very well that it was all true what Jane was saying. Jeff Campbell was beginning now to see very clearly. He was beginning to feel very sick inside him. He knew now many things Melanctha had not yet taught him. He felt very sick and his heart was very heavy, and Melanctha certainly did seem very ugly to him. Jeff was at last beginning to know what it was to have deep feeling. He took care a little longer of Jane Harden, and then he went to his other patients, and then he went home to his room, and he sat down and at last he had stopped thinking. He was very sick and his heart was very heavy in him. He was very tired and all the world was very dreary to him, and he knew very well now at last, he was really feeling. He knew it now from the way it hurt him. He knew very well that now at last he was beginning to really have understanding. The next day he had arranged to spend, long and happy, all alone in the spring fields with Melanctha, wandering. He wrote her a note and said he could not go, he had a sick patient and would have to stay home with him. For three days after, he made no sign to Melanctha. He was very sick all these days, and his heart was very heavy in him, and he knew very well that now at last he had learned what it was to have deep feeling.

At last one day he got a letter from Melanctha. "I certainly don't rightly understand what you are doing now to me Jeff Campbell," wrote Melanctha Herbert. "I certainly don't rightly understand Jeff Campbell why you ain't all these days been near me, but I certainly do suppose it's just another one of the queer kind of ways you have to be good, and repenting of yourself all of a sudden. I certainly don't say to you Jeff Campbell I admire very much the way you take to be good Jeff Campbell. I am sorry Dr. Campbell, but I certainly am afraid I can't stand it no more from you the way you have been just acting. I certainly can't stand it any more the way you act when you have been as if you thought I was always good enough for anybody to have with them, and then you act as if I was a bad one and you always just despise me. I certainly am afraid Dr. Campbell I can't stand it any more like that. I certainly can't stand it any more the way you are always changing. I certainly am afraid Dr. Campbell you ain't man enough to deserve to have anybody care so much to be always with you. I certainly am awful afraid Dr. Campbell I don't ever any more

want to really see you. Good-by Dr. Campbell I wish you always to be
real happy."

Jeff Campbell sat in his room, very quiet, a long time, after he got
through reading this letter. He sat very still and first he was very angry.
As if he, too, did not know very badly what it was to suffer keenly. As
if he had not been very strong to stay with Melanctha when he never
knew what it was that she really wanted. He knew he was very right to
be angry, he knew he really had not been a coward. He knew Melanctha had done many things it was very hard for him to forgive her. He
knew very well he had done his best to be kind, and to trust her, and to
be loyal to her, and now; — and then Jeff suddenly remembered how
one night Melanctha had been so strong to suffer, and he felt come
back to him the sweetness in her, and then Jeff knew that really, he
always forgave her, and that really, it all was that he was so sorry he
had hurt her, and he wanted to go straight away and be a comfort to
her. Jeff knew very well, that what Jane Harden had told him about
Melanctha and her bad ways, had been a true story, and yet he wanted
very badly to be with Melanctha. Perhaps she could teach him to really
understand it better. Perhaps she could teach him how it could be all
true, and yet how he could be right to believe in her and to trust her.

Jeff sat down and began his answer to her. "Dear Melanctha," Jeff
wrote to her. "I certainly don't think you got it all just right in the letter, I just been reading, that you just wrote me. I certainly don't think
you are just fair or very understanding to all I have to suffer to keep
straight on to really always to believe in you and trust you. I certainly
don't think you always are fair to remember right how hard it is for a
man, who thinks like I was always thinking, not to think you do things
very bad very often. I certainly don't think, Melanctha, I ain't right
when I was so angry when I got your letter to me. I know very well,
Melanctha, that with you, I never have been a coward. I find it very
hard, and I never said it any different, it is hard to me to be understanding, and to know really what it is you wanted, and what it is you
are meaning by what you are always saying to me. I don't say ever, it
ain't very hard for you to be standing that I ain't very quick to be following whichever way that you are always leading. You know very
well, Melanctha, it hurts me very bad and way inside me when I have
to hurt you, but I always got to be real honest with you. There ain't no
other way for me to be, with you, and I know very well it hurts me too,
a whole lot, when I can't follow so quick as you would have me. I
don't like to be a coward to you, Melanctha, and I don't like to say
what I ain't meaning to you. And if you don't want me to do things

honest, Melanctha, why I can't ever talk to you, and you are right when you say, you never again want to see me, but if you got any real sense of what I always been feeling with you, and if you got any right sense, Melanctha, of how hard I been trying to think and to feel right for you, I will be very glad to come and see you, and to begin again with you. I don't say anything now, Melanctha, about how bad I been this week, since I saw you, Melanctha. It don't ever do any good to talk such things over. All I know is I do my best, Melanctha, to you, and I don't say, no, never, I can do any different than just to be honest and come as fast as I think it's right for me to be going in the ways you teach me to be really understanding. So don't talk any more foolishness, Melanctha, about my always changing. I don't change, never, and I got to do what I think is right and honest to me, and I never told you any different, and you always knew it very well that I always would do just so. If you like me to come and see you to-morrow, and go out with you, I will be very glad to, Melanctha. Let me know right away, what it is you want me to be doing for you, Melanctha.

Very truly yours,
JEFFERSON CAMPBELL

"Please come to me, Jeff." Melanctha wrote back for her answer. Jeff went very slowly to Melanctha, glad as he was, still to be going to her. Melanctha came, very quick, to meet him, when she saw him from where she had been watching for him. They went into the house together. They were very glad to be together. They were very good to one another.

"I certainly did think, Melanctha, this time almost really, you never did want me to come to you at all any more to see you," said Jeff Campbell to her, when they had begun again with their talking to each other. "You certainly did make me think, perhaps really this time, Melanctha, it was all over, my being with you ever, and I was very mad, and very sorry, too, Melanctha."

"Well you certainly was very bad to me, Jeff Campbell," said Melanctha, fondly.

"I certainly never do say any more you ain't always right, Melanctha," Jeff answered and he was very ready now with cheerful laughing, "I certainly never do say that any more, Melanctha, if I know it, but still, really, Melanctha, honest, I think perhaps I wasn't real bad to you any more than you just needed from me."

Jeff held Melanctha in his arms and kissed her. He sighed then and was very silent with her. "Well, Melanctha," he said at last, with some more laughing, "well, Melanctha, any way you can't say ever it ain't, if we are ever friends good and really, you can't say, no, never, but that we certainly have worked right hard to get both of us together for it, so we shall sure deserve it then, if we can ever really get it." "We certainly have worked real hard, Jeff, I can't say that ain't all right the way you say it," said Melanctha. "I certainly never can deny it, Jeff, when I feel so worn with all the trouble you been making for me, you bad boy, Jeff," and then Melanctha smiled and then she sighed, and then she was very silent with him.

At last Jeff was to go away. They stood there on the steps for a long time trying to say good-by to each other. At last Jeff made himself really say it. At last he made himself, that he went down the steps and went away.

On the next Sunday they arranged, they were to have the long happy day of wandering that they had lost last time by Jane Harden's talking. Not that Melanctha Herbert had heard yet of Jane Harden's talking.

Jeff saw Melanctha every day now. Jeff was a little uncertain all this time inside him, for he had never yet told to Melanctha what it was that had so nearly made him really want to leave her. Jeff knew that for him, it was not right he should not tell her. He knew they could only have real peace between them when he had been honest, and had really told her. On this long Sunday Jeff was certain that he would really tell her.

They were very happy all that day in their wandering. They had taken things along to eat together. They sat in the bright fields and they were happy, they wandered in the woods and they were happy. Jeff always loved in this way to wander. Jeff always loved to watch everything as it was growing, and he loved all the colors in the trees and on the ground, and the little, new, bright colored bugs he found in the moist ground and in the grass he loved to lie on and in which he was always so busy searching. Jeff loved everything that moved and that was still, and that had color, and beauty, and real being.

Jeff loved very much this day while they were wandering. He almost forgot that he had any trouble with him still inside him. Jeff loved to be there with Melanctha Herbert. She was always so sympathetic to him for the way she listened to everything he found and told her, the way she felt his joy in all this being, the way she never said she wanted anything different from the way they had it. It was certainly a busy and a happy day, this their first long day of really wandering.

Later they were tired, and Melanctha sat down on the ground, and Jeff threw himself his full length beside her. Jeff lay there, very quiet, and then he pressed her hand and kissed it and murmured to her, "You certainly are very good to me, Melanctha." Melanctha felt it very deep and did not answer. Jeff lay there a long time, looking up above him. He was counting all the little leaves he saw above him. He was following all the little clouds with his eyes as they sailed past him. He watched all the birds that flew high beyond him, and all the time Jeff knew he must tell to Melanctha what it was he knew now, that which Jane Harden, just a week ago, had told him. He knew very well that for him it was certain that he had to say it. It was hard, but for Jeff Campbell the only way to lose it was to say it, the only way to know Melanctha really, was to tell her all the struggle he had made to know her, to tell her so she could help him to understand his trouble better, to help him so that never again he could have any way to doubt her.

Jeff lay there a long time, very quiet, always looking up above him, and yet feeling very close now to Melanctha. At last he turned a little toward her, took her hands closer in his to make him feel it stronger, and then very slowly, for the words came very hard for him, slowly he began his talk to her.

"Melanctha," began Jeff, very slowly, "Melanctha, it ain't right I shouldn't tell you why I went away last week and almost never got the chance again to see you. Jane Harden was sick, and I went in to take care of her. She began to tell everything she ever knew about you. She didn't know how well now I know you. I didn't tell her not to go on talking. I listened while she told me everything about you. I certainly found it very hard with what she told me. I know she was talking truth in everything she said about you. I knew you had been free in your ways, Melanctha, I knew you liked to get excitement the way I always hate to see the colored people take it. I didn't know, till I heard Jane Harden say it, you had done things so bad, Melanctha. When Jane Harden told me, I got very sick, Melanctha. I couldn't bear hardly, to think, perhaps I was just another like them to you, Melanctha. I was wrong not to trust you perhaps, Melanctha, but it did make things very ugly to me. I try to be honest to you, Melanctha, the way you say you really want it from me."

Melanctha drew her hands from Jeff Campbell. She sat there, and there was deep scorn in her anger.

"If you wasn't all through just selfish and nothing else, Jeff Campbell, you would take care you wouldn't have to tell me things like this, Jeff Campbell."

Jeff was silent a little, and he waited before he gave his answer. It was not the power of Melanctha's words that held him, for, for them, he had his answer, it was the power of the mood that filled Melanctha, and for that he had no answer. At last he broke through this awe, with his slow fighting resolution, and he began to give his answer.

"I don't say ever, Melanctha," he began, "it wouldn't have been more right for me to stop Jane Harden in her talking and to come to you to have you tell me what you were when I never knew you. I don't say it, no never to you, that that would not have been the right way for me to do, Melanctha. But I certainly am without any kind of doubting, I certainly do know for sure, I had a good right to know about what you were and your ways and your trying to use your understanding, every kind of way you could to get your learning. I certainly did have a right to know things like that about you, Melanctha. I don't say it ever, Melanctha, and I say it very often, I don't say ever I shouldn't have stopped Jane Harden in her talking and come to you and asked you yourself to tell me all about it, but I guess I wanted to keep myself from how much it would hurt me more, to have you yourself say it to me. Perhaps it was I wanted to keep you from having it hurt you so much more, having you to have to tell it to me. I don't know, I don't say it was to help you from being hurt most, or to help me. Perhaps I was a coward to let Jane Harden tell me 'stead of coming straight to you, to have you tell me, but I certainly am sure, Melanctha, I certainly had a right to know such things about you. I don't say it ever, ever, Melanctha, I hadn't the just right to know those things about you."
Melanctha laughed her harsh laugh. "You needn't have been under no kind of worry, Jeff Campbell, about whether you should have asked me. You could have asked, it wouldn't have hurt nothing. I certainly never would have told you nothing." "I am not so sure of that, Melanctha," said Jeff Campbell. "I certainly do think you would have told me. I certainly do think I could make you feel it right to tell me. I certainly do think all I did wrong was to let Jane Harden tell me. I certainly do know I never did wrong, to learn what she told me. I certainly know very well, Melanctha, if I had come here to you, you would have told it all to me, Melanctha."
He was silent, and this struggle lay there, strong, between them. It was a struggle, sure to be going on always between them. It was a struggle that was as sure always to be going on between them, as their minds and hearts always were to have different ways of working.
At last Melanctha took his hand, leaned over him and kissed him. "I sure am very fond of you, Jeff Campbell," Melanctha whispered to him.

Now for a little time there was not any kind of trouble between Jeff Campbell and Melanctha Herbert. They were always together now for long times, and very often. They got much joy now, both of them, from being all the time together. It was summer now, and they had warm sunshine to wander. It was summer now, and Jeff Campbell had more time to wander, for colored people never get sick much in summer. It was summer now, and there was a lovely silence everywhere, and all the noises, too, that they heard around them were lovely ones, and added to the joy, in these warm days, they loved so much to be together.

They talked some to each other in these days, did Jeff Campbell and Melanctha Herbert, but always in these days their talking more and more was like it always is with real lovers. Jeff did not talk so much now about what he before always had been thinking. Sometimes Jeff would be, as if he was just waking from himself to be with Melanctha, and then he would find he had been really all the long time with her, and he had really never needed to be doing any thinking.

It was sometimes pure joy Jeff would be talking to Melanctha, in these warm days he loved so much to wander with her. Sometimes Jeff would lose all himself in a strong feeling. Very often now, and always with more joy in his feeling, he would find himself, he did not know how or what it was he had been thinking. And Melanctha always loved very well to make him feel it. She always now laughed a little at him, and went back a little in him to his before, always thinking, and she teased him with his always now being so good with her in his feeling, and then she would so well and freely, and with her pure, strong ways of reaching, she would give him all the love she knew now very well, how much he always wanted to be sure he really had it.

And Jeff took it straight now, and he loved it, and he felt, strong, the joy of all this being, and it swelled out full inside him, and he poured it all out back to her in freedom, in tender kindness, and in joy, and in gentle brother fondling. And Melanctha loved him for it always, her Jeff Campbell now, who never did things ugly, for her, like all the men she always knew before always had been doing to her. And they loved it always, more and more, together, with this new feeling they had now, in these long summer days so warm; they, always together now, just these two so dear, more and more to each other always, and the summer evenings when they wandered, and the noises in the full streets, and the music of the organs, and the dancing, and the warm smell of the people, and of dogs and of the horses, and all the joy of the strong, sweet pungent, dirty, moist, warm negro southern summer.

Every day now, Jeff seemed to be coming nearer, to be really loving. Every day now, Melanctha poured it all out to him, with more freedom. Every day now, they seemed to be having more and more, both together, of this strong, right feeling. More and more every day now they seemed to know more really, what it was each other one always feeling. More and more now every day Jeff found in himself, he felt more trusting. More and more every day now, he did not think anything in words about what he was always doing. Every day now more and more Melanctha would let out to Jeff her real, strong feeling.

One day there had been much joy between them, more than they ever yet had had with their new feeling. All the day they had lost themselves in warm wandering. Now they were lying there and resting, with a green, bright, light-flecked world around them.

What was it that now really happened to them? What was it that Melanctha did, that made everything get all ugly for them? What was it that Melanctha felt then, that made Jeff remember all the feeling he had had in him when Jane Harden told him how Melanctha had learned to be so very understanding? Jeff did not know how it was that it had happened to him. It was all green, and warm, and very lovely to him, and now Melanctha somehow had made it all so ugly for him. What was it Melanctha was now doing with him? What was it he used to be thinking was the right way for him and all the colored people to be always trying to make it right, the way they should be always living? Why was Melanctha Herbert now all so ugly for him?

Melanctha Herbert somehow had made him feel deeply just then, what very more it was that she wanted from him. Jeff Campbell now felt in him what everybody always had needed to make them really understanding, to him. Jeff felt a strong disgust inside him; not for Melanctha herself, to him, not for himself really, in him, not for what it was that everybody wanted, in them; he only had disgust because he never could know really in him, what it was he wanted, to be really right in understanding, for him, he only had disgust because he never could know really what it was really right to him to be always doing, in the things he had before believed in, the things he before had believed in for himself and for all the colored people, the living regular, and the never wanting to be always having new things, just to keep on, always being in excitements. All the old thinking now came up very strong inside him. He sort of turned away then, and threw Melanctha from him.

Jeff never, even now, knew what it was that moved him. He never, even now, was ever sure, he really knew what Melanctha was, when she was real herself, and honest. He thought he knew, and then there came to him some moment, just like this one, when she really woke him up to be strong in him. Then he really knew he could know nothing. He knew then, he never could know what it was he really wanted with him. He knew then he never could know really what it was he felt inside him. It was all so mixed up inside him. All he knew was he wanted very badly Melanctha should be there beside him, and he wanted very badly, too, always to throw her from him. What was it really that Melanctha wanted with him? What was it really, he, Jeff Campbell, wanted she should give him? "I certainly did think now," Jeff Campbell groaned inside him, "I certainly did think now I really was knowing all right, what I wanted. I certainly did really think now I was knowing how to be trusting with Melanctha. I certainly did think it was like that now with me sure, after all I've been through all this time with her. And now I certainly do know I don't know anything that's very real about her. Oh the good Lord help and keep me!" and Jeff groaned hard inside him, and he buried his face deep in the green grass underneath him, and Melanctha Herbert was very silent there beside him.

Then Jeff turned to look and see her. She was lying very still there by him, and the bitter water[5] on her face was biting. Jeff was so very sorry then, all over and inside him, the way he always was when Melanctha had been deep hurt by him. "I didn't mean to be so bad again to you, Melanctha, dear one," and he was very tender to her. "I certainly didn't never mean to go to be so bad to you, Melanctha, darling. I certainly don't know, Melanctha, darling, what it is makes me act so to you sometimes, when I certainly ain't meaning anything like I want to hurt you. I certainly don't mean to be so bad, Melanctha, only it comes so quick on me before I know what I am acting to you. I certainly am all sorry, hard, to be so bad to you, Melanctha, darling." "I suppose, Jeff," said Melanctha, very low and bitter, "I suppose you are always thinking, Jeff, somebody had ought to be ashamed with us two together, and you certainly do think you don't see any way to it, Jeff, for me to be feeling that way ever, so you certainly don't see any way to it, only to do it just so often for me. That certainly is the way always with you, Jeff Campbell, if I understand you right the way you are always acting to me. That certainly is right the way I am saying it to

---

[5] *bitter water*: Tears — perhaps of frustration as much as sadness.

you now, Jeff Campbell. You certainly didn't anyway trust me now no more, did you, when you just acted so bad to me. I certainly am right the way I say it Jeff now to you. I certainly am right when I ask you for it now, to tell me what I ask you, about not trusting me more then again, Jeff, just like you never really knew me. You certainly never did trust me just then, Jeff, you hear me?" "Yes, Melanctha," Jeff answered slowly. Melanctha paused. "I guess I certainly never can forgive you this time, Jeff Campbell," she said firmly. Jeff paused too, and thought a little. "I certainly am afraid you never can no more now again, Melanctha," he said sadly.

They lay there very quiet now a long time, each one thinking very hard on their own trouble. At last Jeff began again to tell Melanctha what it was he was always thinking with her. "I certainly do know, Melanctha, you certainly now don't want any more to be hearing me just talking, but you see, Melanctha, really, it's just like this way always with me. You see, Melanctha, its like this way now all the time with me. You remember, Melanctha, what I was once telling to you, when I didn't know you very long together, about how I certainly never did know more than just two kinds of ways of loving, one way the way it is good to be in families and the other kind of way, like animals are all the time just with each other, and how I didn't ever like that last kind of way much for any of the colored people. You see Melanctha, it's like this way with me. I got a new feeling now, you been teaching to me, just like I told you once, just like a new religion to me, and I see perhaps what really loving is like, like really having everything together, new things, little pieces all different, like I always before been thinking was bad to be having, all go together like, to make one good big feeling. You see, Melanctha, it's certainly like that you make me been seeing, like I never know before any way there was of all kinds of loving to come together to make one way really truly lovely. I see that now, sometimes, the way you certainly been teaching me, Melanctha, really, and then I love you those times, Melanctha, like a real religion, and then it comes over me all sudden, I don't know anything real about you Melanctha, dear one, and then it comes over me sudden, perhaps I certainly am wrong now, thinking all this way so lovely, and not thinking now any more the old way I always before was always thinking, about what was the right way for me, to live regular and all the colored people, and then I think, perhaps, Melanctha you are really just a bad one, and I think, perhaps I certainly am doing it so because I just am too anxious to be just having all the time excitements, like I don't ever like really to be doing when I know it, and then

I always get so bad to you, Melanctha, and I can't help it with myself then, never, for I want to be always right really in the ways, I have to do them. I certainly do very badly want to be right, Melanctha, the only way I know is right Melanctha really, and I don't know any way, Melanctha, to find out really, whether my old way, the way I always used to be thinking, or the new way, you make so like a real religion to me sometimes, Melanctha, which way certainly is the real right way for me to be always thinking, and then I certainly am awful good and sorry, Melanctha, I always give you so much trouble, hurting you with the bad ways I am acting. Can't you help me to any way, to make it all straight for me, Melanctha, so I know right and real what it is I should be acting. You see, Melanctha, I don't want always to be a coward with you, if I only could know certain what was the right way for me to be acting. I certainly am real sure, Melanctha, that would be the way I would be acting, if I only knew it sure for certain now, Melanctha. Can't you help me any way to find out real and true, Melanctha, dear one. I certainly do badly want to know always, the way I should be acting."

"No, Jeff, dear, I certainly can't help you much in that kind of trouble you are always having. All I can do now, Jeff, is to just keep certainly with my believing you are good always, Jeff, and though you certainly do hurt me bad, I always got strong faith in you, Jeff, more in you certainly, than you seem to be having in your acting to me, always so bad, Jeff."

"You certainly are very good to me, Melanctha, dear one," Jeff said, after a long, tender silence. "You certainly are very good to me, Melanctha, darling, and me so bad to you always, in my acting. Do you love me good, and right, Melanctha, always?" "Always and always, you be sure of that now you have me. Oh you Jeff, you always be so stupid." "I certainly never can say now you ain't right, when you say that to me so, Melanctha," Jeff answered. "Oh, Jeff dear, I love you always, you know that now, all right, for certain. If you don't know it right now, Jeff, really, I prove it to you now, for good and always." And they lay there a long time in their loving, and then Jeff began again with his happy free enjoying.

"I sure am a good boy to be learning all the time the right way you are teaching me, Melanctha, darling," began Jeff Campbell, laughing, "You can't say no, never, I ain't a good scholar for you to be teaching now, Melanctha, and I am always so ready to come to you every day, and never playing hooky ever from you. You can't say ever, Melanctha, now can you, I ain't a real good boy to be always studying to be learning to be real bright, just like my teacher. You can't say ever to me, I ain't a good boy to you now, Melanctha." "Not near so good, Jeff Campbell, as such a good, patient kind of teacher, like me, who never teaches any ways it ain't good her scholars should be knowing, ought to be really having, Jeff, you hear me? I certainly don't think I am right for you, to be forgiving always, when you are so bad, and I so patient, with all this hard teaching always." "But you do forgive me always, sure, Melanctha, always?" "Always and always, you be sure Jeff, and I certainly am afraid I never can stop with my forgiving, you always are going to be so bad to me, and I always going to have to be so good with my forgiving." "Oh! Oh!" cried Jeff Campbell, laughing, "I ain't going to be so bad for always, sure I ain't, Melanctha, my own darling. And sure you do forgive me really, and sure you love me true and really, sure, Melanctha?" "Sure, sure, Jeff, boy, sure now and always, sure now you believe me, sure you do, Jeff, always." "I sure hope I does, with all my heart, Melanctha, darling." "I sure do that same, Jeff, dear boy, now you really know what it is to be loving, and I prove it to you now so, Jeff, you never can be forgetting. You see now, Jeff, good and certain, what I always before been saying to you, Jeff, now." "Yes, Melanctha, darling," murmured Jeff, and he was very happy in it, and so the two of them now in the warm air of the sultry, southern, negro sunshine, lay there for a long time just resting.

And now for a real long time there was no open trouble any more between Jeff Campbell and Melanctha Herbert. Then it came that Jeff knew he could not say out any more, what it was he wanted, he could not say out any more, what it was, he wanted to know about, what Melanctha wanted.

Melanctha sometimes now, when she was tired with being all the time so much excited, when Jeff would talk a long time to her about what was right for them both to be always doing, would be, as if she gave way in her head, and lost herself in a bad feeling. Sometimes when they had been strong in their loving, and Jeff would have rise inside him some strange feeling, and Melanctha felt it in him as it would soon be coming, she would lose herself then in this bad feeling that made her head act as if she never knew what it was they were doing. And slowly now, Jeff soon always came to be feeling that his Melanctha would be hurt very much in her head in the ways he never liked to think of, if she would ever now again have to listen to his trouble, when he was telling about what it was he still was wanting to make things for himself really understanding.

Now Jeff began to have always a strong feeling that Melanctha could no longer stand it, with all her bad suffering, to let him fight out with himself what was right for him to be doing. Now he felt he must not, when she was there with him, keep on, with this kind of fighting that was always going on inside him. Jeff Campbell never knew yet, what he thought was the right way, for himself and for all the colored people to be living. Jeff was coming always each time closer to be really understanding, but now Melanctha was so bad in her suffering with him, that he knew she could not any longer have him with her while he was always showing that he never really yet was sure what it was, the right way, for them to be really loving.

Jeff saw now he had to go so fast, so that Melanctha never would have to wait any to get from him always all that she ever wanted. He never could be honest now, he never could be now, any more, trying to be really understanding, for always every moment now he felt it to be a strong thing in him, how very much it was Melanctha Herbert always suffered.

Jeff did not know very well these days, what it was, was really happening to him. All he knew every now and then, when they were getting strong to get excited, the way they used to when he gave his feeling out so that he could be always honest, that Melanctha somehow never seemed to hear him, she just looked at him and looked as if her head hurt with him, and then Jeff had to keep himself from being honest, and he had to go so fast, and to do everything Melanctha ever wanted from him.

Jeff did not like it very well these days, in his true feeling. He knew now very well Melanctha was not strong enough inside her to stand any more of his slow way of doing. And yet now he knew he was not honest in his feeling. Now he always had to show more to Melanctha than he was ever feeling. Now she made him go so fast, and he knew it was not real with his feeling, and yet he could not make her suffer so any more because he always was so slow with his feeling.

It was very hard for Jeff Campbell to make all this way of doing, right, inside him. If Jeff Campbell could not be straight out, and real honest, he never could be very strong inside him. Now Melanctha, with her making him feel, always, how good she was and how very much she suffered in him, made him always go so fast then, he could not be strong then, to feel things out straight then inside him. Always now when he was with her, he was being more, than he could already yet, be feeling for her. Always now, with her, he had something inside him always holding in him, always now, with her, he was far ahead of his own feeling.

Jeff Campbell never knew very well these days what it was that was going on inside him. All he knew was, he was uneasy now always to be with Melanctha. All he knew was, that he was always uneasy when he was with Melanctha, not the way he used to be from just not being very understanding, but now, because he never could be honest with her, because he was now always feeling her strong suffering, in her, because he knew now he was having a straight, good feeling with her, but she went so fast, and he was so slow to her; Jeff knew his right feeling never got a chance to show itself as strong, to her.

All this was always getting harder for Jeff Campbell. He was very proud to hold himself to be strong, was Jeff Campbell. He was very tender not to hurt Melanctha, when he knew she would be sure to feel it badly in her head a long time after, he hated that he could not now be honest with her, he wanted to stay away to work it out all alone, without her, he was afraid she would feel it to suffer, if he kept away now from her. He was uneasy always, with her, he was uneasy when he thought about her, he knew now he had a good, straight, strong feeling of right loving for her, and yet now he never could use it to be good and honest with her.

Jeff Campbell did not know, these days, anything he could do to make it better for her. He did not know anything he could do, to set himself really right in his acting and his thinking toward her. She pulled him so fast with her, and he did not dare to hurt her, and he could not come right, so fast, the way she always needed he should be doing it now, for her.

These day were not very joyful ones now any more, to Jeff Campbell, with Melanctha. He did not think it out to himself now, in words, about her. He did not know enough, what was his real trouble, with her.

Sometimes now and again with them, and with all this trouble for a little while well forgotten by him, Jeff, and Melanctha with him, would be very happy in a strong, sweet loving. Sometimes then, Jeff would find himself to be soaring very high in his true loving. Sometimes Jeff would find then, in his loving, his soul swelling out full inside him. Always Jeff felt now in himself, deep feeling.

Always now Jeff had to go so much faster than was real with his feeling. Yet always Jeff knew now he had a right, strong feeling. Always now when Jeff was wondering, it was Melanctha he was doubting, in the loving. Now he would often ask her, was she real now to him, in her loving. He would ask her often, feeling something queer about it all inside him, though yet he was never really strong in his doubting, and always Melanctha would answer to him, "Yes Jeff, sure, you know it, always," and always Jeff felt a doubt now, in her loving.

Always now Jeff felt in himself, deep loving. Always now he did not know really, if Melanctha was true in her loving.

All these days Jeff was uncertain in him, and he was uneasy about which way he should act so as not to be wrong and put them both into bad trouble. Always now he was, as if he must feel deep into Melanctha to see if it was real loving he would find she now had in her, and always he would stop himself, with her, for always he was afraid now that he might badly hurt her.

Always now he liked it better when he was detained when he had to go and see her. Always now he never liked to go to be with her, although he never wanted really, not to be always with her. Always now he never felt really at ease with her, even when they were good friends together. Always now he felt, with her, he could not be really honest to her. And Jeff never could be happy with her when he could not feel strong to tell all his feeling to her. Always now every day he found it harder to make the time pass, with her, and not let his feeling come so that he would quarrel with her.

And so one evening, late, he was to go to her. He waited a little long, before he went to her. He was afraid, in himself, to-night, he would surely hurt her. He never wanted to go when he might quarrel with her.

Melanctha sat there looking very angry, when he came in to her. Jeff took off his hat and coat and then sat down by the fire with her.

"If you come in much later to me just now, Jeff Campbell, I certainly never would have seen you no more never to speak to you, 'thout your apologising real humble to me." "Apologising Melanctha," and Jeff laughed and was scornful to her, "Apologising, Melanctha, I ain't proud that kind of way, Melanctha, I don't mind apologising to you, Melanctha, all I mind, Melanctha is to be doing of things wrong, to you." "That's easy, to say things that way, Jeff to me. But you never was very proud Jeff, to be courageous to me." "I don't know about that Melanctha. I got courage to say some things hard, when I mean them, to you." "Oh, yes, Jeff, I know all about that, Jeff, to me. But I mean real courage, to run around and not care nothing about what happens, and always to be game in any kind of trouble. That's what I mean by real courage, to me, Jeff, if you want to know it." "Oh, yes, Melanctha, I know all that kind of courage. I see plenty of it all the time with some kind of colored men and with some girls like you Melanctha, and Jane Harden. I know all about how you are always making a fuss to be proud because you don't holler so much when you run in to where you ain't got any business to be, and so you get hurt, the way you ought to. And then, you kind of people are very brave then, sure, with all your kinds of suffering, but the way I see it, going round with all my patients, that kind of courage makes all kind of trouble, for them who ain't so noble with their courage, and then they got it, always to be bearing it, when the end comes, to be hurt the hardest. It's like running around and being game to spend all your money always, and then a man's wife and children are the ones do all the starving and they don't ever get a name for being brave, and they don't ever want to be doing all that suffering, and they got to stand it and say nothing. That's the way I see it a good deal now with all that kind of braveness in some of the colored people. They always make a lot of noise to show they are so brave not to holler, when they got so much suffering they always bring all on themselves, just by doing things they got no business to be doing. I don't say, never, Melanctha, they ain't got good courage not to holler, but I never did see much in looking for that kind of trouble just to show you ain't going to holler. No its all right being brave every day, just living regular and not having new ways all the time just to get excitements, the way I hate to see it in all the colored people. No I don't see much, Melanctha, in being brave just to get it good, where you've got no business. I ain't ashamed Melanctha, right here to tell you, I ain't ashamed ever to say I ain't got no longing to be brave, just to go around and look for trouble. "Yes that's just like you always, Jeff, you never understand things right, the way you are always feeling in you. You ain't got no way to understand right, how it depends what way somebody goes to look for new things, the way it makes it right for them to get excited." "No Melanctha, I certainly never do say I understand much anybody's got a right to think they won't have real bad trouble, if they go and look hard where they are certain sure to find it. No Melanctha, it certainly does sound very pretty all this talking about danger and being game and never hollering, and all that way of talking, but when two men are just fighting, the strong man mostly gets on top with doing good hard pounding, and the man that's getting all that pounding, he mostly never likes it so far as I have been able yet to see it, and I don't see much difference what kind of noble way they are made of when they ain't got any kind of business to get together there to be fighting. That certainly is the only way I ever see it happen right, Melanctha, whenever I happen to be anywhere I can be looking." "That's because you never can see anything that ain't just so simple, Jeff, with everybody, the way you always think it. It do make all the difference the kind of way anybody is made to do things game Jeff Campbell." "Maybe Melanctha, I certainly never say no you ain't right, Melanctha. I just been telling it to you all straight, Melanctha, the way I always see it. Perhaps if you run around where you ain't got any business, and you stand up very straight and say, I am so brave, nothing can ever ever hurt me, maybe nothing will ever hurt you then Melanctha. I never have seen it do so. I never can say truly any differently to you Melanctha, but I always am ready to be learning from you, Melanctha. And perhaps when somebody cuts into you real hard, with a brick he is throwing, perhaps you never will do any hollering then, Melanctha. I certainly don't ever say no, Melanctha to you, I only say that ain't the way yet I ever see it happen when I had a chance to be there looking."

They sat there together, quiet by the fire, and they did not seem to feel very loving.

"I certainly do wonder," Melanctha said dreamily, at last breaking into their long unloving silence. "I certainly do wonder why always it happens to me I care for anybody who ain't no ways good enough for me ever to be thinking to respect him."

Jeff looked at Melanctha. Jeff got up then and walked a little up and down the room, and then he came back, and his face was set and dark and he was very quiet to her.

"Oh dear, Jeff, sure, why you look so solemn now to me. Sure Jeff I never am meaning anything real by what I just been saying. What was I just been saying Jeff to you. I only certainly was just thinking how everything always was just happening to me."

Jeff Campbell sat very still and dark, and made no answer.

"Seems to me, Jeff you might be good to me a little to-night when my head hurts so, and I am so tired with all the hard work I have been doing, thinking, and I always got so many things to be a trouble to me, living like I do with nobody ever who can help me. Seems to me you might be good to me Jeff to-night, and not get angry, every little thing I am ever saying to you."

"I certainly would not get angry ever with you, Melanctha, just because you say things to me. But now I certainly been thinking you really mean what you have been just then saying to me." "But you say all the time to me Jeff, you ain't no ways good enough in your loving to me, you certainly say to me all the time you ain't no ways good or understanding to me." "That certainly is what I say to you always, just the way I feel it to you Melanctha always, and I got it right in me to say it, and I have got a right in me to be very strong and feel it, and to be always sure to believe it, but it ain't right for you Melanctha to feel it. When you feel it so Melanctha, it does certainly make everything all wrong with our loving. It makes it so I certainly never can bear to have it."

They sat there then a long time by the fire, very silent, and not loving, and never looking to each other for it. Melanctha was moving and twitching herself and very nervous with it. Jeff was heavy and sullen and dark and very serious in it.

"Oh why can't you forget I said it to you Jeff now, and I certainly am so tired, and my head and all now with it."

Jeff stirred, "All right Melanctha, don't you go make yourself sick now in your head, feeling so bad with it," and Jeff made himself do it, and he was a patient doctor again now with Melanctha when he felt her really having her head hurt with it. "It's all right now Melanctha darling, sure it is now I tell you. You just lie down now a little, dear one, and I sit here by the fire and just read awhile and just watch with you so I will be here ready, if you need me to give you something to help you resting." And then Jeff was a good doctor to her, and very sweet and tender with her, and Melanctha loved him to be there to help her, and then Melanctha fell asleep a little, and Jeff waited there beside her until he saw she was really sleeping, and then he went back and sat down by the fire.

And Jeff tried to begin again with his thinking, and he could not make it come clear to himself, with all his thinking, and he felt everything all thick and heavy and bad, now inside him, everything that he could not understand right, with all the hard work he made, with his thinking. And then he moved himself a little, and took a book to forget his thinking, and then as always, he loved it when he was reading, and then soon he was deep in his reading, and so he forgot now for a little while that he could seem to be very understanding.

And so Jeff forgot himself for awhile in his reading, and Melanctha was sleeping. And then Melanctha woke up and she was screaming. "Oh, Jeff, I thought you gone away for always from me. Oh, Jeff, never now go away no more from me. Oh, Jeff, sure, sure, always be just so good to me."

There was a weight in Jeff Campbell from now on, always with him, that he could never lift out from him, to feel easy. He always was trying not to have it in him and he always was trying not to let

Melanctha feel it, with him, but it was always there inside him. Now Jeff Campbell always was serious, and dark, and heavy, and sullen, and he would often sit a long time with Melanctha without moving. "You certainly never have forgiven to me, what I said to you that night, Jeff, now have you?" Melanctha asked him after a long silence, late one evening with him. "It ain't ever with me a question like forgiving, Melanctha, I got in me. It's just only what you are feeling for me, makes any difference to me. I ain't ever seen anything since in you, makes me think you didn't mean it right, what you said about not thinking now any more I was good, to make it right for you to be really caring so very much to love me."

"I certainly never did see no man like you, Jeff. You always wanting to have it all clear out in words always, what everybody is always feeling. I certainly don't see a reason, why I should always be explaining to you what I mean by what I am just saying. And you ain't got no feeling ever for me, to ask me what I meant, by what I was saying when I was so tired, that night. I never know anything right I was saying." "But you don't ever tell me now, Melanctha, so I really hear you say it, you don't mean it the same way, the way you said it to me." "Oh Jeff, you so stupid always to me and always just bothering with your always asking to me. And I don't never any way remember ever anything I been saying to you, and I am always my head, so it hurts me it half kills me, and my heart jumps so, sometimes I think I die so when it hurts me, and I am so blue always, I think sometimes I take something to just kill me, and I got so much to bother thinking always and doing, and I got so much to worry, and all that, and then you come and ask me what I mean by what I was just saying to you. I certainly don't know, Jeff, when you ask me. Seems to me, Jeff, sometimes you might have some kind of a right feeling to be careful to me." "You ain't got no right Melanctha Herbert," flashed out Jeff through his dark, frowning anger, "you certainly ain't got no right always to be using your being hurt and being sick, and having pain, like a weapon, so as to make me do things it ain't never right for me to be doing for you. You certainly ain't got no right to be always holding your pain out to show me." "What do you mean by them words, Jeff Campbell." "I certainly do mean them just like I am saying them, Melanctha. You act always, like I been responsible all myself for all our loving one another. And if its anything anyway that ever hurts you, you act like as if it was me made you just begin it all with me. I ain't no coward, you hear me, Melanctha? I never put my trouble back on anybody, thinking that they made me. I certainly am right ready always,

Melanctha, you certainly had ought to know me, to stand all my own trouble for me, but I tell you straight now, the way I think it Melanctha, I ain't going to be as if I was the reason why you wanted to be loving, and to be suffering so now with me." "But ain't you certainly ought to be feeling it so, to be right, Jeff Campbell. Did I ever do anything but just let you do everything you wanted to me. Did I ever try to .make you be loving to me. Did I ever do nothing except just sit there ready to endure your loving with me. But I certainly never, Jeff Campbell, did make any kind of way as if I wanted really to be having you for me."

Jeff stared at Melanctha. "So that's the way you say it when you are thinking right about it all, Melanctha. Well I certainly ain't got a word to say ever to you any more, Melanctha, if that's the way its straight out to you now, Melanctha." And Jeff almost laughed out to her, and he turned to take his hat and coat, and go away now forever from her.

Melanctha dropped her head on her arms, and she trembled all over and inside her. Jeff stopped a little and looked very sadly at her. Jeff could not so quickly make it right for himself, to leave her.

"Oh, I certainly shall go crazy now, I certainly know that," Melanctha moaned as she sat there, all fallen and miserable and weak together.

Jeff came and took her in his arms, and held her. Jeff was very good then to her, but they neither of them felt inside all right, as they once did, to be together.

From now on, Jeff had real torment in him.

Was it true what Melanctha had said that night to him? Was it true that he was the one had made all this trouble for them? Was it true, he was the only one, who always had had wrong ways in him? Waking or sleeping Jeff now always had this torment going on inside him.

Jeff did not know now any more, what to feel within him. He did not know how to begin thinking out this trouble that must always now be bad inside him. He just felt a confused struggle and resentment always in him, a knowing, no, Melanctha was not right in what she had said that night to him, and then a feeling, perhaps he always had been wrong in the way he never could be understanding. And then would come strong to him, a sense of the deep sweetness in Melanctha's loving and a hating the cold slow way he always had to feel things in him.

Always Jeff knew, sure, Melanctha was wrong in what she had said that night to him, but always Melanctha had had deep feeling with him, always he was poor and slow in the only way he knew how to have any feeling. Jeff knew Melanctha was wrong, and yet he always had a deep doubt in him. What could he know, who had such slow feeling in him? What could he ever know, who always had to find his way with just thinking. What could he know, who had to be taught such a long time to learn about what was really loving? Jeff now always had this torment in him.

Melanctha was now always making him feel her way, strong whenever she was with him. Did she go on to do it just to show him, did she do it so now because she was no longer loving, did she do it so because that was her way to make him be really loving. Jeff never did know how it was that it all happened so to him.

Melanctha acted now the way she had said it always had been with them. Now it was always Jeff who had to do the asking. Now it was always Jeff who had to ask when would be the next time he should come to see her. Now always she was good and patient to him, and now always she was kind and loving with him, and always Jeff felt it was, that she was good to give him anything he ever asked or wanted, but never now any more for her own sake to make her happy in him. Now she did these things, as if it was just to please her Jeff Campbell who needed she should now have kindness for him. Always now he was the beggar, with them. Always now Melanctha gave it, not of her need, but from her bounty to him. Always now Jeff found it getting harder for him.

Sometimes Jeff wanted to tear things away from before him, always now he wanted to fight things and be angry with them, and always now Melanctha was so patient to him.

Now, deep inside him, there was always a doubt with Jeff, of Melanctha's loving. It was not a doubt yet to make him really doubting, for with that, Jeff never could be really loving, but always now he knew that something, and that not in him, something was wrong with their loving. Jeff Campbell could not know any right way to think out what was inside Melanctha with her loving, he could not use any way now to reach inside her to find if she was true in her loving, but now something had gone wrong between them, and now he never felt sure in him, the way once she had made him, that now at last he really had got to be understanding.

Melanctha was too many[6] for him. He was helpless to find out the way she really felt now for him. Often Jeff would ask her, did she

---

[6] *too many:* Too complex, complicated; he could not understand her various emotions.

really love him. Always she said, "Yes Jeff, sure, you know that," and now instead of a full sweet strong love with it, Jeff only felt a patient, kind endurance in it.

Jeff did not know. If he was right in such a feeling, he certainly never any more did want to have Melanctha Herbert with him. Jeff Campbell hated badly to think Melanctha ever would give him love, just for his sake, and not because she needed it herself, to be with him. Such a way of loving would be very hard for Jeff to be enduring.

"Jeff what makes you act so funny to me. Jeff you certainly now are jealous to me. Sure Jeff, now I don't see ever why you be so foolish to look so to me." "Don't you ever think I can be jealous of anybody ever Melanctha, you hear me. It's just, you certainly don't ever understand me. It's just this way with me always now Melanctha. You love me, and I don't care anything what you do or what you ever been to anybody. You don't love me, then I don't care any more about what you ever do or what you ever be to anybody. But I never want you to be being good Melanctha to me, when it ain't your loving makes you need it. I certainly don't ever want to be having any of your kind of kindness to me. If you don't love me, I can stand it. All I never want to have is your being good to me from kindness. If you don't love me, then you and I certainly do quit right here Melanctha, all strong feeling, to be always living to each other. It certainly never is anybody I ever am thinking about when I am thinking with you Melanctha darling. That's the true way I am telling you Melanctha, always. It's only your loving me ever gives me anything to bother me Melanctha, so all you got to do, if you don't really love me, is just certainly to say so to me. I won't bother you more then than I can help to keep from it Melanctha. You certainly need never to be in any worry, never, about me Melanctha. You just tell me straight out Melanctha, real, the way you feel it. I certainly can stand it all right, I tell you true Melanctha. And I never will care to know why or nothing Melanctha. Loving is just living Melanctha to me, and if you don't really feel it now Melanctha to me, there ain't ever nothing between us then Melanctha, is there? That's straight and honest just the way I always feel it to you now Melanctha. Oh Melanctha, darling, do you love me? Oh Melanctha, please, please, tell me honest, tell me, do you really love me?"

"Oh you so stupid Jeff boy, of course I always love you. Always and always Jeff and I always just so good to you. Oh you so stupid Jeff and don't know when you got it good with me. Oh dear, Jeff I certainly am so tired Jeff to-night, don't you go be a bother to me. Yes I love you Jeff, how often you want me to tell you. Oh you so stupid Jeff, but yes

I love you. Now I won't say it no more now tonight Jeff, you hear me. You just be good Jeff now to me or else I certainly get awful angry with you. Yes I love you, sure, Jeff, though you don't any way deserve it from me. Yes, yes I love you. Yes Jeff I say it till I certainly am very sleepy. Yes I love you now Jeff, and you certainly must stop asking me to tell you. Oh you great silly boy Jeff Campbell, sure I love you, oh you silly stupid, my own boy Jeff Campbell. Yes I love you and I certainly never won't say it one more time to-night Jeff, now you hear me."

Yes Jeff Campbell heard her, and he tried hard to believe her. He did not really doubt her but somehow it was wrong now, the way Melanctha said it. Jeff always now felt baffled with Melanctha. Something, he knew, was not right now in her. Something in her always now was making stronger the torment that was tearing every minute at the joy he once always had had with her.

Always now Jeff wondered did Melanctha love him. Always now he was wondering, was Melanctha right when she said, it was he had made all their beginning. Was Melanctha right when she said, it was he had the real responsibility for all the trouble they had and still were having now between them. If she was right, what a brute he always had been in his acting. If she was right, how good she had been to endure the pain he had made so bad so often for her. But no, surely she had made herself to bear it, for her own sake, not for his to make him happy. Surely he was not so twisted in all his long thinking. Surely he could remember right what it was had happened every day in their long loving. Surely he was not so poor a coward as Melanctha always seemed to be thinking. Surely, surely, and then the torment would get worse every minute in him.

One night Jeff Campbell was lying in his bed with his thinking, and night after night now he could not do any sleeping for his thinking. To-night suddenly he sat up in his bed, and it all came clear to him, and he pounded his pillow with his fist, and he almost shouted out alone there to him, "I ain't a brute the way Melanctha has been saying. Its all wrong the way I been worried thinking. We did begin fair, each not for the other but for ourselves, what we were wanting. Melanctha Herbert did it just like I did it, because she liked it bad enough to want to stand it. It's all wrong in me to think it any way except the way we really did it. I certainly don't know now whether she is now real and true in her loving. I ain't got any way ever to find out if she is real and true now always to me. All I know is I didn't ever make her to begin to be with me. Melanctha has got to stand for her own trouble, just like I got to stand for my own trouble. Each man has got to do it for himself when he is in real trouble. Melanctha, she certainly don't remember right when she says I made her begin and then I made her trouble. No by God, I ain't no coward nor a brute either ever to her. I been the way I felt it honest, and that certainly is all about it now between us, and everybody always has just got to stand for their own trouble. I certainly am right this time the way I see it." And Jeff lay down now, at last in comfort, and he slept, and he was free from his long doubting torment.

"You know Melanctha," Jeff Campbell began, the next time he was alone to talk a long time to Melanctha. "You know Melanctha, sometimes I think a whole lot about what you like to say so much about being game and never doing any hollering. Seems to me Melanctha, I certainly don't understand right what you mean by not hollering. Seems to me it certainly ain't only what comes right away when one is hit, that counts to be brave to be bearing, but all that comes later from your getting sick from the shock of being hurt once in a fight, and all that, and all the being taken care of for years after, and the suffering of your family, and all that, you certainly must stand and not holler, to be certainly really brave the way I understand it." "What you mean Jeff by your talking." "I mean, seems to me really not to holler, is to be strong not to show you ever have been hurt. Seems to me, to get your head hurt from your trouble and to show it, ain't certainly no braver than to say, oh, oh, how bad you hurt me, please don't hurt me mister. It just certainly seems to me, like many people think themselves so game just to stand what we all of us always just got to be standing, and everybody stands it, and we don't certainly none of us like it, and yet we don't ever most of us think we are so much being game, just because we got to stand it."

"I know what you mean now by what you are saying to me now Jeff Campbell. You make a fuss now to me, because I certainly just have stopped standing everything you like to be always doing so cruel to me. But that's just the way always with you Jeff Campbell, if you want to know it. You ain't got no kind of right feeling for all I always been forgiving to you." "I said it once for fun, Melanctha, but now I certainly do mean it, you think you got a right to go where you got no business, and you say, I am so brave nothing can hurt me, and then something, like always, it happens to hurt you, and you show your hurt always so everybody can see it, and you say, I am so brave nothing did hurt me except he certainly didn't have any right to, and see how bad I suffer, but you never hear me make a holler, though certainly anybody got any feeling, to see me suffer, would certainly never touch me except to take good care of me. Sometimes I certainly don't rightly see Melanctha, how much more game that is than just the ordinary kind of holler." "No, Jeff Campbell, and made the way you is you certainly ain't likely ever to be much more understanding." "No, Melanctha, nor you neither. You think always, you are the only one who ever can do any way to really suffer." "Well, and ain't I certainly always been the only person knows how to bear it. No, Jeff Campbell, I certainly be glad to love anybody really worthy, but I made so, I never seem to be able in this world to find him." "No, and your kind of way of thinking, you certainly Melanctha never going to any way be able ever to be finding of him. Can't you understand Melanctha, ever, how no man certainly ever really can hold your love for long times together. You certainly Melanctha, you ain't got down deep loyal feeling, true inside you, and when you ain't just that moment quick with feeling, then you certainly ain't ever got anything more there to keep you. You see Melanctha, it certainly is this way with you, it is, that you ain't ever got any way to remember right what you been doing, or anybody else that has been feeling with you. You certainly Melanctha, never can remember right, when it comes what you have done and what you think happens to you." "It certainly is all easy for you Jeff Campbell to be talking. You remember right, because you don't remember nothing till you get home with your thinking everything all over, but I certainly don't think much ever of that kind of way of remembering right, Jeff Campbell. I certainly do call it remembering right Jeff Campbell, to remember right just when it happens to you, so you have a right kind of feeling not to act the way you always been doing to me, and then you go home Jeff Campbell, and you begin with your thinking, and then it certainly is very easy for you to be good and forgiving with it. No, that ain't to me, the way of remembering Jeff Campbell, not as I can see it not to make people always suffer, waiting for you certainly to get to do it. Seems to me like Jeff Campbell, I never could feel so like a man was low and to be scorning of him, like that day in the summer, when you threw me off just because you got one of those fits of your remembering. No, Jeff Campbell, its real feeling every moment when its needed, that certainly does seem to me like real remembering. And that way, certainly, you don't never know nothing like what should be right Jeff Campbell. No Jeff, it's me that always certainly has had to bear it with you. It's always me that certainly has had to suffer, while you go home to remember. No you certainly ain't got no sense yet Jeff, what you need to make you really feeling. No, it certainly is me Jeff Campbell, that always has got to be remembering for us both, always. That's what's the true way with us Jeff Campbell, if you want to know what it is I am always thinking." "You is certainly real modest Melanctha, when you do this kind of talking, you sure is Melanctha," said Jeff Campbell laughing. "I think sometimes Melanctha I am certainly awful conceited, when I think sometimes I am all out doors, and I think I certainly am so bright, and better than most everybody I ever got anything now to do with, but when I hear you talk this way Melanctha, I certainly do think I am a real modest kind of fellow." "Modest!" said Melanctha, angry, "Modest, that certainly is a queer thing for you Jeff to be calling yourself even when you are laughing." "Well it certainly does depend a whole lot what you are thinking with," said Jeff Campbell. "I never did use to think I was so much on being real modest Melanctha, but now I know really I am, when I hear you talking. I see all the time there are many people living just as good as I am, though they are a little different to me. Now with you Melanctha if I understand you right what you are talking, you don't think that way of no other one that you are ever knowing." "I certainly could be real modest too, Jeff Campbell," said Melanctha, "If I could meet somebody once I could keep right on respecting when I got so I was really knowing with them. But I certainly never met anybody like that yet, Jeff Campbell, if you want to know it." "No, Melanctha, and with the way you got of thinking, it certainly don't look like as if you ever will Melanctha, with your never remembering anything only what you just then are feeling in you, and you not understanding what any one else is ever feeling, if they don't holler just the way you are doing. No Melanctha, I certainly don't see any ways you are likely ever to meet one, so good as you are always thinking you be." "No, Jeff Campbell, it certainly ain't that way with me at all the way you say it. It's because I am always knowing what it is I am wanting, when I get it. I certainly don't never have to wait till I have it, and then throw away what I got in me, and then come back and say, that's a mistake I just been making, it ain't that never at all like I understood it, I want to have, bad, what I didn't think it was I wanted. It's that way of knowing right what I am wanting, makes me feel nobody can come right with me, when I am feeling things, Jeff Campbell. I certainly do say Jeff Campbell, I certainly don't think much of the way you always do it, always never knowing what it is you are ever really wanting and everybody always got to suffer. No Jeff, I don't certainly think there is much doubting which is better and the stronger with us two, Jeff Campbell."

"As you will, Melanctha Herbert," cried Jeff Campbell, and he rose up, and he thundered out a black oath, and he was fierce to leave her now forever, and then with the same movement, he took her in his arms and held her.

"What a silly goose boy you are, Jeff Campbell," Melanctha whispered to him fondly.

"Oh yes," said Jeff, very dreary. "I never could keep really mad with anybody, not when I was a little boy and playing. I used most to cry sometimes, I couldn't get real mad and keep on a long time with it, the way everybody always did it. It's certainly no use to me Melanctha, I certainly can't ever keep mad with you Melanctha, my dear one. But don't you ever be thinking it's because I think you right in what you been just saying to me. I don't Melanctha really think it that way, honest, though I certainly can't get mad the way I ought to. No Melanctha, little girl, really truly, you ain't right the way you think it. I certainly do know that Melanctha, honest. You certainly don't do me right Melanctha, the way you say you are thinking. Good-bye Melanctha, though you certainly is my own little girl for always." And then they were very good a little to each other, and then Jeff went away for that evening, from her.

Melanctha had begun now once more to wander. Melanctha did not yet always wander, but a little now she needed to begin to look for others. Now Melanctha Herbert began again to be with some of the better kind of black girls, and with them she sometimes wandered. Melanctha had not yet come again to need to be alone, when she wandered.

Jeff Campbell did not know that Melanctha had begun again to wander. All Jeff knew, was that now he could not be so often with her.

Jeff never knew how it had come to happen to him, but now he never thought to go to see Melanctha Herbert, until he had before, asked her if she could be going to have time then to have him with her. Then Melanctha would think a little, and then she would say to him, "Let me see Jeff, to-morrow, you was just saying to me. I certainly am awful busy you know Jeff just now. It certainly does seem to me this week Jeff, I can't anyways fix it. Sure I want to see you soon Jeff. I certainly Jeff got to do a little more now, I been giving so much time, when I had no business, just to be with you when you asked me. Now I guess Jeff, I certainly can't see you no more this week Jeff, the way I got to do things." "All right Melanctha," Jeff would answer and he would be very angry. "I want to come only just certainly as you want me now Melanctha." "Now Jeff you know I certainly can't be neglecting always to be with everybody just to see you. You come see me next week Tuesday Jeff, you hear me. I don't think Jeff I certainly be so busy, Tuesday." Jeff Campbell would then go away and leave her, and he would be hurt and very angry, for it was hard for a man with a great pride in himself, like Jeff Campbell, to feel himself no better than a beggar. And yet he always came as she said he should, on the day she had fixed for him, and always Jeff Campbell was not sure yet that he really understood what it was Melanctha wanted. Always Melanctha said to him, yes she loved him, sure he knew that. Always Melanctha said to him, she certainly did love him just the same as always, only sure he knew now she certainly did seem to be right busy with all she certainly now had to be doing.

Jeff never knew what Melanctha had to do now, that made her always be so busy, but Jeff Campbell never cared to ask Melanctha such a question. Besides Jeff knew Melanctha Herbert would never, in such a matter, give him any kind of a real answer. Jeff did not know whether it was that Melanctha did not know how to give a simple answer. And then how could he, Jeff, know what was important to her. Jeff Campbell always felt strongly in him, he had no right to interfere with Melanctha in any practical kind of a matter. There they had always, never asked each other any kind of question. There they had felt always in each other, not any right to take care of one another. And Jeff Campbell now felt less than he had ever, any right to claim to know what Melanctha thought it right that she should do in any of her ways of living. All Jeff felt a right in himself to question, was her loving.

Jeff learned every day now, more and more, how much it was that he could really suffer. Sometimes it hurt so in him, when he was alone, it would force some slow tears from him. But every day, now that Jeff Campbell, knew more how it could hurt him, he lost his feeling of deep awe that he once always had had for Melanctha's feeling. Suffering was not so much after all, thought Jeff Campbell, if even he could feel it so it hurt him. It hurt him bad, just the way he knew he once had hurt Melanctha, and yet he too could have it and not make any kind of a loud holler with it.

In tender hearted natures, those that mostly never feel strong passion, suffering often comes to make them harder. When these do not know in themselves what it is to suffer, suffering is then very awful to them and they badly want to help everyone who ever has to suffer, and they have a deep reverence for anybody who knows really how to always suffer. But when it comes to them to really suffer, they soon begin to lose their fear and tenderness and wonder. Why it isn't so very much to suffer, when even I can bear to do it. It isn't very pleasant to be having all the time, to stand it, but they are not so much wiser after all, all the others just because they know too how to bear it.

Passionate natures who have always made themselves, to suffer, that is all the kind of people who have emotions that come to them as sharp as a sensation, they always get more tender-hearted when they suffer, and it always does them good to suffer. Tender-hearted, unpassionate, and comfortable natures always get much harder when they suffer, for so they lose the fear and reverence and wonder they once had for everybody who ever has to suffer, for now they know themselves what it is to suffer and it is not so awful any longer to them when they know too, just as well as all the others, how to have it.

And so it came in these days to Jeff Campbell. Jeff knew now always, way inside him, what it is to really suffer, and now every day with it, he knew how to understand Melanctha better. Jeff Campbell still loved Melanctha Herbert and he still had a real trust in her and he still had a little hope that some day they would once more get together, but slowly, every day, this hope in him would keep growing always weaker. They still were a good deal of time together, but now they never any more were really trusting with each other. In the days when they used to be together, Jeff had felt he did not know much what was inside Melanctha, but he knew very well, how very deep always was his trust in her; now he knew Melanctha Herbert better, but now he never felt a deep trust in her. Now Jeff never could be really honest with her. He never doubted yet, that she was steady only to him, but somehow he could not believe much really in Melanctha's loving.

Melanctha Herbert was a little angry now when Jeff asked her, "I never give nobody before Jeff, ever more than one chance with me, and I certainly been giving you most a hundred Jeff, you hear me." "And why shouldn't you Melanctha, give me a million, if you really love me!" Jeff flashed out very angry. "I certainly don't know as you deserve that anyways from me, Jeff Campbell." "It ain't deserving, I am ever talking about to you Melanctha. Its loving, and if you are really loving to me you won't certainly never any ways call them chances." "Deed Jeff, you certainly are getting awful wise Jeff now, ain't you, to me." "No I ain't Melanctha, and I ain't jealous either to you. I just am doubting from the way you are always acting to me." "Oh yes Jeff, that's what they all say, the same way, when they certainly got jealousy all through them. You ain't got no cause to be jealous with me Jeff, and I am awful tired of all this talking now, you hear me."

Jeff Campbell never asked Melanctha any more if she loved him. Now things were always getting worse between them. Now Jeff was always very silent with Melanctha. Now Jeff never wanted to be honest to her, and now Jeff never had much to say to her.

Now when they were together, it was Melanctha always did most of the talking. Now she often had other girls there with her. Melanctha was always kind to Jeff Campbell but she never seemed to need to be alone now with him. She always treated Jeff, like her best friend, and she always spoke so to him and yet she never seemed now to very often want to see him.

Every day it was getting harder for Jeff Campbell. It was as if now, when he had learned to really love Melanctha, she did not need any more to have him. Jeff began to know this very well inside him.

Jeff Campbell did not know yet that Melanctha had begun again to wander. Jeff was not very quick to suspect Melanctha. All Jeff knew was, that he did not trust her to be now really loving to him.

Jeff was no longer now in any doubt inside him. He knew very well now he really loved Melanctha. He knew now very well she was not any more a real religion to him. Jeff Campbell knew very well too now inside him, he did not really want Melanctha, now if he could no longer trust her, though he loved her hard and really knew now what it was to suffer.

Every day Melanctha Herbert was less and less near to him. She always was very pleasant in her talk and to be with him, but somehow now it never was any comfort to him.

Melanctha Herbert now always had a lot of friends around her. Jeff Campbell never wanted to be with them. Now Melanctha began to find it, he said it often to him, always harder to arrange to be alone now with him. Sometimes she would be late for him. Then Jeff always would try to be patient in his waiting, for Jeff Campbell knew very well how to remember, and he knew it was only right that he should now endure this from her.

Then Melanctha began to manage often not to see him, and once she went away when she had promised to be there to meet him.

Then Jeff Campbell was really filled up with his anger. Now he knew he could never really want her. Now he knew he never any more could really trust her.

Jeff Campbell never knew why Melanctha had not come to meet him. Jeff had heard a little talking now, about how Melanctha Herbert had commenced once more to wander. Jeff Campbell still sometimes saw Jane Harden, who always needed a doctor to be often there to help her. Jane Harden always knew very well what happened to Melanctha. Jeff Campbell never would talk to Jane Harden anything about Melanctha. Jeff was always loyal to Melanctha. Jeff never let Jane Harden say much to him about Melanctha, though he never let her know that now he loved her. But somehow Jeff did know now about Melanctha, and he knew about some men that Melanctha met with Rose Johnson very often.

Jeff Campbell would not let himself really doubt Melanctha, but Jeff began to know now very well, he did not want her. Melanctha Herbert did not love him ever, Jeff knew it now, the way he once had thought that she could feel it. Once she had been greater for him than he had thought he could ever know how to feel it. Now Jeff had come to where he could understand Melanctha Herbert. Jeff was not bitter to her because she could not really love him, he was bitter only that he had let himself have a real illusion in him. He was a little bitter too, that he had lost now, what he had always felt real in the world, that had made it for him always full of beauty, and now he had not got this new religion really, and he had lost what he before had to know what was good and had real beauty.

Jeff Campbell was so angry now in him, because he had begged Melanctha always to be honest to him. Jeff could stand it in her not to love him, he could not stand it in her not to be honest to him.

Jeff Campbell went home from where Melanctha had not met him, and he was sore and full of anger in him.

Jeff Campbell could not be sure what to do, to make it right inside him. Surely he must be strong now and cast this loving from him, and yet, was he sure he now had real wisdom in him. Was he sure that Melanctha Herbert never had had a real deep loving for him. Was he sure Melanctha Herbert never had deserved a reverence from him. Always now Jeff had this torment in him, but always now he felt more that Melanctha never had real greatness for him.

Jeff waited to see if Melanctha would send any word to him. Melanctha Herbert never sent a line to him.

At last Jeff wrote his letter to Melanctha. "Dear Melanctha, I certainly do know you ain't been any way sick this last week when you never met me right the way you promised, and never sent me any word to say why you acted a way you certainly never could think was the right way you should do it to me. Jane Harden said she saw you that day and you went out walking with some people you like now to be with. Don't be misunderstanding me now any more Melanctha. I love you now because that's my slow way to learn what you been teaching, but I know now you certainly never had what seems to me real kind of feeling. I don't love you Melanctha any more now like a real religion, because now I know you are just made like all us others. I know now no man can ever really hold you because no man can ever be real to trust in you, because you mean right Melanctha, but you never can remember, and so you certainly never have got any way to be honest. So please you understand me right now Melanctha, it never is I don't know how to love you. I do know now how to love you, Melanctha, really. You sure do know that, Melanctha, in me. You certainly always can trust me. And so now Melanctha, I can say to you certainly real honest with you, I am better than you are in my right kind of feeling. And so Melanctha, I don't never any more want to be a trouble to you. You certainly make me see things Melanctha, I never any other way could be knowing. You been very good and patient to me, when I was certainly below you in my right feeling. I certainly never have been near so good and patient to you ever any way Melanctha, I certainly know that Melanctha. But Melanctha, with me, it certainly is, always to be good together, two people certainly must be thinking each one as good as the other, to be really loving right Melanctha. And it certainly must never be any kind of feeling, of one only taking, and one only just giving, Melanctha, to me. I know you certainly don't really ever understand me now Melanctha, but that's no matter. I certainly do know what I am feeling now with you real Melanctha. And so good-bye now for good Melanctha. I say I can never ever really trust you real Melanctha, that's only just certainly from your way of not being ever equal in your feeling to anybody real, Melanctha, and your way never to know right how to remember. Many ways I really trust you deep Melanctha, and I certainly do feel deep all the good sweetness you certainly got real in you Melanctha. Its only just in your loving me Melanctha. You never can be equal to me and that way I certainly never can bear any more to have it. And so now Melanctha, I always be your friend, if you need me, and now we never see each other any more to talk to."

And then Jeff Campbell thought and thought, and he could never make any way for him now, to see it different, and so at last he sent this letter to Melanctha.

And now surely it was all over in Jeff Campbell. Surely now he never any more could know Melanctha. And yet, perhaps Melanctha really loved him. And then she would know how much it hurt him never any more, any way, to see her, and perhaps she would write a line to tell him. But that was a foolish way for Jeff ever to be thinking. Of course Melanctha never would write a word to him. It was all over now for always, everything between them, and Jeff felt it a real relief to him. For many days now Jeff Campbell only felt it as a relief in him. Jeff was all locked up and quiet now inside him. It was all settling down heavy in him, and these days when it was sinking so deep in him, it was only the rest and quiet of not fighting that he could really feel inside him. Jeff Campbell could not think now, or feel anything else in him. He had no beauty nor any goodness to see around him. It was a dull, pleasant kind of quiet he now had inside him. Jeff almost began to love this dull quiet in him, for it was more nearly being free for him than anything he had known in him since Melanctha Herbert first had moved him. He did not find it a real rest yet for him, he had not really conquered what had been working so long in him, he had not learned to see beauty and real goodness yet in what had happened to him, but it was rest even if he was sodden now all through him. Jeff Campbell liked it very well, not to have fighting always going on inside him.

And so Jeff went on every day, and he was quiet, and he began again to watch himself in his working; and he did not see any beauty now around him, and it was dull and heavy always now inside him, and yet he was content to have gone so far in keeping steady to what he knew was the right way for him to come back to, to be regular, and see beauty in every kind of quiet way of living, the way he had always wanted it for himself and for all the colored people. He knew he had lost the sense he once had of joy all through him, but he could work, and perhaps he would bring some real belief back into him about the beauty that he could not now any more see around him.

And so Jeff Campbell went on with his working, and he staid home every evening, and he began again with his reading, and he did not do much talking, and he did not seem to himself to have any kind of feeling.

And one day Jeff thought perhaps he really was forgetting, one day he thought he could soon come back and be happy in his old way of regular and quiet living.

Jeff Campbell had never talked to any one of what had been going on inside him. Jeff Campbell liked to talk and he was honest, but it never came out from him, anything he was ever really feeling, it only came out from him, what it was that he was always thinking. Jeff Campbell always was very proud to hide what he was really feeling.

Always he blushed hot to think things he had been feeling. Only to Melanctha Herbert, had it ever come to him, to tell what it was that he was feeling.

And so Jeff Campbell went on with this dull and sodden,[7] heavy, quiet always in him, and he never seemed to be able to have any feeling. Only sometimes he shivered hot with shame when he remembered some things he once had been feeling. And then one day it all woke up, and was sharp in him.

Dr. Campbell was just then staying long times with a sick man who might soon be dying. One day the sick man was resting. Dr. Campbell went to the window to look out a little, while he was waiting. It was very early now in the southern springtime. The trees were just beginning to get the little zigzag crinkles in them, which the young buds always give them. The air was soft and moist and pleasant to them. The earth was wet and rich and smelling for them. The birds were making sharp fresh noises all around them. The wind was very gentle and yet urgent to them. And the buds and the long earthworms, and the negroes, and all the kinds of children, were coming out every minute farther into the new spring, watery, southern sunshine.

Jeff Campbell too began to feel a little his old joy inside him. The sodden quiet began to break up in him. He leaned far out of the window to mix it all up with him. His heart went sharp and then it almost stopped inside him. Was it Melanctha Herbert he had just seen passing by him? Was it Melanctha, or was it just some other girl, who made him feel so bad inside him? Well, it was no matter, Melanctha was there in the world around him, he did certainly always know that in him. Melanctha Herbert was always in the same town with him, and he could never any more feel her near him. What a fool he was to throw her from him. Did he know she did not really love him. Suppose Melanctha was now suffering through him. Suppose she really would be glad to see him. And did anything else he did, really mean anything now to him? What a fool he was to cast her from him. And yet did Melanctha Herbert want him, was she honest to him, had Melanctha ever loved him, and did Melanctha now suffer by him? Oh! Oh! Oh! and the bitter water once more rose up in him.

All that long day, with the warm moist young spring stirring in him, Jeff Campbell worked, and thought, and beat his breast, and wandered, and spoke aloud, and was silent, and was certain, and then in doubt and then keen to surely feel, and then all sodden in him; and he

---

[7] *sodden:* Sluggish, dampened, the opposite state of airy lightness and happiness.

walked, and he sometimes ran fast to lose himself in his rushing, and
he bit his nails to pain and bleeding, and he tore his hair[8] so that he
could be sure he was really feeling, and he never could know what it
was right, he now should be doing. And then late that night he wrote
it all out to Melanctha Herbert, and he made himself quickly send it
without giving himself any time to change it.

"It has come to me strong to-day Melanctha, perhaps I am wrong
the way I now am thinking. Perhaps you do want me badly to be with
you. Perhaps I have hurt you once again the way I used to. I certainly
Melanctha, if I ever think that really, I certainly do want bad not to be
wrong now ever any more to you. If you do feel the way to-day it came
to me strong may-be you are feeling, then say so Melanctha to me, and
I come again to see you. If not, don't say anything any more ever to
me. I don't want ever to be bad to you Melanctha, really. I never want
ever to be a bother to you. I never can stand it to think I am wrong;
really, thinking you don't want me to come to you. Tell me Melanctha,
tell me honest to me, shall I come now any more to see you." "Yes"
came the answer from Melanctha, "I be home Jeff to-night to see
you."

Jeff Campbell went that evening late to see Melanctha Herbert. As
Jeff came nearer to her, he doubted that he wanted really to be with
her, he felt that he did not know what it was he now wanted from her.
Jeff Campbell knew very well now, way inside him, that they could
never talk their trouble out between them. What was it Jeff wanted
now to tell Melanctha Herbert? What was it that Jeff Campbell now
could tell her? Surely he never now could learn to trust her. Surely Jeff
knew very well all that Melanctha always had inside her. And yet it
was awful, never any more to see her.

Jeff Campbell went in to Melanctha, and he kissed her, and he held
her, and then he went away from her and he stood still and looked at
her. "Well Jeff!" "Yes Melanctha!" "Jeff what was it made you act so
to me?" "You know very well Melanctha, it's always I am thinking
you don't love me, and you are acting to me good out of kindness, and
then Melanctha you certainly never did say anything to me why you
never came to meet me, as you certainly did promise to me you would
that day I never saw you!" "Jeff don't you really know for certain, I
always love you?" "No Melanctha, deed I don't know it in me. Deed
and certain sure Melanctha, if I only know that in me, I certainly never
would give you any bother." "Jeff, I certainly do love you more seems

---

[8] *tore his hair:* Pulled hard at his hair, trying to create pain.

to me always, you certainly had ought to feel that in you." "Sure Melanctha?" "Sure Jeff boy, you know that." "But then Melanctha why did you act so to me?" "Oh Jeff you certainly been such a bother to me. I just had to go away that day Jeff, and I certainly didn't mean not to tell you, and then that letter you wrote came to me and something happened to me. I don't know right what it was Jeff, I just kind of fainted, and what could I do Jeff, you said you certainly never any more wanted to come and see me!" "And no matter Melanctha, even if you knew, it was just killing me to act so to you, you never would have said nothing to me?" "No of course, how could I Jeff when you wrote that way to me. I know how you was feeling Jeff to me, but I certainly couldn't say nothing to you." "Well Melanctha, I certainly know I am right proud too in me, but I certainly never could act so to you Melanctha, if I ever knew any way at all you ever really loved me. No Melanctha darling, you and me certainly don't feel much the same way ever. Any way Melanctha, I certainly do love you true Melanctha." "And I love you too Jeff, even though you don't never certainly seem to believe me." "No I certainly don't any way believe you Melanctha, even when you say it to me. I don't know Melanctha how, but sure I certainly do trust you, only I don't believe now ever in your really being loving to me. I certainly do know you trust me always Melanctha, only somehow it ain't ever all right to me. I certainly don't know any way otherwise Melanctha, how I can say it to you." "Well I certainly can't help you no ways any more Jeff Campbell, though you certainly say it right when you say I trust you Jeff now always. You certainly is the best man Jeff Campbell, I ever can know, to me. I never been anyways thinking it can be ever different to me." "Well you trust me then Melanctha, and I certainly love you Melanctha, and seems like to me Melanctha, you and me had ought to be a little better than we certainly ever are doing now to be together. You certainly do think that way, too, Melanctha to me. But may be you do really love me. Tell me, please, real honest now Melanctha darling, tell me so I really always know it in me, do you really truly love me?" "Oh you stupid, stupid boy, Jeff Campbell. Love you, what do you think makes me always to forgive you. If I certainly didn't always love you Jeff, I certainly never would let you be always being all the time such a bother to me the way you certainly Jeff always are to me. Now don't you dass ever any more say words like that ever to me. You hear me now Jeff, or I do something real bad sometime, so I really hurt you. Now Jeff you just be good to me. You know Jeff how bad I need it, now you should always be good to me!"

Jeff Campbell could not make an answer to Melanctha. What was it he should now say to her? What words could help him to make their feeling any better? Jeff Campbell knew that he had learned to love deeply, that, he always knew very well now in him, Melanctha had learned to be strong to be always trusting, that he knew too now inside him, but Melanctha did not really love him, that he felt always too strong for him. That fact always was there in him, and it always thrust itself firm, between them. And so this talk did not make things really better for them.

Jeff Campbell was never any more a torment to Melanctha, he was only silent to her. Jeff often saw Melanctha and he was very friendly with her and he never any more was a bother to her. Jeff never any more now had much chance to be loving with her. Melanctha never was alone now when he saw her.

Melanctha Herbert had just been getting thick in her trouble with Jeff Campbell, when she went to that church where she first met Rose, who later was married regularly to Sam Johnson. Rose was a good-looking, better kind of black girl, and had been brought up quite like their own child by white folks. Rose was living now with colored people. Rose was staying just then with a colored woman, who had known 'Mis' Herbert and her black husband and this girl Melanctha.

Rose soon got to like Melanctha Herbert and Melanctha now always wanted to be with Rose, whenever she could do it. Melanctha Herbert always was doing everything for Rose that she could think of that Rose ever wanted. Rose always liked to be with nice people who would do things for her. Rose had strong common sense and she was lazy. Rose liked Melanctha Herbert, she had such kind of fine ways in her. Then, too, Rose had it in her to be sorry for the subtle, sweet-natured, docile, intelligent Melanctha Herbert who always was so blue sometimes, and always had had so much trouble. Then, too, Rose could scold Melanctha, for Melanctha Herbert never could know how to keep herself from trouble, and Rose was always strong to keep straight, with her simple selfish wisdom.

But why did the subtle, intelligent, attractive, half white girl Melanctha Herbert, with her sweetness and her power and her wisdom, demean herself to do for and to flatter and to be scolded, by this lazy, stupid, ordinary, selfish black girl. This was a queer thing in Melanctha Herbert.

And so now in these new spring days, it was with Rose that Melanctha began again to wander. Rose always knew very well in herself what was the right way to do when you wandered. Rose knew very well, she was not just any common kind of black girl, for she had been raised by white folks, and Rose always saw to it that she was engaged to him when she had any one man with whom she ever always wandered. Rose always had strong in her the sense for proper conduct. Rose always was telling the complex and less sure Melanctha, what was the right way she should do when she wandered.

Rose never knew much about Jeff Campbell with Melanctha Herbert. Rose had not known about Melanctha Herbert when she had been almost all her time with Dr. Campbell.

Jeff Campbell did not like Rose when he saw her with Melanctha. Jeff would never, when he could help it, meet her. Rose did not think much about Dr. Campbell. Melanctha never talked much about him to her. He was not important now to be with her.

Rose did not like Melanctha's old friend Jane Harden when she saw her. Jane despised Rose for an ordinary, stupid, sullen black girl. Jane could not see what Melanctha could find in that black girl, to endure her. It made Jane sick to see her. But then Melanctha had a good mind, but she certainly never did care much to really use it. Jane Harden now really never cared any more to see Melanctha, though Melanctha still always tried to be good to her. And Rose, she hated that stuck up, mean speaking, nasty, drunk thing, Jane Harden. Rose did not see how Melanctha could bear to ever see her, but Melanctha always was so good to everybody, she never would know how to act to people the way they deserved that she should do it.

Rose did not know much about Melanctha, and Jeff Campbell and Jane Harden. All Rose knew about Melanctha was her old life with her mother and her father. Rose was always glad to be good to poor Melanctha, who had had such an awful time with her mother and her father, and now she was alone and had nobody who could help her. "He was a awful black man to you Melanctha, I like to get my hands on him so he certainly could feel it. I just would Melanctha, now you hear me."

Perhaps it was this simple faith and simple anger and simple moral way of doing in Rose, that Melanctha now found such a comfort to her. Rose was selfish and was stupid and was lazy, but she was decent and knew always what was the right way she should do, and what she wanted, and she certainly did admire how bright was her friend Melanctha Herbert, and she certainly did feel how very much it was she always suffered and she scolded her to keep her from more trouble, and she never was angry when she found some of the different ways Melanctha Herbert sometimes had to do it.

And so always Rose and Melanctha were more and more together, and Jeff Campbell could now hardly ever any more be alone with Melanctha.

Once Jeff had to go away to another town to see a sick man. "When I come back Monday Melanctha, I come Monday evening to see you. You be home alone once Melanctha to see me." "Sure Jeff, I be glad to see you!"

When Jeff Campbell came to his house on Monday there was a note there from Melanctha. Could Jeff come day after to-morrow, Wednesday? Melanctha was so sorry she had to go out that evening. She was awful sorry and she hoped Jeff would not be angry.

Jeff was angry and he swore a little, and then he laughed, and then he sighed. "Poor Melanctha, she don't know any way to be real honest, but no matter, I sure do love her and I be good if only she will let me."

Jeff Campbell went Wednesday night to see Melanctha. Jeff Campbell took her in his arms and kissed her. "I certainly am awful sorry not to see you Jeff Monday, the way I promised, but I just couldn't Jeff, no way I could fix it." Jeff looked at her and then he laughed a little at her. "You want me to believe that really now Melanctha. All right I believe it if you want me to Melanctha. I certainly be good to you to-night the way you like it. I believe you certainly did want to see me Melanctha, and there was no way you could fix it." "Oh Jeff dear," said Melanctha, "I sure was wrong to act so to you. It's awful hard for me ever to say it to you, I have been wrong in my acting to you, but I certainly was bad this time Jeff to you. It do certainly come hard to me to say it Jeff, but I certainly was wrong to go away from you the way I did it. Only you always certainly been so bad Jeff, and such a bother to me, and making everything always so hard for me, and I certainly got some way to do it to make it come back sometimes to you. You bad boy Jeff, now you hear me, and this certainly is the first time Jeff I ever yet said it to anybody, I ever been wrong, Jeff, you hear me!" "All right Melanctha, I sure do forgive you, cause it's certainly the first time I ever heard you say you ever did anything wrong the way you shouldn't," and Jeff Campbell laughed and kissed her, and Melanctha laughed and loved him, and they really were happy now for a little time together.

And now they were very happy in each other and then they were silent and then they became a little sadder and then they were very quiet once more with each other.

"Yes I certainly do love you Jeff!" Melanctha said and she was very dreamy. "Sure, Melanctha." "Yes Jeff sure, but not the way you are now ever thinking. I love you more and more seems to me Jeff always, and I certainly do trust you more and more always to me when I know you. I do love you Jeff, sure yes, but not the kind of way of loving you are ever thinking it now Jeff with me. I ain't got certainly no hot passion any more now in me. You certainly have killed all that kind of feeling now Jeff in me. You certainly do know that Jeff, now the way I am always, when I am loving with you. You certainly do know that Jeff, and that's the way you certainly do like it now in me. You certainly don't mind now Jeff, to hear me say this to you."

Jeff Campbell was hurt so that it almost killed him. Yes he certainly did know now what it was to have real hot love in him, and yet Melanctha certainly was right, he did not deserve she should ever give it to him. "All right Melanctha I ain't ever kicking. I always will give you certainly always everything you want that I got in me. I take anything you want now to give me. I don't say never Melanctha it don't hurt me, but I certainly don't say ever Melanctha it ought ever to be any different to me." And the bitter tears rose up in Jeff Campbell, and they came and choked his voice to be silent, and he held himself hard to keep from breaking.

"Good-night Melanctha," and Jeff was very humble to her. "Good-night Jeff, I certainly never did mean any way to hurt you. I do love you, sure Jeff every day more and more, all the time I know you." "I know Melanctha, I know, it's never nothing to me. You can't help it, anybody ever the way they are feeling. It's all right now Melanctha, you believe me, good-night now Melanctha, I got now to leave you, good-by Melanctha, sure don't look so worried to me, sure Melanctha I come again soon to see you." And then Jeff stumbled down the steps, and he went away fast to leave her.

And now the pain came hard and harder in Jeff Campbell, and he groaned, and it hurt him so, he could not bear it. And the tears came, and his heart beat, and he was hot and worn and bitter in him.

Now Jeff knew very well what it was to love Melanctha. Now Jeff Campbell knew he was really understanding. Now Jeff knew what it was to be good to Melanctha. Now Jeff was good to her always.

Slowly Jeff felt it a comfort in him to have it hurt so, and to be good to Melanctha always. Now there was no way Melanctha ever had had to bear things from him, worse than he now had it in him. Now Jeff was strong inside him. Now with all the pain there was peace in him. Now he knew he was understanding, now he knew he had a hot love in him, and he was good always to Melanctha Herbert who was the one had made him have it. Now he knew he could be good, and not cry out for help to her to teach him how to bear it. Every day Jeff felt himself more a strong man, the way he once had thought was his real self, the way he knew it. Now Jeff Campbell had real wisdom in him, and it did not make him bitter when it hurt him, for Jeff knew now all through him that he was really strong to bear it.

And so now Jeff Campbell could see Melanctha often, and he was patient, and always very friendly to her, and every day Jeff Campbell understood Melanctha Herbert better. And always Jeff saw Melanctha could not love him the way he needed she should do it. Melanctha Herbert had no way she ever really could remember.

And now Jeff knew there was a man Melanctha met very often, and perhaps she wanted to try to have this man to be good, for her. Jeff Campbell never saw the man Melanctha Herbert perhaps now wanted. Jeff Campbell only knew very well that there was one. Then there was Rose that Melanctha now always had with her when she wandered.

Jeff Campbell was very quiet to Melanctha. He said to her, now he thought he did not want to come any more especially to see her. When they met, he always would be glad to see her, but now he never would go anywhere any more to meet her. Sure he knew she always would have a deep love in him for her. Sure she knew that. "Yes Jeff, I always trust you Jeff, I certainly do know that all right." Jeff Campbell said, all right he never could say anything to reproach her. She knew always that he really had learned all through him how to love her. "Yes, Jeff, I certainly do know that." She knew now she could always trust him. Jeff always would be loyal to her though now she never was any more to him like a religion, but he never could forget the real sweetness in her. That Jeff must remember always, though now he never can trust her to be really loving to any man for always, she never did have any way she ever could remember. If she ever needed anybody to be good to her, Jeff Campbell always would do anything he could to help her. He never can forget the things she taught him so he could be really understanding, but he never any more wants to see her. He be like a brother to her always, when she needs it, and he always will be a good friend to her. Jeff Campbell certainly was sorry never any more to see her, but it was good that they now knew each other really. "Good-by Jeff you always been very good always to me." "Good-by Melanctha you know you always can trust yourself to me." "Yes, I know, I know Jeff, really." "I certainly got to go now Melanctha, from you. I go this time, Melanctha really," and Jeff Campbell went away and this time he never looked back to her. This time Jeff Campbell just broke away and left her.

Jeff Campbell loved to think now he was strong again to be quiet, and to live regular, and to do everything the way he wanted it to be right for himself and all the colored people. Jeff went away for a little while to another town to work there, and he worked hard, and he was very sad inside him, and sometimes the tears would rise up in him, and then he would work hard, and then he would begin once more to see some beauty in the world around him. Jeff had behaved right and he had learned to have a real love in him. That was very good to have inside him.

Jeff Campbell never could forget the sweetness in Melanctha Herbert, and he was always very friendly to her, but they never any more came close to one another. More and more Jeff Campbell and Melanctha fell away from all knowing of each other, but Jeff never could forget Melanctha. Jeff never could forget the real sweetness she had in her, but Jeff never any more had the sense of a real religion for her. Jeff always had strong in him the meaning of all the new kind of beauty Melanctha Herbert once had shown him, and always more and more it helped him with his working for himself and for all the colored people.

Melanctha Herbert, now that she was all through with Jeff Campbell, was free to be with Rose and the new men she met now.

Rose was always now with Melanctha Herbert. Rose never found any way to get excited. Rose always was telling Melanctha Herbert the right way she should do, so that she would not always be in trouble. But Melanctha Herbert could not help it, always she would find new ways to get excited.

Melanctha was all ready now to find new ways to be in trouble. And yet Melanctha Herbert never wanted not to do right. Always Melanctha Herbert wanted peace and quiet, and always she could only find new ways to get excited.

"Melanctha," Rose would say to her, "Melanctha, I certainly have got to tell you, you ain't right to act so with that kind of feller. You better just had stick to black men now, Melanctha, you hear me what I tell you, just the way you always see me do it. They're real bad men, now I tell you Melanctha true, and you better had hear to me. I been raised by real nice kind of white folks, Melanctha, and I certainly knows awful well, soon as ever I can see 'em acting, what is a white man will act decent to you and the kind it ain't never no good to a colored girl to ever go with. Now you know real Melanctha how I always mean right good to you, and you ain't got no way like me Melanctha, what was raised by white folks, to know right what is the way you

should be acting with men. I don't never want to see you have bad
trouble come hard to you now Melanctha, and so you just hear to me
now Melanctha, what I tell you, for I knows it. I don't say never certainly to you Melanctha, you never had ought to have nothing to do
ever with no white men, though it ain't never to me Melanctha, the
best kind of a way a colored girl can have to be acting, no I never do
say to you Melanctha, you hadn't never ought to be with white men,
though it ain't never the way I feel it ever real right for a decent colored girl to be always doing, but not never Melanctha, now you hear
me, no not never no kind of white men like you been with always now
Melanctha when I see you. You just hear to me Melanctha, you certainly had ought to hear to me Melanctha, I say it just like I knows it
awful well, Melanctha, and I knows you don't know no better,
Melanctha, how to act so, the ways I seen it with them kind of white
fellers, them as never can know what to do right by a decent girl they
have ever got to be with them. Now you hear to me Melanctha, what I
tell you. "

And so it was Melanctha Herbert found new ways to be in trouble.
But it was not very bad this trouble, for these white men Rose never
wanted she should be with, never meant very much to Melanctha. It
was only that she liked it to be with them, and they knew all about fine
horses, and it was just good to Melanctha, now a little, to feel real
reckless with them. But mostly it was Rose and other better kind of
colored girls and colored men with whom Melanctha Herbert now
always wandered.

It was summer now and the colored people came out into the sunshine, full blown with the flowers. And they shone in the streets and in
the fields with their warm joy, and they glistened in their black heat,
and they flung themselves free in their wide abandonment of shouting
laughter.

It was very pleasant in some ways, the life Melanctha Herbert now
led with Rose and all the others. It was not always that Rose had to
scold her.

There was not anybody of all these colored people, excepting only
Rose, who ever meant much to Melanctha Herbert. But they all liked
Melanctha, and the men all liked to see her do things, she was so game
always to do anything anybody ever could do, and then she was good
and sweet to do anything anybody ever wanted from her.

These were pleasant days then, in the hot southern negro sunshine,
with many simple jokes and always wide abandonment of laughter.
"Just look at that Melanctha there a running. Don't she just go like a

bird when she is flying. Hey Melanctha there, I come and catch you, hey Melanctha, I put salt on your tail to catch you," and then the man would try to catch her, and he would fall full on the earth and roll in an agony of wide-mouthed shouting laughter. And this was the kind of way Rose always liked to have Melanctha do it, to be engaged to him, and to have a good warm nigger time with colored men, not to go about with that kind of white man, never could know how to act right, to any decent kind of girl they could ever get to be with them.

Rose, always more and more, liked Melanctha Herbert better. Rose often had to scold Melanctha Herbert, but that only made her like Melanctha better. And then Melanctha always listened to her, and always acted every way she could to please her. And then Rose was so sorry for Melanctha, when she was so blue sometimes, and wanted somebody should come and kill her.

And Melanctha Herbert clung to Rose in the hope that Rose could save her. Melanctha felt the power of Rose's selfish, decent kind of nature. It was so solid, simple, certain to her. Melanctha clung to Rose, she loved to have her scold her, she always wanted to be with her. She always felt a solid safety in her. Rose always was, in her way, very good to let Melanctha be loving to her. Melanctha never had any way she could really be a trouble to her. Melanctha never had any way that she could ever get real power, to come close inside to her. Melanctha was always very humble to her. Melanctha was always ready to do anything Rose wanted from her. Melanctha needed badly to have Rose always willing to let Melanctha cling to her. Rose was a simple, sullen, selfish, black girl, but she had a solid power in her. Rose had strong the sense of decent conduct, she had strong the sense for decent comfort. Rose always knew very well what it was she wanted, and she knew very well what was the right way to do to get everything she wanted, and she never had any kind of trouble to perplex her. And so the subtle intelligent attractive half white girl Melanctha Herbert loved and did for, and demeaned herself in service to this coarse, decent, sullen, ordinary, black, childish Rose and now this unmoral promiscuous shiftless Rose was to be married to a good man of the negroes, while Melanctha Herbert with her white blood and attraction and her desire for a right position was perhaps never to be really regularly married. Sometimes the thought of how all her world was made filled the complex, desiring Melanctha with despair. She wondered often how she could go on living when she was so blue. Sometimes Melanctha thought she would just kill herself, for sometimes she thought this would be really the best thing for her to do.

Rose was now to be married to a decent good man of the negroes. His name was Sam Johnson, and he worked as a deck-hand on a coasting steamer, and he was very steady, and he got good wages. Rose first met Sam Johnson at church, the same place where she had met Melanctha Herbert. Rose liked Sam when she saw him, she knew he was a good man and worked hard and got good wages, and Rose thought it would be very nice and very good now in her position to get really, regularly married.

Sam Johnson liked Rose very well and he always was ready to do anything she wanted. Sam was a tall, square shouldered, decent, a serious, straightforward, simple, kindly, colored workman. They got on very well together, Sam and Rose, when they were married. Rose was lazy, but not dirty, and Sam was careful but not fussy. Sam was a kindly, simple, earnest, steady workman, and Rose had good common decent sense in her, of how to live regular, and not to have excitements, and to be saving so you could be always sure to have money, so as to have everything you wanted.

It was not very long that Rose knew Sam Johnson, before they were regularly married. Sometimes Sam went into the country with all the other young church people, and then he would be a great deal with Rose and with her Melanctha Herbert. Sam did not care much about Melanctha Herbert. He liked Rose's ways of doing, always better. Melanctha's mystery had no charm for Sam ever. Sam wanted a nice little house to come to when he was tired from his working, and a little baby all his own he could be good to. Sam Johnson was ready to marry as soon as ever Rose wanted he should do it. And so Sam Johnson and Rose one day had a grand real wedding and were married. Then they furnished completely, a little red brick house and then Sam went back to his work as deck hand on a coasting steamer.

Rose had often talked to Sam about how good Melanctha was and how much she always suffered. Sam Johnson never really cared about Melanctha Herbert, but he always did almost everything Rose ever wanted, and he was a gentle, kindly creature, and so he was very good to Rose's friend Melanctha. Melanctha Herbert knew very well Sam did not like her, and so she was very quiet, and always let Rose do the talking for her. She only was very good to always help Rose, and to do anything she ever wanted from her, and to be very good and listen and be quiet whenever Sam had anything to say to her. Melanctha liked Sam Johnson, and all her life Melanctha loved and wanted good and kind and considerate people, and always Melanctha loved and wanted people to be gentle to her, and always she wanted to be regular, and to have peace and quiet in her, and always Melanctha could only find new ways to be in trouble. And Melanctha needed badly to have Rose, to believe her, and to let her cling to her. Rose was the only steady thing Melanctha had to cling to and so Melanctha demeaned herself to be like a servant, to wait on, and always to be scolded, by this ordinary, sullen, black, stupid, childish woman.

Rose was always telling Sam he must be good to poor Melanctha. "You know Sam," Rose said very often to him, "You certainly had ought to be very good to poor Melanctha, she always do have so much trouble with her. You know Sam how I told you she had such a bad time always with that father, and he was awful mean to her always that awful black man, and he never took no kind of care ever to her, and he never helped her when her mother died so hard, that poor Melanctha. Melanctha's ma you know Sam, always was just real religious. One day Melanctha was real little, and she heard her ma say to her pa, it was awful sad to her, Melanctha had not been the one the Lord had took from them stead of the little brother who was dead in the house there from fever. That hurt Melanctha awful when she heard her ma say it. She never could feel it right, and I don't no ways blame Melanctha, Sam, for not feeling better to her ma always after, though Melanctha, just like always she is, always was real good to her ma after, when she was so sick, and died so hard, and nobody never to help Melanctha do it, and she just all alone to do everything without no help come to her no way, and that ugly awful black man she have for a father never all the time come near her. But that's always the way Melanctha is just doing Sam, the way I been telling to you. She always is being just so good to everybody and nobody ever there to thank her for it. I never did see nobody ever Sam, have such bad luck, seems to me always with them, like that poor Melanctha always has it, and she always so good with it, and never no murmur in her, and never no complaining from her, and just never saying nothing with it. You be real good to her Sam, now you hear me, now you and me is married right together. He certainly was an awful black man to her Sam, that father she had, acting always just like a brute to her and she so game and never to tell anybody how it hurt her. And she so sweet and good always to do anything anybody ever can be wanting. I don't see Sam how some men can be to act so awful. I told you Sam, how once Melanctha broke her arm bad and she was so sick and it hurt her awful and he never would let no doctor come near to her and he do some things so awful to her, she don't never want to tell nobody how bad he hurt her. That's just the way Sam with Melanctha always, you never can know how bad it is, it hurts her. You hear me Sam, you always be real good to her now you and me is married right to each other."

And so Rose and Sam Johnson were regularly married, and Rose sat at home and bragged to all her friends how nice it was to be married really to a husband.

Rose did not have Melanctha to live with her, now Rose was married. Melanctha was with Rose almost as much as ever but it was a little different now their being together.

Rose Johnson never asked Melanctha to live with her in the house, now Rose was married. Rose liked to have Melanctha come all the time to help her, Rose liked Melanctha to be almost always with her, but Rose was shrewd in her simple selfish nature, she did not ever think to ask Melanctha to live with her.

Rose was hard headed, she was decent, and she always knew what it was she needed. Rose needed Melanctha to be with her, she liked to have her help her, the quick, good Melanctha to do for the slow, lazy, selfish, black girl, but Rose could have Melanctha to do for her and she did not need her to live with her.

Sam never asked Rose why she did not have her. Sam always took what Rose wanted should be done for Melanctha, as the right way he should act toward her.

It could never come to Melanctha to ask Rose to let her. It never could come to Melanctha to think that Rose would ask her. It would never ever come to Melanctha to want it, if Rose should ask her, but Melanctha would have done it for the safety she always felt when she was near her. Melanctha Herbert wanted badly to be safe now, but this living with her, that, Rose would never give her. Rose had strong the sense for decent comfort, Rose had strong the sense for proper conduct, Rose had strong the sense to get straight always what she wanted, and she always knew what was the best thing she needed, and always Rose got what she wanted.

And so Rose had Melanctha Herbert always there to help her, and she sat and was lazy and she bragged and she complained a little and she told Melanctha how she ought to do, to get good what she wanted like she Rose always did it, and always Melanctha was doing everything Rose ever needed. "Don't you bother so, doing that Melanctha, I do it or Sam when he comes home to help me. Sure you don't mind lifting it Melanctha? You is very good Melanctha to do it, and when you go out Melanctha, you stop and get some rice to bring me tomorrow when you come in. Sure you won't forget Melanctha. I never

see anybody like you Melanctha to always do things so nice for
me." And then Melanctha would do some more for Rose, and then
very late Melanctha would go home to the colored woman where she
lived now.

And so though Melanctha still was so much with Rose Johnson, she
had times when she could not stay there. Melanctha now could not
really cling there. Rose had Sam, and Melanctha more and more lost
the hold she had had there.

Melanctha Herbert began to feel she must begin again to look and
see if she could find what it was she had always wanted. Now Rose
Johnson could no longer help her.

And so Melanctha Herbert began once more to wander and with
men Rose never thought it was right she should be with.

One day Melanctha had been very busy with the different kinds of
ways she wandered. It was a pleasant late afternoon at the end of a
long summer. Melanctha was walking along, and she was free and
excited. Melanctha had just parted from a white man and she had a
bunch of flowers he had left with her. A young buck, a mulatto, passed
by and snatched them from her. "It certainly is real sweet in you sister,
to be giving me them pretty flowers," he said to her.

"I don't see no way it can make them sweeter to have with you,"
said Melanctha. "What one man gives, another man had certainly just
as much good right to be taking." "Keep your old flowers then, I certainly don't never want to have them." Melanctha Herbert laughed at
him and took them. "No, I didn't nohow think you really did want to
have them. Thank you kindly mister, for them. I certainly always do
admire to see a man always so kind of real polite to people." The man
laughed, "You ain't nobody's fool I can say for you, but you certainly
are a damned pretty kind of girl, now I look at you. Want men to be
polite to you? All right, I can love you, that's real polite now, want to
see me try it." "I certainly ain't got no time this evening just only left
to·thank you. I certainly got to be real busy now, but I certainly always
will admire to see you." The man tried to catch and stop her, Melanctha Herbert laughed and dodged so that he could not touch her.
Melanctha went quickly down a side street near her and so the man
for that time lost her.

For some days Melanctha did not see any more of her mulatto. One
day Melanctha was with a white man and they saw him. The white
man stopped to speak to him. Afterwards Melanctha left the white
man and she then soon met him. Melanctha stopped to talk to him.
Melanctha Herbert soon began to like him.

Jem Richards, the new man Melanctha had begun to know now, was a dashing kind of fellow, who had to do with fine horses and with racing. Sometimes Jem Richards would be betting and would be good and lucky, and be making lots of money. Sometimes Jem would be betting badly, and then he would not be having any money. Jem Richards was a straight man. Jem Richards always knew that by and by he would win again and pay it, and so Jem mostly did win again, and then he always paid it.

Jem Richards was a man other men always trusted. Men gave him money when he lost all his, for they all knew Jem Richards would win again, and when he did win they knew, and they were right, that he would pay it.

Melanctha Herbert all her life had always loved to be with horses. Melanctha liked it that Jem knew all about fine horses. He was a reckless man was Jem Richards. He knew how to win out, and always all her life, Melanctha Herbert loved successful power.

Melanctha Herbert always liked Jem Richards better. Things soon began to be very strong between them.

Jem was more game even than Melanctha. Jem always had known what it was to have real wisdom. Jem had always all his life been understanding.

Jem Richards made Melanctha Herbert come fast with him. He never gave her any time with waiting. Soon Melanctha always had Jem with her. Melanctha did not want anything better. Now in Jem Richards, Melanctha found everything she had ever needed to content her.

Melanctha was now less and less with Rose Johnson. Rose did not think much of the way Melanctha now was going. Jem Richards was all right, only Melanctha never had no sense of the right kind of way she should be doing. Rose often was telling Sam now, she did not like the fast way Melanctha was going. Rose told it to Sam, and to all the girls and men, when she saw them. But Rose was nothing just then to Melanctha. Melanctha Herbert now only needed Jem Richards to be with her.

And things were always getting stronger between Jem Richards and Melanctha Herbert. Jem Richards began to talk now as if he wanted to get married to her. Jem was deep in his love now for her. And as for Melanctha, Jem was all the world now to her. And so Jem gave her a ring, like white folks, to show he was engaged to her, and would by and by be married to her. And Melanctha was filled full with joy to have Jem so good to her.

Melanctha always loved to go with Jem to the races. Jem had been lucky lately with his betting, and he had a swell turn-out[9] to drive in, and Melanctha looked very handsome there beside him.

Melanctha was very proud to have Jem Richards want her. Melanctha loved it the way Jem knew how to do it. Melanctha loved Jem and loved that he should want her. She loved it too, that he wanted to be married to her. Jem Richards was a straight decent man, whom other men always looked up to and trusted. Melanctha needed badly a man to content her.

Melanctha's joy made her foolish. Melanctha told everybody about how Jem Richards, that swell man who owned all those fine horses and was so game, nothing ever scared him, was engaged to be married to her, and that was the ring he gave her.

Melanctha let out her joy very often to Rose Johnson. Melanctha had begun again now to go there.

Melanctha's love for Jem made her foolish. Melanctha had to have some one always now to talk to and so she went often to Rose Johnson.

Melanctha put all herself into Jem Richards. She was mad and foolish in the joy she had there.

Rose never liked the way Melanctha did it. "No Sam I don't say never Melanctha ain't engaged to Jem Richards the way she always says it, and Jem he is all right for that kind of a man he is, though he do think himself so smart and like he owns the earth and everything he can get with it, and he sure gave Melanctha a ring like he really meant he should be married right soon with it, only Sam, I don't ever like it the way Melanctha is going. When she is engaged to him Sam, she ain't not right to take on so excited. That ain't no decent kind of a way a girl ever should be acting. There ain't no kind of a man going stand that, not like I knows men Sam, and I sure does know them. I knows them white and I knows them colored, for I was raised by white folks, and they don't none of them like a girl to act so. That's all right to be so when you is just only loving, but it ain't no ways right to be acting so when you is engaged to him, and when he says, all right he get really regularly married to you. You see Sam I am right like I am always and I knows it. Jem Richards, he ain't going to the last to get real married, not if I knows it right, the way Melanctha now is acting to him. Rings or anything ain't nothing to them, and they don't never do no good for them, when a girl acts foolish like Melanctha always now is acting. I

[9] *a swell turn-out:* Buggy or carriage of the style used by the middle or upper class.

certainly will be right sorry Sam, if Melanctha has real bad trouble come now to her, but I certainly don't no ways like it Sam the kind of way Melanctha is acting to him. I don't never say nothing to her Sam. I just listens to what she is saying always, and I thinks it out like I am telling to you Sam but I don't never say nothing no more now to Melanctha. Melanctha didn't say nothing to me about that Jem Richards till she was all like finished with him, and I never did like it Sam, much, the way she was acting, not coming here never when she first ran with those men and met him. And I didn't never say nothing to her, Sam, about it, and it ain't nothing ever to me, only I don't never no more want to say nothing to her, so I just listens to what she got to tell like she wants it. No Sam, I don't never want to say nothing to her. Melanctha just got to go her own way, not as I want to see her have bad trouble ever come hard to her, only it ain't in me never Sam, after Melanctha did so, ever to say nothing more to her how she should be acting. You just see Sam like I tell you, what way Jem Richards will act to her, you see Sam I just am right like I always am when I knows it."

Melanctha Herbert never thought she could ever again be in trouble. Melanctha's joy had made her foolish.

And now Jem Richards had some bad trouble with his betting. Melanctha sometimes felt now when she was with him that there was something wrong inside him. Melanctha knew he had had trouble with his betting but Melanctha never felt that that could make any difference to them.

Melanctha once had told Jem, sure he knew she always would love to be with him, if he was in jail or only just a beggar. Now Melanctha said to him, "Sure you know Jem that it don't never make any kind of difference you're having any kind of trouble, you just try me Jem and be game, don't look so worried to me. Jem sure I know you love me like I love you always, and its all I ever could be wanting Jem to me, just your wanting me always to be with you. I get married Jem to you soon ever as you can want me, if you once say it Jem to me. It ain't nothing to me ever, anything like having any money Jem, why you look so worried to me."

Melanctha Herbert's love had surely made her mad and foolish. She thrust it always deep into Jem Richards and now that he had trouble with his betting, Jem had no way that he ever wanted to be made to feel it. Jem Richards never could want to marry any girl while he had trouble. That was no way a man like him should do it. Melanctha's love had made her mad and foolish, she should be silent now and let him do it. Jem Richards was not a kind of man to want a woman to be strong to him, when he was in trouble with his betting. That was not the kind of a time when a man like him needed to have it.

Melanctha needed so badly to have it, this love which she had always wanted, she did not know what she should do to save it. Melanctha saw now, Jem Richards always had something wrong inside him. Melanctha soon dared not ask him. Jem was busy now, he had to sell things and see men to raise money. Jem could not meet Melanctha now so often.

It was lucky for Melanctha Herbert that Rose Johnson was coming now to have her baby. It had always been understood between them, Rose should come and stay then in the house where Melanctha lived with an old colored woman, so that Rose could have the Doctor from the hospital near by to help her, and Melanctha there to take care of her the way Melanctha always used to do it.

Melanctha was very good now to Rose Johnson. Melanctha did everything that any woman could, she tended Rose, and she was patient, submissive, soothing and untiring, while the sullen, childish, cowardly, black Rosie grumbled, and fussed, and howled, and made herself to be an abomination and like a simple beast.

All this time Melanctha was always being every now and then with Jem Richards. Melanctha was beginning to be stronger with Jem Richards. Melanctha was never so strong and sweet and in her nature as when she was deep in trouble, when she was fighting so with all she had, she could not do any foolish thing with her nature.

Always now Melanctha Herbert came back again to be nearer to Rose Johnson. Always now Melanctha would tell all about her troubles to Rose Johnson. Rose had begun now a little again to advise her.

Melanctha always told Rose now about the talks she had with Jem Richards, talks where they neither of them liked very well what the other one was saying. Melanctha did not know what it was Jem Richards wanted. All Melanctha knew was, he did not like it when she wanted to be good friends and get really married, and then when Melanctha would say, "all right, I never wear your ring no more Jem, we ain't not any more to meet ever like we ever going to get really regular married," then Jem did not like it either. What was it Jem Richards really wanted?

Melanctha stopped wearing Jem's ring on her finger. Poor Melanctha, she wore it on a string she tied around her neck so that she could always feel it, but Melanctha was strong now with Jem Richards, and he never saw it. And sometimes Jem seemed to be awful sorry for it, and sometimes he seemed kind of glad of it. Melanctha never could make out really what it was Jem Richards wanted.

There was no other woman yet to Jem, that Melanctha knew, and so she always trusted that Jem would come back to her, deep in his love, the way once he had had it and had made all the world like she once had never believed anybody could really make it. But Jem Richards was more game than Melanctha Herbert. He knew how to fight to win out, better. Melanctha really had already lost it, in not keeping quiet and waiting for Jem to do it.

Jem Richards was not yet having better luck in his betting. He never before had had such a long time without some good coming to him in his betting. Sometimes Jem talked as if he wanted to go off on a trip somewhere and try some other place for luck with his betting. Jem Richards never talked as if he wanted to take Melanctha with him.

And so Melanctha sometimes was really trusting, and sometimes she was all sick inside her with her doubting. What was it Jem really wanted to do with her? He did not have any other woman, in that Melanctha could be really trusting, and when she said no to him, no she never would come near him, now he did not want to have her, then Jem would change and swear, yes sure he did want her, now and always right here near him, but he never now any more said he wanted to be married soon to her. But then Jem Richards never would marry a girl, he said that very often, when he was in this kind of trouble, and now he did not see any way he could get out of his trouble. But Melanctha ought to wear his ring, sure she knew he never had loved any kind of woman like he loved her. Melanctha would wear the ring a little while, and then they would have some more trouble, and then she would say to him, no she certainly never would any more wear anything he gave her, and then she would wear it on the string so nobody could see it but she could always feel it on her.

Poor Melanctha, surely her love had made her mad and foolish.

And now Melanctha needed always more and more to be with Rose Johnson, and Rose had commenced again to advise her, but Rose could not help her. There was no way now that anybody could advise her. The time when Melanctha could have changed it with Jem Richards was now all past for her. Rose knew it, and Melanctha too, she knew it, and it almost killed her to let herself believe it.

The only comfort Melanctha ever had now was waiting on Rose till she was so tired she could hardly stand it. Always Melanctha did everything Rose ever wanted. Sam Johnson began now to be very gentle and a little tender to Melanctha. She was so good to Rose and Sam was so glad to have her there to help Rose and to do things and to be a comfort to her.

Rose had a hard time to bring her baby to its birth and Melanctha did everything that any woman could.

The baby though it was healthy after it was born did not live long. Rose Johnson was careless and negligent and selfish and when Melanctha had to leave for a few days the baby died. Rose Johnson had liked her baby well enough and perhaps she just forgot it for a while, anyway the child was dead and Rose and Sam were very sorry, but then these things came so often in the negro world in Bridgepoint that they neither of them thought about it very long. When Rose had become strong again she went back to her house with Sam. And Sam Johnson was always now very gentle and kind and good to Melanctha who had been so good to Rose in her bad trouble.

Melanctha Herbert's troubles with Jem Richards were never getting any better. Jem always now had less and less time to be with her. When Jem was with Melanctha now he was good enough to her. Jem Richards was worried with his betting. Never since Jem had first begun to make a living had he ever had so much trouble for such a long time together with his betting. Jem Richards was good enough now to Melanctha but he had not much strength to give her. Melanctha could never any more now make him quarrel with her. Melanctha never now could complain of his treatment of her, for surely, he said it always by his actions to her, surely she must know how a man was when he had trouble on his mind with trying to make things go a little better.

Sometimes Jem and Melanctha had long talks when they neither of them liked very well what the other one was saying, but mostly now Melanctha could not make Jem Richards quarrel with her, and more and more, Melanctha could not find any way to make it right to blame him for the trouble she now always had inside her. Jem was good to her, and she knew, for he told her, that he had trouble all the time now with his betting. Melanctha knew very well that for her it was all wrong inside Jem Richards, but Melanctha had now no way that she could really reach him.

Things between Melanctha and Jem Richards were now never getting any better. Melanctha now more and more needed to be with Rose Johnson. Rose still liked to have Melanctha come to her house and do things for her, and Rose liked to grumble to her and to scold her and to tell Melanctha what was the way Melanctha always should be doing so she could make things come out better and not always be so much in trouble. Sam Johnson in these days was always very good and gentle to Melanctha. Sam was now beginning to be very sorry for her. Jem Richards never made things any better for Melanctha. Often Jem would talk so as to make Melanctha almost certain that he never any more wanted to have her. Then Melanctha would get very blue, and she would say to Rose, sure she would kill herself, for that certainly now was the best way she could do.

Rose Johnson never saw it the least bit that way. "I don't see Melanctha why you should talk like you would kill yourself just because you're blue. I'd never kill myself Melanctha cause I was blue. I'd maybe kill somebody else but I'd never kill myself. If I ever killed myself, Melanctha it'd be by accident and if I ever killed myself by accident, Melanctha, I'd be awful sorry. And that certainly is the way you should feel it Melanctha, now you hear me, not just talking foolish like you always do. It certainly is only your way just always being foolish makes you all that trouble to come to you always now, Melanctha, and I certainly right well knows that. You certainly never can learn no way Melanctha ever with all I certainly been telling to you, ever since I know you good, that it ain't never no way like you do always is the right way you be acting ever and talking, the way I certainly always have seen you do so Melanctha always. I certainly am right Melanctha about them ways you have to do it, and I knows it; but you certainly never can noways learn to act right Melanctha, I certainly do know that, I certainly do my best Melanctha to help you with it only you certainly never do act right Melanctha, not to nobody ever, I can see it. You never act right by me Melanctha no more than by everybody. I never say nothing to you Melanctha when you do so, for I certainly never do like it when I just got to say it to you, but you just certainly done with that Jem Richards you always say wanted real bad to be married to you, just like I always said to Sam you certainly was going to do it. And I certainly am real kind of sorry like for you Melanctha, but you certainly had ought to have come to see me to talk to you, when you first was engaged to him so I could show you, and now you got all this trouble come to you Melanctha like I certainly know you always catch it. It certainly ain't never Melanctha I ain't real sorry to see trouble come so hard to you, but I certainly can see Melanctha it all is always just the way you always be having it in you not never to do right. And now you always talk like you just kill yourself because you are so blue, that certainly never is Melanctha, no kind of a way for any decent kind of a girl to do."

Rose had begun to be strong now to scold Melanctha and she was impatient very often with her, but Rose could now never any more be a help to her. Melanctha Herbert never could know now what it was right she should do. Melanctha always wanted to have Jem Richards with her and now he never seemed to want her, and what could Melanctha do. Surely she was right now when she said she would just kill herself, for that was the only way now she could do.

Sam Johnson always, more and more, was good and gentle to Melanctha. Poor Melanctha, she was so good and sweet to do anything anybody ever wanted, and Melanctha always liked it if she could have peace and quiet, and always she could only find new ways to be in trouble. Sam often said this now to Rose about Melanctha.

"I certainly don't never want Sam to say bad things about Melanctha, for she certainly always do have most awful kind of trouble come hard to her, but I never can say I like it real right Sam the way Melanctha always has to do it. Its now just the same with her like it is always she has got to do it, now the way she is with that Jem Richards. He certainly now don't never want to have her but Melanctha she ain't got no right kind of spirit. No Sam I don't never like the way any more Melanctha is acting to him, and then Sam, she ain't never real right honest, the way she always should do it. She certainly just don't kind of never Sam tell right what way she is doing with it. I don't never like to say nothing Sam no more to her about the way she always has to be acting. She always say, yes all right Rose, I do the way you say it, and then Sam she don't never noways do it. She certainly is right sweet and good, Sam, is Melanctha, nobody ever can hear me say she ain't always ready to do things for everybody anyway she ever can see to do it, only Sam some ways she never does act real right ever, and some ways, Sam, she ain't ever real honest with it. And Sam sometimes I hear awful kind of things she been doing, some girls know about her how she does it, and sometimes they tell me what kind of ways she has to do it, and Sam it certainly do seem to me like more and more I certainly am awful afraid Melanctha never will come to any good. And then Sam, some-times, you hear it, she always talk like she kill herself all the time she is so blue, and Sam that certainly never is no kind of way any decent girl ever had ought to do. You see Sam, how I am right like I always is when I knows it. You just be careful, Sam, now you hear me, you be careful Sam sure, I tell you, Melanctha more and more I see her I certainly do feel Melanctha no way is really honest. You be careful, Sam now, like I tell you, for I knows it, now you hear to me, Sam, what I tell you, for I certainly always is right, Sam, when I knows it."

At first Sam tried a little to defend Melanctha, and Sam always was good and gentle to her, and Sam liked the ways Melanctha had to be quiet to him, and to always listen as if she was learning, when she was there and heard him talking, and then Sam liked the sweet way she always did everything so nicely for him; but Sam never liked to fight with anybody ever, and surely Rose knew best about Melanctha and anyway Sam never did really care much about Melanctha. Her mystery never had had any interest for him. Sam liked it that she was sweet to him and that she always did everything Rose ever wanted that she should be doing, but Melanctha never could be important to him. All Sam ever wanted was to have a little house and to live regular and to work hard and to come home to his dinner, when he was tired with his working and by and by he wanted to have some children all his own to be good to, and so Sam was real sorry for Melanctha, she was so good and so sweet always to them, and Jem Richards was a bad man to behave so to her, but that was always the way a girl got it when she liked that kind of a fast fellow. Anyhow Melanctha was Rose's friend, and Sam never cared to have anything to do with the kind of trouble always came to women, when they wanted to have men, who never could know how to behave good and steady to their women.

And so Sam never said much to Rose about Melanctha. Sam was always very gentle to her, but now he began less and less to see her. Soon Melanctha never came any more to the house to see Rose and Sam never asked Rose anything about her.

Melanctha Herbert was beginning now to come less and less to the house to be with Rose Johnson. This was because Rose seemed always less and less now to want her, and Rose would not let Melanctha now do things for her. Melanctha was always humble to her and Melanctha always wanted in every way she could to do things for her. Rose said no, she guessed she do that herself like she likes to have it better. Melanctha is real good to stay so long to help her, but Rose guessed perhaps Melanctha better go home now, Rose don't need nobody to help her now, she is feeling real strong, not like just after she had all that trouble with the baby, and then Sam, when he comes home for his dinner he likes it when Rose is all alone there just to give him his dinner. Sam always is so tired now, like he always is in the summer, so many people always on the steamer, and they make so much work so Sam is real tired now, and he likes just to eat his dinner and never have people in the house to be a trouble to him.

Each day Rose treated Melanctha more and more as if she never wanted Melanctha any more to come there to the house to see her. Melanctha dared not ask Rose why she acted in this way to her. Melanctha badly needed to have Rose always there to save her. Melanctha wanted badly to cling to her and Rose had always been so solid for her. Melanctha did not dare to ask Rose if she now no longer wanted her to come and see her.

Melanctha now never any more had Sam to be gentle to her. Rose always sent Melanctha away from her before it was time for Sam to come home to her. One day Melanctha had stayed a little longer, for Rose that day had been good to let Melanctha begin to do things for her. Melanctha then left her and Melanctha met Sam Johnson who stopped a minute to speak kindly to her.

The next day Rose Johnson would not let Melanctha come in to her. Rose stood on the steps, and there she told Melanctha what she thought now of her.

"I guess Melanctha it certainly ain't no ways right for you to come here no more just to see me. I certainly don't Melanctha no ways like to be a trouble to you. I certainly think Melanctha I get along better now when I don't have nobody like you are, always here to help me, and Sam he do so good now with his working, he pay a little girl something to come every day to help me. I certainly do think Melanctha I don't never want you no more to come here just to see me." "Why Rose, what I ever done to you, I certainly don't think you is right Rose to be so bad now to me." "I certainly don't no ways Melanctha Herbert think you got any right ever to be complaining the way I been acting to you. I certainly never do think Melanctha Herbert, you hear to me, nobody ever been more patient to you than I always been to like you, only Melanctha, I hear more things now so awful bad about you, everybody always is telling to me what kind of a way you always have been doing so much, and me always so good to you, and you never no ways, knowing how to be honest to me. No Melanctha it ain't ever in me, not to want you to have good luck come to you, and I like it real well Melanctha when you some time learn how to act the way it is decent and right for a girl to be doing, but I don't no ways ever like it the kind of things everybody tell me now about you. No Melanctha, I can't never any more trust you. I certainly am real sorry to have never any more to see you, but there ain't no other way, I ever can be acting to you. That's all I ever got any more to say to you now Melanctha." "But Rose, deed; I certainly don't know, no more than the dead, nothing I ever done to make you act so to me. Anybody say anything bad about me Rose, to you, they just a pack of liars to you, they certainly is Rose, I tell you true. I certainly never done nothing I ever been

ashamed to tell you. Why you act so bad to me Rose. Sam he certainly
don't think ever like you do, and Rose I always do everything I can,
you ever want me to do for you." "It ain't never no use standing there
talking, Melanctha Herbert. I just can tell it to you, and Sam, he don't
know nothing about women ever the way they can be acting. I certainly am very sorry Melanctha, to have to act so now to you, but I
certainly can't do no other way with you, when you do things always
so bad, and everybody is talking so about you. It ain't no use to you to
stand there and say it different to me Melanctha. I certainly am always
right Melanctha Herbert, the way I certainly always have been when I
knows it, to you. No Melanctha, it just is, you never can have no kind
of a way to act right, the way a decent girl has to do, and I done my
best always to be telling it to you Melanctha Herbert, but it don't
never do no good to tell nobody how to act right; they certainly never
can learn when they ain't got no sense right to know it, and you never
have no sense right Melanctha to be honest, and I ain't never wishing
no harm to you ever Melanctha Herbert, only I don't never want any
more to see you come here. I just say to you now, like I always been
saying to you, you don't know never the right way, any kind of decent
girl has to be acting, and so Melanctha Herbert, me and Sam, we don't
never any more want you to be setting your foot in my house here
Melanctha Herbert, I just tell you. And so you just go along now,
Melanctha Herbert, you hear me, and I don't never wish no harm to
come to you."

Rose Johnson went into her house and closed the door behind her.
Melanctha stood like one dazed, she did not know how to bear this
blow that almost killed her. Slowly then Melanctha went away without even turning to look behind her.

Melanctha Herbert was all sore and bruised inside her. Melanctha
had needed Rose always to believe her, Melanctha needed Rose always
to let her cling to her, Melanctha wanted badly to have somebody who
could make her always feel a little safe inside her, and now Rose had
sent her from her. Melanctha wanted Rose more than she had ever
wanted all the others. Rose always was so simple, solid, decent, for
her. And now Rose had cast her from her. Melanctha was lost, and all
the world went whirling in a mad weary dance around her.

Melanctha Herbert never had any strength alone ever to feel safe
inside her. And now Rose Johnson had cast her from her, and Melanctha could never any more be near her. Melanctha Herbert knew now,
way inside her, that she was lost, and nothing any more could ever
help her.

Melanctha went that night to meet Jem Richards who had promised to be at the old place to meet her. Jem Richards was absent in his manner to her. By and by he began to talk to her, about the trip he was going to take soon, to see if he could get some luck back in his betting. Melanctha trembled, was Jem too now going to leave her. Jem Richards talked some more then to her, about the bad luck he always had now, and how he needed to go away to see if he could make it come out any better.

Then Jem stopped, and then he looked straight at Melanctha.

"Tell me Melanctha right and true, you don't care really nothing more about me now Melanctha," he said to her.

"Why you ask me that, Jem Richards," said Melanctha.

"Why I ask you that Melanctha, God Almighty, because I just don't give a damn now for you any more Melanctha. That the reason I was asking."

Melanctha never could have for this an answer. Jem Richards waited and then he went away and left her.

Melanctha Herbert never again saw Jem Richards. Melanctha never again saw Rose Johnson, and it was hard to Melanctha never any more to see her. Rose Johnson had worked in to be the deepest of all Melanctha's emotions.

"No, I don't never see Melanctha Herbert no more now," Rose would say to anybody who asked her about Melanctha. "No, Melanctha she never comes here no more now, after we had all that trouble with her acting so bad with them kind of men she liked so much to be with. She don't never come to no good Melanctha Herbert don't, and me and Sam don't want no more to see her. She didn't do right ever the way I told her. Melanctha just wouldn't, and I always said it to her, if she don't be more kind of careful, the way she always had to be acting, I never did want no more she should come here in my house no more to see me. I ain't no ways ever against any girl having any kind of a way, to have a good time like she wants it, but not that kind of a way Melanctha always had to do it. I expect some day Melanctha kill herself, when she act so bad like she do always, and then she get so awful blue. Melanctha always says that's the only way she ever can think it a easy way for her to do. No, I always am real sorry for Melanctha, she never was no just common kind of nigger, but she don't never know not with all the time I always was telling it to her, no she never no way could learn, what was the right way she should do. I certainly don't never want no kind of harm to come bad to Melanctha, but I certainly do think she will most kill herself some time, the way she always say it would be easy way for her to do. I never see nobody ever could be so awful blue."

But Melanctha Herbert never really killed herself because she was so blue, though often she thought this would be really the best way for her to do. Melanctha never killed herself, she only got a bad fever and went into the hospital where they took good care of her and cured her. When Melanctha was well again, she took a place and began to work and to live regular. Then Melanctha got very sick again, she began to cough and sweat and be so weak she could not stand to do her work.

Melanctha went back to the hospital, and there the Doctor told her she had the consumption,[10] and before long she would surely die. They sent her where she would be taken care of, a home for poor consumptives, and there Melanctha stayed until she died.

---

[10] *consumption:* Tuberculosis, at this period in history, often incurable.

THE GENTLE LENA

Lena was patient, gentle, sweet and german.[1] She had been a servant for four years and had liked it very well.

Lena had been brought from Germany to Bridgepoint by a cousin and had been in the same place there for four years.

This place Lena had found very good. There was a pleasant, unexacting[2] mistress and her children, and they all liked Lena very well.

There was a cook there who scolded Lena a great deal but Lena's german patience held no suffering and the good incessant woman really only scolded so for Lena's good.

Lena's german voice when she knocked and called the family in the morning was as awakening, as soothing, and as appealing, as a delicate soft breeze in midday, summer. She stood in the hallway every morning a long time in her unexpectant and unsuffering german patience calling to the young ones to get up. She would call and wait a long time and then call again, always even, gentle, patient, while the young ones fell back often into that precious, tense, last bit of sleeping that gives a strength of joyous vigor in the young, over them that have come to the readiness of middle age, in their awakening.

Lena had good hard work all morning, and on the pleasant, sunny afternoons she was sent out into the park to sit and watch the little two year old girl baby of the family.

The other girls, all them that make the pleasant, lazy crowd, that watch the children in the sunny afternoons out in the park, all liked the simple, gentle, german Lena very well. They all, too, liked very well to tease her, for it was so easy to make her mixed and troubled, and all helpless, for she could never learn to know just what the other quicker girls meant by the queer things they said.

The two or three of these girls, the ones that Lena always sat with, always worked together to confuse her. Still it was pleasant, all this life for Lena.

The little girl fell down sometimes and cried, and then Lena had to soothe her. When the little girl would drop her hat, Lena had to pick it up and hold it. When the little girl was bad and threw away her play-

---

[1] *german:* Using a lowercase form of a national marker, Stein persists in this work in describing characters by their country of origin. In her schema, women of German origin are likely to be passive, somber, and easily dominated.

[2] *unexacting:* A word coined by Stein, signifying a person who is easy to please.

Working-class family, circa 1912. Courtesy Corbis/Hulton-Deutsch Collection.

things, Lena told her she could not have them and took them from her to hold until the little girl should need them.

It was all a peaceful life for Lena, almost as peaceful as a pleasant leisure. The other girls, of course, did tease her, but then that only made a gentle stir within her.

Lena was a brown and pleasant creature, brown as blonde races often have them brown, brown, not with the yellow or the red or the chocolate brown of sun burned countries, but brown with the clear color laid flat on the light toned skin beneath, the plain, spare brown that makes it right to have been made with hazel eyes, and not too abundant straight, brown hair, hair that only later deepens itself into brown from the straw yellow of a german childhood.

Lena had the flat chest, straight back and forward falling shoulders of the patient and enduring working woman, though her body was now still in its milder girlhood and work had not yet made these lines too clear.

The rarer feeling that there was with Lena, showed in all the even quiet of her body movements, but in all it was the strongest in the patient, old-world ignorance, and earth made pureness of her brown, flat, soft featured face. Lena had eyebrows that were a wondrous thickness. They were black, and spread, and very cool, with their dark color and their beauty, and beneath them were her hazel eyes, simple and human, with the earth patience of the working, gentle, german woman.

Yes it was all a peaceful life for Lena. The other girls, of course, did tease her, but then that only made a gentle stir within her.

"What you got on your finger Lena," Mary, one of the girls she always sat with, one day asked her. Mary was good natured, quick, intelligent and Irish.

Lena had just picked up the fancy paper made accordion that the little girl had dropped beside her, and was making it squeak sadly as she pulled it with her brown, strong, awkward finger.

"Why, what is it, Mary, paint?" said Lena, putting her finger to her mouth to taste the dirt spot.

"That's awful poison Lena, don't you know?" said Mary, "that green paint that you just tasted."

Lena had sucked a good deal of the green paint from her finger. She stopped and looked hard at the finger. She did not know just how much Mary meant by what she said.

"Ain't it poison, Nellie, that green paint, that Lena sucked just now," said Mary. "Sure it is Lena, its real poison, I ain't foolin' this time anyhow."

Lena was a little troubled. She looked hard at her finger where the paint was, and she wondered if she had really sucked it.

It was still a little wet on the edges and she rubbed it off a long time on the inside of her dress, and in between she wondered and looked at the finger and thought, was it really poison that she had just tasted.

"Ain't it too bad, Nellie, Lena should have sucked that," Mary said.

Nellie smiled and did not answer. Nellie was dark and thin, and looked Italian. She had a big mass of black hair that she wore high up on her head, and that made her face look very fine.

Nellie always smiled and did not say much, and then she would look at Lena to perplex her.

And so they all three sat with their little charges in the pleasant sunshine a long time. And Lena would often look at her finger and wonder if it was really poison that she had just tasted and then she would rub her finger on her dress a little harder.

Mary laughed at her and teased her and Nellie smiled little and looked queerly at her.

Then it came time, for it was growing cooler, for them to drag together the little ones, who had begun to wander, and to take each one back to its own mother. And Lena never knew for certain whether it was really poison, that green stuff that she had tasted.

During these four years of service, Lena always spent her Sundays out at the house of her aunt, who had brought her four years before to Bridgepoint.

This aunt, who had brought Lena, four years before, to Bridgepoint, was a hard, ambitious, well meaning, german woman. Her husband was a grocer in the town, and they were very well to do. Mrs. Haydon, Lena's aunt, had two daughters who were just beginning as young ladies, and she had a little boy who was not honest and who was very hard to manage.

Mrs. Haydon was a short, stout, hard built, german woman. She always hit the ground very firmly and compactly as she walked. Mrs. Haydon was all a compact and well hardened mass, even to her face, reddish and darkened from its early blonde, with its hearty, shiny, cheeks, and doubled chin well covered over with the up-roll from her short, square neck.

The two daughters, who were fourteen and fifteen, looked like unkneaded, unformed mounds of flesh beside her.

The elder girl, Mathilda, was blonde, and slow, and simple, and quite fat. The younger, Bertha, who was almost as tall as her sister, was dark, and quicker, and she was heavy, too, but not really fat.

These two girls the mother had brought up very firmly. They were well taught for their position. They were always both well dressed, in the same kinds of hats and dresses, as is becoming in two german sisters. The mother liked to have them dressed in red. Their best clothes were red dresses, made of good heavy cloth, and strongly trimmed with braid of a glistening black. They had stiff, red felt hats, trimmed with black velvet ribbon, and a bird. The mother dressed matronly, in a bonnet and in black, always sat between her two big daughters, firm, directing, and repressed.

The only weak spot in this good german woman's conduct was the way she spoiled her boy, who was not honest and who was very hard to manage.

The father of this family was a decent, quiet, heavy, and uninterfering german man. He tried to cure the boy of his bad ways, and make him honest, but the mother could not make herself let the father manage, and so the boy was brought up very badly.

Mrs. Haydon's girls were now only just beginning as young ladies, and so to get her niece, Lena, married, was just then the most important thing that Mrs. Haydon had to do.

Mrs. Haydon had four years before gone to Germany to see her parents, and had taken the girls with her. This visit had been for Mrs. Haydon most successful, though her children had not liked it very well.

Mrs. Haydon was a good and generous woman, and she patronized her parents grandly, and all the cousins who came from all about to see her. Mrs. Haydon's people were of the middling class of farmers. They were not peasants, and they lived in a town of some pretension, but it all seemed very poor and smelly to Mrs. Haydon's american born daughters.

Mrs. Haydon liked it all. It was familiar, and then here she was so wealthy and important. She listened and decided, and advised all of her relations how to do things better. She arranged their present and their future for them, and showed them how in the past they had been wrong in all their methods.

Mrs. Haydon's only trouble was with her two daughters, whom she could not make behave well to her parents. The two girls were very nasty to all their numerous relations. Their mother could hardly make them kiss their grandparents, and every day the girls would get a scolding. But then Mrs. Haydon was so very busy that she did not have time to really manage her stubborn daughters.

These hard working, earth-rough german cousins were to these american born children, ugly and dirty, and as far below them as were italian or negro[3] workmen, and they could not see how their mother could ever bear to touch them, and then all the women dressed so funny, and were worked all rough and different.

The two girls stuck up their noses at them all, and always talked in English to each other about how they hated all these people and how they wished their mother would not do so. The girls could talk some German, but they never chose to use it.

It was her eldest brother's family that most interested Mrs. Haydon. Here there were eight children, and out of the eight, five of them were girls.

Mrs. Haydon thought it would be a fine thing to take one of these girls back with her to Bridgepoint and get her well started. Everybody

---

[3] *italian or negro:* Class or economic positions are often determined by race or place of origin. This pairing suggests equality between these groups.

liked that she should do so, and they were all willing that it should be Lena.

Lena was the second girl in her large family. She was at this time just seventeen years old. Lena was not an important daughter in the family. She was always sort of dreamy and not there. She worked hard and went very regularly at it, but even good work never seemed to bring her near.

Lena's age just suited Mrs. Haydon's purpose. Lena could first go out to service, and learn how to do things, and then, when she was a little older, Mrs. Haydon could get her a good husband. And then Lena was so still and docile, she would never want to do things her own way. And then, too, Mrs. Haydon, with all her hardness had wisdom, and she could feel the rarer strain there was in Lena.

Lena was willing to go with Mrs. Haydon. Lena did not like her german life very well. It was not the hard work but the roughness that disturbed her. The people were not gentle, and the men when they were glad were very boisterous, and would lay hold of her and roughly tease her. They were good people enough around her, but it was all harsh and dreary for her.

Lena did not really know that she did not like it. She did not know that she was always dreamy and not there. She did not think whether it would be different for her away off there in Bridgepoint. Mrs. Haydon took her and got her different kinds of dresses, and then took her with them to the steamer. Lena did not really know what it was that had happened to her.

Mrs. Haydon, and her daughters, and Lena traveled second class on the steamer. Mrs. Haydon's daughters hated that their mother should take Lena. They hated to have a cousin, who was to them, little better than a nigger, and then everybody on the steamer there would see her. Mrs. Haydon's daughters said things like this to their mother, but she never stopped to hear them, and the girls did not dare to make their meaning very clear. And so they could only go on hating Lena hard, together. They could not stop her from going back with them to Bridgepoint.

Lena was very sick on the voyage. She thought, surely before it was over that she would die. She was so sick she could not even wish that she had not started. She could not eat, she could not moan, she was just blank and scared, and sure that every minute she would die. She could not hold herself in, nor help herself in her trouble. She just staid where she had been put, pale, and scared, and weak, and sick, and sure that she was going to die.

Mathilda and Bertha Haydon had no trouble from having Lena for a cousin on the voyage, until the last day that they were on the ship, and by that time they had made their friends and could explain.

Mrs. Haydon went down every day to Lena, gave her things to make her better, held her head when it was needful, and generally was good and did her duty by her.

Poor Lena had no power to be strong in such trouble. She did not know how to yield to her sickness nor endure. She lost all her little sense of being in her suffering. She was so scared, and then at her best, Lena, who was patient, sweet and quiet, had not self-control, nor any active courage.

Poor Lena was so scared and weak, and every minute she was sure that she would die.

After Lena was on land again a little while, she forgot all her bad suffering. Mrs. Haydon got her the good place, with the pleasant unexacting mistress, and her children, and Lena began to learn some English and soon was very happy and content.

All her Sundays out Lena spent at Mrs. Haydon's house. Lena would have liked much better to spend her Sundays with the girls she always sat with, and who often asked her, and who teased her and made a gentle stir within her, but it never came to Lena's unexpectant and unsuffering german nature to do something different from what was expected of her, just because she would like it that way better. Mrs. Haydon had said that Lena was to come to her house every other Sunday, and so Lena always went there.

Mrs. Haydon was the only one of her family who took any interest in Lena. Mr. Haydon did not think much of her. She was his wife's cousin and he was good to her, but she was for him stupid, and a little simple, and very dull, and sure some day to need help and to be in trouble. All young poor relations, who were brought from Germany to Bridgepoint were sure, before long, to need help and to be in trouble.

The little Haydon boy was always very nasty to her. He was a hard child for any one to manage, and his mother spoiled him very badly. Mrs. Haydon's daughters as they grew older did not learn to like Lena any better. Lena never knew that she did not like them either. She did not know that she was only happy with the other quicker girls, she always sat with in the park, and who laughed at her and always teased her.

Mathilda Haydon, the simple, fat, blonde, older daughter felt very badly that she had to say that this was her cousin Lena, this Lena who was little better for her than a nigger. Mathilda was an overgrown, slow, flabby, blonde, stupid, fat girl, just beginning as a woman; thick in her speech and dull and simple in her mind, and very jealous of all her family and of other girls, and proud that she could have good dresses and new hats and learn music, and hating very badly to have a cousin who was a common servant. And then Mathilda remembered very strongly that dirty nasty place that Lena came from and that Mathilda had so turned up her nose at, and where she had been made so angry because her mother scolded her and liked all those rough cow-smelly people.

Then, too, Mathilda would get very mad when her mother had Lena at their parties, and when she talked about how good Lena was, to certain german mothers in whose sons, perhaps, Mrs. Haydon might find Lena a good husband. All this would make the dull, blonde, fat Mathilda very angry. Sometimes she would get so angry that she would, in her thick, slow way, and with jealous anger blazing in her light blue eyes, tell her mother that she did not see how she could like that nasty Lena; and then her mother would scold Mathilda, and tell her that she knew her cousin Lena was poor and Mathilda must be good to poor people.

Mathilda Haydon did not like relations to be poor. She told all her girl friends what she thought of Lena, and so the girls would never talk to Lena at Mrs. Haydon's parties. But Lena in her unsuffering and unexpectant patience never really knew that she was slighted. When Mathilda was with her girls in the street or in the park and would see Lena, she always turned up her nose and barely nodded to her, and then she would tell her friends how funny her mother was to take care of people like that Lena, and how, back in Germany, all Lena's people lived just like pigs.

The younger daughter, the dark, large, but not fat, Bertha Haydon, who was very quick in her mind, and in her ways, and who was the favorite with her father, did not like Lena, either. She did not like her because for her Lena was a fool and so stupid, and she would let those Irish and Italian girls laugh at her and tease her, and everybody always made fun of Lena, and Lena never got mad, or even had sense enough to know that they were all making an awful fool of her.

Bertha Haydon hated people to be fools. Her father, too, thought Lena was a fool, and so neither the father nor the daughter ever paid any attention to Lena, although she came to their house every other Sunday.

Lena did not know how all the Haydons felt. She came to her aunt's house all her Sunday afternoons that she had out, because Mrs. Haydon had told her she must do so. In the same way Lena always saved all of her wages. She never thought of any way to spend it. The german cook, the good woman who always scolded Lena, helped her to put it in the bank each month, as soon as she got it. Sometimes before it got into the bank to be taken care of, somebody would ask Lena for it. The little Haydon boy sometimes asked and would get it, and sometimes some of the girls, the ones Lena always sat with, needed some more money; but the german cook, who always scolded Lena, saw to it that this did not happen very often. When it did happen she would scold Lena very sharply, and for the next few months she would not let Lena touch her wages, but put it in the bank for her on the same day that Lena got it.

So Lena always saved her wages, for she never thought to spend them, and she always went to her aunt's house for her Sundays because she did not know that she could do anything different.

Mrs. Haydon felt more and more every year that she had done right to bring Lena back with her, for it was all coming out just as she had expected. Lena was good and never wanted her own way,[4] she was learning English, and saving all her wages, and soon Mrs. Haydon would get her a good husband.

All these four years Mrs. Haydon was busy looking around among all the german people that she knew for the right man to be Lena's husband, and now at last she was quite decided.

The man Mrs. Haydon wanted for Lena was a young german-american tailor, who worked with his father. He was good and all the family were very saving, and Mrs. Haydon was sure that this would be just right for Lena, and then too, this young tailor always did whatever his father and his mother wanted.

This old german tailor and his wife, the father and the mother of Herman Kreder, who was to marry Lena Mainz, were very thrifty, careful people. Herman was the only child they had left with them, and he always did everything they wanted. Herman was now twenty-eight years old, but he had never stopped being scolded and directed by his father and his mother. And now they wanted to see him married.

Herman Kreder did not care much to get married. He was a gentle soul and a little fearful. He had a sullen temper, too. He was obedient to his father and his mother. He always did his work well. He often

---

[4] *never wanted her own way:* Mrs. Haydon preferred that her niece be docile; notice that Herman shares this behavior, up to a point, within his family.

went out on Saturday nights and on Sundays, with other men. He liked it with them but he never became really joyous. He liked to be with men and he hated to have women with them. He was obedient to his mother, but he did not care much to get married.

Mrs. Haydon and the elder Kreders had often talked the marriage over. They all three liked it very well. Lena would do anything that Mrs. Haydon wanted, and Herman was always obedient in everything to his father and his mother. Both Lena and Herman were saving and good workers and neither of them ever wanted their own way.

The elder Kreders, everybody knew, had saved up all their money, and they were hard, good german people, and Mrs. Haydon was sure that with these people Lena would never be in any trouble. Mr. Haydon would not say anything about it. He knew old Kreder had a lot of money and owned some good houses, and he did not care what his wife did with that simple, stupid Lena, so long as she would be sure never to need help or to be in trouble.

Lena did not care much to get married. She liked her life very well where she was working. She did not think much about Herman Kreder. She thought he was a good man and she always found him very quiet. Neither of them ever spoke much to the other. Lena did not care much just then about getting married.

Mrs. Haydon spoke to Lena about it very often. Lena never answered anything at all. Mrs. Haydon thought, perhaps Lena did not like Herman Kreder. Mrs. Haydon could not believe that any girl not even Lena, really had no feeling about getting married.

Mrs. Haydon spoke to Lena very often about Herman. Mrs. Haydon sometimes got very angry with Lena. She was afraid that Lena, for once, was going to be stubborn, now when it was all fixed right for her to be married.

"Why you stand there so stupid, why don't you answer, Lena," said Mrs. Haydon one Sunday, at the end of a long talking that she was giving Lena about Herman Kreder, and about Lena's getting married to him.

"Yes ma'am," said Lena, and then Mrs. Haydon was furious with this stupid Lena. "Why don't you answer with some sense, Lena, when I ask you if you don't like Herman Kreder. You stand there so stupid and don't answer just like you ain't heard a word what I been saying to you. I never see anybody like you, Lena. If you going to burst out at all, why don't you burst out sudden instead of standing there so silly and don't answer. And here I am so good to you, and find you a good husband so you can have a place to live in all your own. Answer me,

Lena, don't you like Herman Kreder? He is a fine young fellow, almost too good for you, Lena, when you stand there so stupid and don't make no answer. There ain't many poor girls that get the chance you got now to get married."

"Why, I do anything you say, Aunt Mathilda. Yes, I like him. He don't say much to me, but I guess he is a good man, and I do anything you say for me to do."

"Well then Lena, why you stand there so silly all the time and not answer when I asked you."

"I didn't hear you say you wanted I should say anything to you. I didn't know you wanted me to say nothing. I do whatever you tell me it's right for me to do. I marry Herman Kreder, if you want me."

And so for Lena Mainz the match was made.

Old Mrs. Kreder did not discuss the matter with her Herman. She never thought that she needed to talk such things over with him. She just told him about getting married to Lena Mainz who was a good worker and very saving and never wanted her own way, and Herman made his usual little grunt in answer to her.

Mrs. Kreder and Mrs. Haydon fixed the day and made all the arrangements for the wedding and invited everybody who ought to be there to see them married.

In three months Lena Mainz and Herman Kreder were to be married.

Mrs. Haydon attended to Lena's getting all the things that she needed. Lena had to help a good deal with the sewing. Lena did not sew very well. Mrs. Haydon scolded because Lena did not do it better, but then she was very good to Lena, and she hired a girl to come and help her. Lena still stayed on with her pleasant mistress, but she spent all her evenings and her Sundays with her aunt and all the sewing.

Mrs. Haydon got Lena some nice dresses. Lena liked that very well. Lena liked having new hats even better, and Mrs. Haydon had some made for her by a real milliner who made them very pretty.

Lena was nervous these days, but she did not think much about getting married. She did not know really what it was, that, which was always coming nearer.

Lena liked the place where she was with the pleasant mistress and the good cook, who always scolded, and she liked the girls she always sat with. She did not ask if she would like being married any better. She always did whatever her aunt said and expected, but she was always nervous when she saw the Kreders with their Herman. She was excited and she liked her new hats, and everybody teased her and every day her marrying was coming nearer, and yet she did not really know what it was, this that was about to happen to her.

Herman Kreder knew more what it meant to be married and he did not like it very well. He did not like to see girls and he did not want to have to have one always near him. Herman always did everything that his father and his mother wanted and now they wanted that he should be married.

Herman had a sullen temper; he was gentle and he never said much. He liked to go out with other men, but he never wanted that there should be any women with them. The men all teased him about getting married. Herman did not mind the teasing but he did not like very well the getting married and having a girl always with him.

Three days before the wedding day, Herman went away to the country to be gone over Sunday. He and Lena were to be married Tuesday afternoon. When the day came Herman had not been seen or heard from.

The old Kreder couple had not worried much about it. Herman always did everything they wanted and he would surely come back in time to get married. But when Monday night came, and there was no Herman, they went to Mrs. Haydon to tell her what had happened.

Mrs. Haydon got very much excited. It was hard enough to work so as to get everything all ready, and then to have that silly Herman go off that way, so no one could tell what was going to happen. Here was Lena and everything all ready, and now they would have to make the wedding later so that they would know that Herman would be sure to be there.

Mrs. Haydon was very much excited, and then she could not say much to the old Kreder couple. She did not want to make them angry, for she wanted very badly now that Lena should be married to their Herman.

At last it was decided that the wedding should be put off a week longer. Old Mr. Kreder would go to New York to find Herman, for it was very likely that Herman had gone there to his married sister.

Mrs. Haydon sent word around, about waiting until a week from that Tuesday, to everybody that had been invited, and then Tuesday morning she sent for Lena to come down to see her.

Mrs. Haydon was very angry with poor Lena when she saw her. She scolded her hard because she was so foolish, and now Herman had gone off and nobody could tell where he had gone to, and all because Lena always was so dumb and silly. And Mrs. Haydon was just like a mother to her, and Lena always stood there so stupid and did not answer what anybody asked her, and Herman was so silly too, and now his father had to go and find him. Mrs. Haydon did not think that any old people should be good to their children. Their children always were so thankless, and never paid any attention, and older people were always doing things for their good. Did Lena think it gave Mrs. Haydon any pleasure, to work so hard to make Lena happy, and get her a good husband, and then Lena was so thankless and never did anything that anybody wanted. It was a lesson to poor Mrs. Haydon not to do things any more for anybody. Let everybody take care of themselves and never come to her with any troubles; she knew better now than to meddle to make other people happy. It just made trouble for her and her husband did not like it. He always said she was too good, and nobody ever thanked her for it, and there Lena was always standing stupid and not answering anything anybody wanted. Lena could always talk enough to those silly girls she liked so much, and always sat with, but who never did anything for her except to take away her money, and here was her aunt who tried so hard and was so good to her and treated her just like one of her own children and Lena stood there, and never made any answer and never tried to please her aunt, or to do anything that her aunt wanted. "No, it ain't no use your standin' there and cryin', now, Lena. Its too late now to care about that Herman. You should have cared some before, and then you wouldn't have to stand and cry now, and be a disappointment to me, and then I get scolded by my husband for taking care of everybody, and nobody ever thankful. I am glad you got the sense to feel sorry now, Lena, anyway, and I try to do what I can to help you out in your trouble, only you don't deserve to have anybody take any trouble for you. But perhaps you know better next time. You go home now and take care you don't spoil your clothes and that new hat, you had no business to be wearin' that this morning, but you ain't got no sense at all, Lena. I never in my life see anybody be so stupid."

Mrs. Haydon stopped and poor Lena stood there in her hat, all trimmed with pretty flowers, and the tears coming out of her eyes, and Lena did not know what it was that she had done, only she was not going to be married and it was a disgrace for a girl to be left by a man on the very day she was to be married.

Lena went home all alone, and cried in the street car.

Poor Lena cried very hard all alone in the street car. She almost spoiled her new hat with her hitting it against the window in her crying. Then she remembered that she must not do so.

The conductor was a kind man and he was very sorry when he saw her crying. "Don't feel so bad, you get another feller, you are such a nice girl," he said to make her cheerful. "But Aunt Mathilda said now, I never get married," poor Lena sobbed out for her answer. "Why you really got trouble like that," said the conductor, "I just said that now to josh[5] you. I didn't ever think you really was left by a feller. He must be a stupid feller. But don't you worry, he wasn't much good if he could go away and leave you, lookin' to be such a nice girl. You just tell all your trouble to me, and I help you." The car was empty and the conductor sat down beside her to put his arm around her, and to be a comfort to her. Lena suddenly remembered where she was, and if she did things like that her aunt would scold her. She moved away from the man into the corner. He laughed, "Don't be scared," he said, "I wasn't going to hurt you. But you just keep up your spirit. You are a real nice girl, and you'll be sure to get a real good husband. Don't you let nobody fool you. You're all right and I don't want to scare you."

The conductor went back to his platform to help a passenger get on the car. All the time Lena stayed in the street car, he would come in every little while and reassure her, about her not to feel so bad about a man who hadn't no more sense than to go away and leave her. She'd be sure yet to get a good man, she needn't be so worried, he frequently assured her.

He chatted with the other passenger who had just come in, a very well dressed old man, and then with another who came in later, a good sort of a working man, and then another who came in, a nice lady, and he told them all about Lena's having trouble, and it was too bad there were men who treated a poor girl so badly. And everybody in the car was sorry for poor Lena and the workman tried to cheer her, and the old man looked sharply at her, and said she looked like a good girl, but she ought to be more careful and not to be so careless, and things like that would not happen to her, and the nice lady went and sat beside her and Lena liked it, though she shrank away from being near her.

So Lena was feeling a little better when she got off the car, and the conductor helped her, and he called out to her, "You be sure you keep up a good heart now. He wasn't no good that feller and you were lucky for to lose him. You'll get a real man yet, one that will be better for you. Don't you be worried, you're a real nice girl as I ever see in such trouble," and the conductor shook his head and went back into his car to talk it over with the other passengers he had there.

[5] *josh:* Tease.

The german cook, who always scolded Lena, was very angry when she heard the story. She never did think Mrs. Haydon would do so much for Lena, though she was always talking so grand about what she could do for everybody. The good german cook always had been a little distrustful of her. People who always thought they were so much never did really do things right for anybody. Not that Mrs. Haydon wasn't a good woman. Mrs. Haydon was a real, good, german woman, and she did really mean to do well by her niece Lena. The cook knew that very well, and she had always said so, and she always had liked and respected Mrs. Haydon, who always acted very proper to her, and Lena was so backward, when there was a man to talk to, Mrs. Haydon did have hard work when she tried to marry Lena. Mrs. Haydon was a good woman, only she did talk sometimes too grand. Perhaps this trouble would make her see it wasn't always so easy to do, to make everybody do everything just like she wanted. The cook was very sorry now for Mrs. Haydon. All this must be such a disappointment, and such a worry to her, and she really had always been very good to Lena. But Lena had better go and put on her other clothes and stop with all that crying. That wouldn't do nothing now to help her, and if Lena would be a good girl, and just be real patient, her aunt would make it all come out right yet for her. "I just tell Mrs. Aldrich, Lena, you stay here yet a little longer. You know she is always so good to you, Lena, and I know she let you, and I tell her all about that stupid Herman Kreder. I got no patience, Lena, with anybody who can be so stupid. You just stop now with your crying, Lena, and take off them good clothes and put them away so you don't spoil them when you need them, and you can help me with the dishes and everything will come off better for you. You see if I ain't right by what I tell you. You just stop crying now Lena quick, or else I scold you."

Lena still choked a little and was very miserable inside her but she did everything just as the cook told her.

The girls Lena always sat with were very sorry to see her look so sad with her trouble. Mary the Irish girl sometimes got very angry with her. Mary was always very hot when she talked of Lena's aunt Mathilda, who thought she was so grand, and had such stupid, stuck up daughters. Mary wouldn't be a fat fool like that ugly tempered Mathilda Haydon, not for anything anybody could ever give her. How Lena could keep on going there so much when they all always acted as if she was just dirt to them, Mary never could see. But Lena never had any sense of how she should make people stand round for her, and that was always all the trouble with her. And poor Lena, she was so stupid to be sorry for losing that gawky fool who didn't ever know what he wanted and just said "ja" to his mamma and his papa, like a baby, and was scared to look at a girl straight, and then sneaked away the last day like as if somebody was going to do something to him. Disgrace, Lena talking about disgrace! It was a disgrace for a girl to be seen with the likes of him, let alone to be married to him. But that poor Lena, she never did know how to show herself off for what she was really. Disgrace to have him go away and leave her. Mary would just like to get a chance to show him. If Lena wasn't worth fifteen like Herman Kreder, Mary would just eat her own head all up. It was a good riddance Lena had of that Herman Kreder and his stingy, dirty parents, and if Lena didn't stop crying about it, — Mary would just naturally despise her.

Poor Lena, she knew very well how Mary meant it all, this she was always saying to her. But Lena was very miserable inside her. She felt the disgrace it was for a decent german girl that a man should go away and leave her. Lena knew very well that her aunt was right when she said the way Herman had acted to her was a disgrace to everyone that knew her. Mary and Nellie and the other girls she always sat with were always very good to Lena but that did not make her trouble any better. It was a disgrace the way Lena had been left, to any decent family, and that could never be made any different to her.

And so the slow days wore on, and Lena never saw her Aunt Mathilda. At last on Sunday she got word by a boy to go and see her aunt Mathilda. Lena's heart beat quick for she was very nervous now with all this that had happened to her. She went just as quickly as she could to see her Aunt Mathilda.

Mrs. Haydon quick, as soon as she saw Lena, began to scold her for keeping her aunt waiting so long for her, and for not coming in all the week to see her, to see if her aunt should need her, and so her aunt had to send a boy to tell her. But it was easy, even for Lena, to see that her aunt was not really angry with her. It wasn't Lena's fault, went on Mrs. Haydon, that everything was going to happen all right for her. Mrs. Haydon was very tired taking all this trouble for her, and when Lena couldn't even take trouble to come and see her aunt, to see if she needed anything to tell her. But Mrs. Haydon really never minded things like that when she could do things for anybody. She was tired now, all the trouble she had been taking to make things right for Lena, but perhaps now Lena heard it she would learn a little to be thankful to her. "You get all ready to be married Tuesday, Lena, you hear me," said Mrs. Haydon to her. "You come here Tuesday morning and I have everything all ready for you. You wear your new dress I got you, and your hat with all them flowers on it, and you be very careful coming you don't get your things all dirty, you so careless all the time, Lena, and not thinking, and you act sometimes you never got no head at all on you. You go home now, and you tell your Mrs. Aldrich that you leave her Tuesday. Don't you go forgetting now, Lena, anything I ever told you what you should do to be careful. You be a good girl, now Lena. You get married Tuesday to Herman Kreder." And that was all Lena ever knew of what had happened all this week to Herman Kreder. Lena forgot there was anything to know about it. She was really to be married Tuesday, and her Aunt Mathilda said she was a good girl, and now there was no disgrace left upon her.

Lena now fell back into the way she always had of being always dreamy and not there, the way she always had been, except for the few days she was so excited, because she had been left by a man the very day she was to have been married. Lena was a little nervous all these last days, but she did not think much about what it meant for her to be married.

Herman Kreder was not so content about it. He was quiet and was sullen and he knew he could not help it. He knew now he just had to let himself get married. It was not that Herman did not like Lena Mainz. She was as good as any other girl could be for him. She was a little better perhaps than other girls he saw, she was so very quiet, but Herman did not like to always have to have a girl around him. Herman had always done everything that his mother and his father wanted. His father had found him in New York, where Herman had gone to be with his married sister.

Herman's father when he had found him coaxed Herman a long time and went on whole days with his complaining to him, always troubled but gentle and quite patient with him, and always he was worrying to Herman about what was the right way his boy Herman should always do, always whatever it was his mother ever wanted from him, and always Herman never made him any answer.

Old Mr. Kreder kept on saying to him, he did not see how Herman could think now, it could be any different. When you make a bargain you just got to stick right to it, that was the only way old Mr. Kreder could ever see it, and saying you would get married to a girl and she got everything all ready, that was a bargain just like one you make in business and Herman he had made it, and now Herman he would just have to do it, old Mr. Kreder didn't see there was any other way a good boy like his Herman had, to do it. And then too that Lena Mainz was such a nice girl and Herman hadn't ought to really give his father so much trouble and make him pay out all that money, to come all the way to New York just to find him, and they both lose all that time from their working, when all Herman had to do was just to stand up, for an hour, and then he would be all right married, and it would be all over for him, and then everything at home would never be any different to him.

And his father went on; there was his poor mother saying always how her Herman always did everything before she ever wanted, and now just because he got notions in him, and wanted to show people how he could be stubborn, he was making all this trouble for her, and making them pay all that money just to run around and find him. "You got no idea Herman, how bad mama is feeling about the way you been acting Herman," said old Mr. Kreder to him. "She says she never can understand how you can be so thankless Herman. It hurts her very much you been so stubborn, and she find you such a nice girl for you, like Lena Mainz who is always just so quiet and always saves up all her wages, and she never wanting her own way at all like some girls are always all the time to have it, and your mama trying so hard, just so you could be comfortable Herman to be married, and then you act so stubborn Herman. You like all young people Herman, you think only about yourself, and what you are just wanting, and your mama she is thinking only what is good for you to have, for you in the future. Do you think your mama wants to have girl around to be a bother, for herself, Herman. Its just for you Herman she is always thinking, and she talks always about how happy she will be, when she sees her Herman married to a nice girl, and then when she fixed it all up so good for you, so it never would be any bother to you, just the way she wanted you should like it, and you say yes all right, I do it, and then you go away like this and act stubborn, and make all this trouble everybody to take for you, and we spend money, and I got to travel all round to find you. You come home now with me Herman and get married, and I tell your mama she better not say anything to you about how much it cost me to come all the way to look for you — Hey Herman," said his father coaxing, "Hey, you come home now and get married. All you got to do Herman is just to stand up for an hour Herman, and then you don't never to have any more bother to it — Hey Herman! — you come home with me to-morrow and get married. Hey Herman."

Herman's married sister liked her brother Herman, and she had always tried to help him, when there was anything she knew he wanted. She liked it that he was so good and always did everything that their father and their mother wanted, but still she wished it could be that he could have more his own way, if there was anything he ever wanted.

But now she thought Herman with his girl was very funny. She wanted that Herman should be married. She thought it would do him lots of good to get married. She laughed at Herman when she heard the story. Until his father came to find him, she did not know why it was Herman had come just then to New York to see her. When she heard the story she laughed a good deal at her brother Herman and teased him a good deal about his running away, because he didn't want to have a girl to be all the time around him.

Herman's married sister liked her brother Herman, and she did not want him not to like to be with women. He was good, her brother Herman, and it would surely do him good to get married. It would make him stand up for himself stronger. Herman's sister always laughed at him and always she would try to reassure him. "Such a nice man as my brother Herman acting like as if he was afraid of women. Why the girls all like a man like you Herman, if you didn't always run away when you saw them. It do you good really Herman to get married, and then you got somebody you can boss around when you want to. It do you good Herman to get married, you see if you don't like it, when you really done it. You go along home now with papa, Herman and get married to that Lena. You don't know how nice you like it Herman when you try once how you can do it. You just don't be afraid of nothing, Herman. You good enough for any girl to marry, Herman. Any girl be glad to have a man like you to be always with them Herman. You just go along home with papa and try it what I say, Herman. Oh you so funny Herman, when you sit there, and then run away and leave your girl behind you. I know she is crying like anything Herman for to lose you. Don't be bad to her Herman. You go along home with papa now and get married Herman. I'd be awful ashamed Herman, to really have a brother didn't have spirit enough to get married, when a girl is just dying for to have him. You always like me to be with you Herman. I don't see why you say you don't want a girl to be all the time around you. You always been good to me Herman, and I know you always be good to that Lena, and you soon feel just like as if she had always been there with you. Don't act like as if you wasn't a nice strong man, Herman. Really I laugh at you Herman, but you know I like awful well to see you real happy. You go home and get married to that Lena, Herman. She is a real pretty girl and real nice and good and quiet and she make my brother Herman very happy. You just stop your fussing now with Herman, papa. He go with you to-morrow papa, and you see he like it so much to be married, he make everybody laugh just to see him be so happy. Really truly, that's the way it will be with you Herman. You just listen to me what I tell you Herman." And so his sister laughed at him and reassured him, and his father kept on telling what the mother always said about her Herman, and he coaxed him and Herman never said anything in answer, and his sister packed his things up and was very cheerful with him, and she kissed him, and then she laughed and then she kissed him, and his father went and bought the tickets for the train, and at last late on Sunday he brought Herman back to Bridgepoint with him.

It was always very hard to keep Mrs. Kreder from saying what she thought, to her Herman, but her daughter had written her a letter, so as to warn her not to say anything about what he had been doing, to him, and her husband came in with Herman and said, "Here we are come home mama, Herman and me, and we are very tired it was so crowded coming," and then he whispered to her. "You be good to Herman, mama, he didn't mean to make us so much trouble," and so old Mrs. Kreder, held in what she felt was so strong in her to say to her Herman. She just said very stiffly to him, "I'm glad to see you come home to-day, Herman." Then she went to arrange it all with Mrs. Haydon.

Herman was now again just like he always had been, sullen and very good, and very quiet, and always ready to do whatever his mother and his father wanted. Tuesday morning came, Herman got his new clothes on and went with his father and his mother to stand up for an hour and get married. Lena was there in her new dress, and her hat with all the pretty flowers, and she was very nervous for now she knew she was really very soon to be married. Mrs. Haydon had everything all ready. Everybody was there just as they should be and very soon Herman Kreder and Lena Mainz were married.

When everything was really over, they went back to the Kreder house together. They were all now to live together, Lena and Herman and the old father and the old mother, in the house where Mr. Kreder had worked so many years as a tailor, with his son Herman always there to help him.

Irish Mary had often said to Lena she never did see how Lena could ever want to have anything to do with Herman Kreder and his dirty stingy parents. The old Kreders were to an Irish nature, a stingy, dirty couple. They had not the free-hearted, thoughtless, fighting, mud bespattered, ragged, peat-smoked cabin dirt that irish Mary knew and could forgive and love. Theirs was the german dirt of saving, of being dowdy and loose and foul in your clothes so as to save them and yourself in washing, having your hair greasy to save it in the soap and drying, having your clothes dirty, not in freedom, but because so it was cheaper, keeping the house close and smelly because so it cost less to get it heated, living so poorly not only so as to save money but so they should never even know themselves that they had it, working all the time not only because from their nature they just had to and because it made them money but also that they never could be put in any way to make them spend their money.

This was the place Lena now had for her home and to her it was very different than it could be for an irish Mary. She too was german and was thrifty, though she was always so dreamy and not there. Lena was always careful with things and she always saved her money, for that was the only way she knew how to do it. She never had taken care of her own money and she never had thought how to use it.

Lena Mainz had been, before she was Mrs. Herman Kreder, always clean and decent in her clothes and in her person, but it was not because she ever thought about it or really needed so to have it, it was the way her people did in the german country where she came from, and her Aunt Mathilda and the good german cook who always scolded, had kept her on and made her, with their scoldings, always more careful to keep clean and to wash real often. But there was no deep need in all this for Lena and so, though Lena did not like the old Kreders, though she really did not know that, she did not think about their being stingy dirty people.

Herman Kreder was cleaner than the old people, just because it was his nature to keep cleaner, but he was used to his mother and his father, and he never thought that they should keep things cleaner. And Herman too always saved all his money, except for that little beer he drank when he went out with other men of an evening the way he always liked to do it, and he never thought of any other way to spend it. His father had always kept all the money for them and he always was doing business with it. And then too Herman really had no money, for he always had worked for his father, and his father had never thought to pay him.

And so they began all four to live in the Kreder house together, and Lena began soon with it to look careless and a little dirty, and to be more lifeless with it, and nobody ever noticed much what Lena wanted, and she never really knew herself what she needed.

The only real trouble that came to Lena with their living all four there together, was the way old Mrs. Kreder scolded. Lena had always been used to being scolded, but this scolding of old Mrs. Kreder was very different from the way she ever before had had to endure it. Herman, now he was married to her, really liked Lena very well. He did not care very much about her but she never was a bother to him being there around him, only when his mother worried and was nasty to them because Lena was so careless, and did not know how to save things right for them with their eating, and all the other ways with money, that the old woman had to save it,

Herman Kreder had always done everything his mother and his father wanted but he did not really love his parents very deeply. With Herman it was always only that he hated to have any struggle. It was all always all right with him when he could just go along and do the same thing over every day with his working, and not to hear things, and not to have people make him listen to their anger. And now his marriage, and he just knew it would, was making trouble for him. It made him hear more what his mother was always saying, with her scolding. He had to really hear it now because Lena was there, and she was so scared and dull always when she heard it. Herman knew very well with his mother, it was all right if one ate very little and worked hard all day and did not hear her when she scolded, the way Herman always had done before they were so foolish about his getting married and having a girl there to be all the time around him, and now he had to help her so the girl could learn too, not to hear it when his mother scolded, and not to look so scared, and not to eat much, and always to be sure to save it.

Herman really did not know very well what he could do to help Lena to understand it. He could never answer his mother back to help Lena, that never would make things any better for her, and he never could feel in himself any way to comfort Lena, to make her strong not to hear his mother, in all the awful ways she always scolded. It just worried Herman to have it like that all the time around him. Herman did not know much about how a man could make a struggle with a mother, to do much to keep her quiet, and indeed Herman never knew much how to make a struggle against anyone who really wanted to have anything very badly. Herman all his life never wanted anything so badly, that he would really make a struggle against any one to get it. Herman all his life only wanted to live regular and quiet, and not talk much and to do the same way every day like every other with his working. And now his mother had made him get married to this Lena and now with his mother making all that scolding, he had all this trouble and this worry always on him.

Mrs. Haydon did not see Lena now very often. She had not lost her interest in her niece Lena, but Lena could not come much to her house to see her, it would not be right, now Lena was a married woman. And then too Mrs. Haydon had her hands full just then with her two daughters, for she was getting them ready to find them good husbands, and then too her own husband now worried her very often about her always spoiling that boy of hers, so he would be sure to turn out no good and be a disgrace to a german family, and all because his mother always spoiled him. All these things were very worrying now to Mrs. Haydon, but still she wanted to be good to Lena, though she could not see her very often. She only saw her when Mrs. Haydon went to call on Mrs. Kreder or when Mrs. Kreder came to see Mrs. Haydon, and that never could be very often. Then too these days Mrs. Haydon could not scold Lena, Mrs. Kreder was always there with her, and it would not be right to scold Lena when Mrs. Kreder was there, who had now the real right to do it. And so her aunt always said nice things now to Lena, and though Mrs. Haydon sometimes was a little worried when she saw Lena looking sad and not careful, she did not have time just then to really worry much about it.

Lena now never any more saw the girls she always used to sit with. She had no way now to see them and it was not in Lena's nature to search out ways to see them, nor did she now ever think much of the days when she had been used to see them. They never any of them had come to the Kreder house to see her. Not even Irish Mary had ever thought to come to see her. Lena had been soon forgotten by them. They had soon passed away from Lena and now Lena never thought any more that she had ever known them.

The only one of her old friends who tried to know what Lena liked and what she needed, and who always made Lena come to see her, was the good german cook who had always scolded. She now scolded Lena hard for letting herself go so, and going out when she was looking so untidy. "I know you going to have a baby Lena, but that's no way for you to be looking. I am ashamed most to see you come and sit here in my kitchen, looking so sloppy and like you never used to Lena. I never see anybody like you Lena. Herman is very good to you, you always say so, and he don't treat you bad ever though you don't deserve to have anybody good to you, you so careless all the time, Lena, letting yourself go like you never had anybody tell you what was the right way you should know how to be looking. No, Lena, I don't see no reason you should let yourself go so and look so untidy Lena, so I am ashamed to see you sit there looking so ugly, Lena. No Lena that ain't no way ever I see a woman make things come out better, letting herself go so every way and crying all the time like as if you had real trouble. I never wanted to see you marry Herman Kreder, Lena, I knew what you got to stand with that old woman always, and that old man, he is so stingy too and he don't say things out but he ain't any better in his heart than his wife with her bad ways, I know that Lena, I know they don't hardly give you enough to eat, Lena, I am real sorry for you Lena, you know that Lena, but that ain't any way to be going round so untidy Lena, even if you have got all that trouble. You never see me do like that Lena, though sometimes I got a headache so I can't see to stand to be working hardly, and nothing comes right with all my cooking, but I always see Lena, I look decent. That's the only way a german girl can make things come out right Lena. You hear me what I am saying to you Lena. Now you eat something nice Lena, I got it all ready for you, and you wash up and be careful Lena and the baby will come all right to you, and then I make your Aunt Mathilda see that you live in a house soon all alone with Herman and your baby, and then everything go better for you. You hear me what I say to you Lena. Now don't let me ever see you come looking like this any more Lena, and you just stop with that always crying. You ain't got no reason to be sitting there now with all that crying, I never see anybody have trouble it did them any good to do the way you are doing, Lena. You hear me Lena. You go home now and you be good the way I tell you Lena, and I see what I can do. I make your Aunt Mathilda make old Mrs. Kreder let you be till you get your baby all right. Now don't you be scared and so silly Lena. I don't like to see you act so Lena when really you got a nice man and so many things really any girl should be grateful to be having. Now you go home Lena to-day and you do the way I say, to you, and I see what I can do to help you."

"Yes Mrs. Aldrich" said the good german woman to her mistress later, "Yes Mrs. Aldrich that's the way it is with them girls when they want so to get married. They dont know when they got it good Mrs. Aldrich. They never know what it is they're really wanting when they got it, Mrs. Aldrich. There's that poor Lena, she just been here crying and looking so careless so I scold her, but that was no good that marrying for that poor Lena, Mrs. Aldrich. She do look so pale and sad now Mrs. Aldrich, it just break my heart to see her. She was a good girl was Lena, Mrs. Aldrich, and I never had no trouble with her like I got with so many young girls nowadays, Mrs. Aldrich, and I never see any

girl any better to work right than our Lena, and now she got to stand
it all the time with that old woman Mrs. Kreder. My! Mrs. Aldrich, she
is a bad old woman to her. I never see Mrs. Aldrich how old people can
be so bad to young girls and not have no kind of patience with them. If
Lena could only live with her Herman, he ain't so bad the way men
are, Mrs. Aldrich, but he is just the way always his mother wants him,
he ain't got no spirit in him, and so I don't really see no help for that
poor Lena. I know her aunt, Mrs. Haydon, meant it all right for her
Mrs. Aldrich, but poor Lena, it would be better for her if her Herman
had stayed there in New York that time he went away to leave her. I
don't like it the way Lena is looking now, Mrs. Aldrich. She looks like
as if she don't have no life left in her hardly, Mrs. Aldrich, she just
drags around and looks so dirty and after all the pains I always took to
teach her and to keep her nice in her ways and looking. It don't do no
good to them, for them girls to get married Mrs. Aldrich, they are
much better when they only know it, to stay in a good place when they
got it, and keep on regular with their working. I don't like it the way
Lena looks now Mrs. Aldrich. I wish I knew some way to help that
poor Lena, Mrs. Aldrich, but she she is a bad old woman, that old
Mrs. Kreder, Herman's mother. I speak to Mrs. Haydon real soon,
Mrs. Aldrich, I see what we can do now to help that poor Lena."

These were really bad days for poor Lena. Herman always was real
good to her and now he even sometimes tried to stop his mother from
scolding Lena. "She ain't well now mama, you let her be now you hear
me. You tell me what it is you want she should be doing, I tell her. I see
she does it right just the way you want it mama. You let be, I say now
mama, with that always scolding Lena. You let be, I say now, you wait
till she is feeling better." Herman was getting really strong to struggle,
for he could see that Lena with that baby working hard inside her,
really could not stand it any longer with his mother and the awful
ways she always scolded.

It was a new feeling Herman now had inside him that made him feel
he was strong to make a struggle. It was new for Herman Kreder really
to be wanting something, but Herman wanted strongly now to be a
father, and he wanted badly that his baby should be a boy and healthy.
Herman never had cared really very much about his father and his
mother, though always, all his life, he had done everything just as they
wanted, and he had never really cared much about his wife, Lena,
though he always had been very good to her, and had always tried to
keep his mother off her, with the awful way she always scolded, but to
be really a father of a little baby, that feeling took hold of Herman very

deeply. He was almost ready, so as to save his baby from all trouble, to really make a strong struggle with his mother and with his father, too, if he would not help him to control his mother.

Sometimes Herman even went to Mrs. Haydon to talk all this trouble over. They decided then together, it was better to wait there all four together for the baby, and Herman could make Mrs. Kreder stop a little with her scolding, and then when Lena was a little stronger, Herman should have his own house for her, next door to his father, so he could always be there to help him in his working, but so they could eat and sleep in a house where the old woman could not control them and they could not hear her awful scolding.

And so things went on, the same way, a little longer. Poor Lena was not feeling any joy to have a baby. She was scared the way she had been when she was so sick on the water. She was scared now every time when anything would hurt her. She was scared and still and life-less, and sure that every minute she would die. Lena had no power to be strong in this kind of trouble, she could only sit still and be scared, and dull, and lifeless, and sure that every minute she would die.

Before very long, Lena had her baby. He was a good, healthy little boy, the baby. Herman cared very much to have the baby. When Lena was a little stronger he took a house next door to the old couple, so he and his own family could eat and sleep and do the way they wanted. This did not seem to make much change now for Lena. She was just the same as when she was waiting with her baby. She just dragged around and was careless with her clothes and all lifeless, and she acted always and lived on just as if she had no feeling. She always did every-thing regular with the work, the way she always had had to do it, but she never got back any spirit in her. Herman was always good and kind, and always helped her with her working. He did everything he knew to help her. He always did all the active new things in the house and for the baby. Lena did what she had to do the way she always had been taught it. She always just kept going now with her working, and she was always careless, and dirty, and a little dazed, and lifeless. Lena never got any better in herself of this way of being that she had had ever since she had been married.

Mrs. Haydon never saw any more of her niece, Lena. Mrs. Haydon had now so much trouble with her own house, and her daughters get-ting married, and her boy, who was growing up, and who always was getting so much worse to manage. She knew she had done right by Lena. Herman Kreder was a good man, she would be glad to get one so good, sometime, for her own daughters, and now they had a home to live in together, separate from the old people, who had made their trouble for them. Mrs. Haydon felt she had done very well by her niece, Lena, and she never thought now she needed any more to go and see her. Lena would do very well now without her aunt to trouble herself any more about her.

The good german cook who had always scolded, still tried to do her duty like a mother to poor Lena. It was very hard now to do right by Lena. Lena never seemed to hear now what anyone was saying to her. Herman was always doing everything he could to help her. Herman always, when he was home, took good care of the baby. Herman loved to take care of his baby. Lena never thought to take him out or to do anything she didn't have to.

The good cook sometimes made Lena come to see her. Lena would come with her baby and sit there in the kitchen, and watch the good woman cooking, and listen to her sometimes a little, the way she used to, while the good german woman scolded her for going around looking so careless when now she had no trouble, and sitting there so dull, and always being just so thankless. Sometimes Lena would wake up a little and get back into her face her old, gentle, patient, and unsuffering sweetness, but mostly Lena did not seem to hear much when the good german woman scolded. Lena always liked it when Mrs. Aldrich her good mistress spoke to her kindly, and then Lena would seem to go back and feel herself to be like she was when she had been in service. But mostly Lena just lived along and was careless in her clothes, and dull, and lifeless.

By and by Lena had two more little babies. Lena was not so much scared now when she had the babies. She did not seem to notice very much when they hurt her, and she never seemed to feel very much now about anything that happened to her.

They were very nice babies, all these three that Lena had, and Herman took good care of them always. Herman never really cared much about his wife, Lena. The only things Herman ever really cared for were his babies. Herman always was very good to his children. He always had a gentle, tender way when he held them. He learned to be very handy with them. He spent all the time he was not working, with them. By and by he began to work all day in his own home so that he could have his children always in the same room with him.

Lena always was more and more lifeless and Herman now mostly never thought about her. He more and more took all the care of their three children. He saw to their eating right and their washing, and he dressed them every morning, and he taught them the right way to do things, and he put them to their sleeping, and he was now always every minute with them. Then there was to come to them, a fourth baby. Lena went to the hospital near by to have the baby. Lena seemed to be going to have much trouble with it. When the baby was come out at last, it was like its mother lifeless. While it was coming, Lena had grown very pale and sicker. When it was all over Lena had died, too, and nobody knew just how it had happened to her.

The good german cook who had always scolded Lena, and had always to the last day tried to help her, was the only one who ever missed her. She remembered how nice Lena had looked all the time she was in service with her, and how her voice had been so gentle and sweet-sounding, and how she always was a good girl, and how she never had to have any trouble with her, the way she always had with all the other girls who had been taken into the house to help her. The good cook sometimes spoke so of Lena when she had time to have a talk with Mrs. Aldrich, and this was all the remembering there now ever was of Lena.

Herman Kreder now always lived very happy, very gentle, very quiet, very well content alone with his three children. He never had a woman any more to be all the time around him. He always did all his own work in his house, when he was through every day with the work he was always doing for his father. Herman always was alone, and he always worked alone, until his little ones were big enough to help him. Herman Kreder was very well content now and he always lived very regular and peaceful, and with every day just like the next one, always alone now with his three good, gentle children.

FINIS

Part Two

Three Lives
Cultural Contexts

Gertrude Stein with fellow students from Harvard and Radcliffe, circa 1896. Reproduced by permission of the Beinecke Library American Literature Collection, Yale University.

1

The Woman Question:
Rights, the Vote, Education, Health

In order for Gertrude Stein to become an author, she had to make a great journey: she had to change from being a woman who led a largely private life to one leading — at least in part — a public one. The dichotomy of *public* and *private* in the consideration of women's lives reflected the same kind of ideology as the concept of the separation of spheres. Just as women were to live cordoned off from the real business of running the United States, watching their male relations — fathers, brothers, or spouses — mix it up in finance, trade, and professional life, they were also expected to show that watchful public world a blank visage. Woman was allowed to be a personality in private, but her public self was to be so bland as to be indistinguishable from others of her sex.

Sitting away from the center of action with an impassive face was not the role Gertrude Stein wanted to play. As she wrote in an 1890s paper, which she presented to an elite female audience in Baltimore, the condition of women was a matter for serious consideration. She asked her audience to think about "the inadequacy of the present system of training" for women and, as a corollary, "the advantages of a college training in helping to prepare one adequately for the complexities of this nineteenth century existence" ("Value" 1). That Stein made the change from private to public with such enthusiasm as well as diligence suggests that she was, by temperament, truly modern. But that she was able to make the change at all was possible largely because of the nineteenth-century women's movement in the United States.

Most late-nineteenth-century American women were frustrated that they remained without the vote — bereft in the midst of their country's democratic blessings. (African American men were franchised after the Civil War; but no woman — regardless of color — could vote.) Modeling themselves after British writer and intellectual Mary Wollstonecraft, writer of the much revered *A Vindication of the Rights of Women* (1792), literate women began expounding their right to vote. Elizabeth Cady Stanton's 1860 address paralleled her famous remarks at the 1848 Seneca Falls convention, where Stanton, Susan B. Anthony, and others became leaders of what was to be known as the National Women's Suffrage Association (NWSA) (see pp. 223–26).

Growth of interest in the women's movement led to a market, and a forum, for women's writing about issues of interest to the female sex. The painful depictions of working-class women's lives (in both Fanny Fern's "The Working-Girls of New York" [see pp. 226–29] and the anonymous column from an Illinois newspaper [see pp. 233–36]) did much to educate better-classed women readers. Without this increased awareness, Charlotte Perkins Stetson Gilman would have had less demand for her journalism and her books about economics — the excerpt here is a famous one from her 1898 *Women & Economics* (see pp. 236–46); Stein quotes from this book several times in her Baltimore speech advocating women's education. Similarly, without the existence of the women's movement to focus intelligent women on demanding their legal and financial rights, Anna Julia Haywood Cooper, as a southern black educator, would have had difficulty publishing her essays on education (see pp. 229–33).

Both those documents that describe the struggles for women's attainment of power — legal, financial, professional — and those that describe the lives of lower-class women contribute to the reader's understanding of the milieu in which Gertrude Stein matured. From her need to be considered a bright and capable daughter (even if getting her father to accept her brightness meant she sometimes had to think of herself as a son) to her relentless scheduling of her reading and studying time (even if she did not do the household duties her father and her sister Bertha expected her to take on), Stein battled the same forces these women writers document. For the subjects of *Three Lives* — Anna, Lena, and Melanctha — that battle was much more difficult because they were lower-class women with seriously diminished opportunities for achievement.

Women's trying to gain independence and earning power through education was a strategy that fit well with universal demands for increased educational opportunities late in the nineteenth century. As the devotees of the separate spheres saw the problem of women's education, the real thrust of the matter was the education of children. Education for women received its greatest impetus with social appreciation for the place of, and the development of, children. Once U.S. culture valued the child as a person, it became clear that a better means of educating him or her had to be found. Women affluent enough to stay at home would probably be the child's primary caregiver and teacher, at least for the years pertaining to grammar school; that woman, then, must be educated herself. Stein speaks to this in her Baltimore essay:

> From the minute the child is born the modern mother's troubles begin; indeed we may say with Holmes that the education of the child begins a hundred years before it is born. . . . we must study the nature of the individual child. We must realise the different significance of different acts of naughtiness and act accordingly. . . . when we see what an intelligent woman can do in the direction of developing even very poor material and making something of it we realise how very necessary it is for the modern mother to be intelligent and educated. ("Value" 2)

Probably the most significant area of women's education at the period was that of women's health. The excerpts here from S. Weir Mitchell's 1877 instruction manual, implementing his famous cure for neurasthenic patients (those suffering from the chronic fatigue associated with clinical depression), clearly show the assumed dependence of women on the care, knowledge, and presence of their male physicians (see pp. 249–51). Mitchell's *Fat and Blood: And How To Make Them* includes chapters on seclusion, rest, massage, electricity, dietetics, and therapeutics. The book's purpose, as stated in the introduction, is to enable the physician to renew "the vitality of feeble people by a combination of entire rest and of excessive feeding, made possible by passive exercise obtained through the steady use of massage and electricity" (Mitchell 7). Although Mitchell is best known for treating Jane Addams, Charlotte Perkins Stetson Gilman, and others in this regime, the treatment was not geared entirely toward women — though the introduction continues that the patients are chiefly "nervous women, who as a rule are thin, and lack blood" (Mitchell 7).

The excerpts from Mitchell, as well as those from William Osler (pp. 251–57) and J. Whitridge Williams (pp. 257–59), are of great relevance to Gertrude Stein, because during her four years at Johns

Hopkins Medical School, she was a student of the latter two doctors. Osler was a disciple of Mitchell's, and was himself the most influential woman's medicine specialist at Johns Hopkins. Stein had wanted to work in that field herself; she had witnessed the lingering illness and death of her mother from intestinal (or perhaps ovarian) cancer, and her sister-in-law's tortured pregnancy and postdelivery medical problems. She was, therefore, particularly sensitive to what she saw as the abuses of women's health in the Mitchell-Osler treatment.

More generally, as the excerpts from Osler's 1901 address show, all medical knowledge at the time was seriously limited, but the areas that impacted on women's health — such as venereal diseases — were considered embarrassing. (Venereal diseases were grouped with leprosy.) Compounding this imposition of self-shame on women was the overbearing tendency of male physicians to glorify themselves. One of the more disturbing elements of the Osler address is his emphasis on the curing power of faith, with faith here defined as the patient's belief in her physician: "If a poor lass, paralyzed apparently, helpless, bedridden for years, comes to me, having worn out in mind, body and estate a devoted family; if she in a few weeks or less by faith in me, and faith alone, takes up her bed and walks, the saints of old could not have done more. . . ." (see pp. 256–57).

The brief section from Williams' textbook, *Obstetrics* (the book that now, decades — and many editions — after Williams's death, is still one of the classic obstetrics texts) hints at the racial prejudice that haunts the turn-of-the-century culture. According to this textbook, the pelvis of a black woman is radically different from that of a white patient. When Stein chose to boycott the seminars of both Dr. Williams and Dr. Osler, she was taking on the most powerful men in American women's medicine.

It was one of the fortunate coincidences of Stein's education that she was nurtured and encouraged even as she studied in a male atmosphere: at Harvard, she was frequently paired with the brightest of Harvard's graduate students, even though she was an undergraduate woman. It was not until she attended the Johns Hopkins Medical School that her personal knowledge of being female led to conflicts with the male doctors and established principles of medicine; she gave up the field rather than go against her own convictions. We seldom think of Gertrude Stein as a feisty pioneer, but perhaps we should add that quality to her profile.

ELIZABETH CADY STANTON

From *Address to the New York State Legislature* (1860)

One of the founders of the National Women's Suffrage Association, Elizabeth Cady Stanton (1815–1902) was the daughter of a comfortably successful New York lawyer and judge. Well-educated in literature and the law, Stanton — schooled with her brothers, at home — also studied mathematics, Greek, and philosophy. In 1840 she married Henry Stanton, an outspoken abolitionist; in 1851 she met reform activist Susan B. Anthony, with whom she devoted fifty years of her life to women's rights projects.

The two women founded and edited the magazine *The Revolution.* With Matilda Gage they wrote the six-volume *History of Woman Suffrage,* and were responsible for the early transcriptions of Sojourner Truth's speeches. Stanton's feminist *The Woman's Bible,* a best-seller in 1885, was followed in 1888 by her autobiography, *Eighty Years and More.*

An eloquent speaker, Stanton is remembered for her 1848 "Address Delivered at Seneca Falls" and, among others, for this "Address to the New York State Legislature" in 1860 (first published in Stanton, Anthony, and Gage's *History of Woman Suffrage* [Rochester, New York: 1881]). Taking on conventional religious and political beliefs, Stanton never flinched from attacking the self-satisfied men — "blinded by custom and self-interest" (see p. 223) — who were adamantly against giving women the vote (along with countless other rights). By the time Gertrude Stein and her women classmates at Radcliffe College were considering these important issues, Stanton and Susan B. Anthony had become either saints or objects of ridicule, depending on the viewer's attitude toward woman suffrage.

You who have read the history of nations, from Moses[1] down to our last election, where have you ever seen one class looking after the interests of another? Any of you can readily see the defects in other governments, and pronounce sentence against those who have sacrificed the masses to themselves; but when we come to our own case, we are blinded by custom and self-interest. Some of you who have no capital can see the injustice which the laborer suffers; some of you who

---

[1] *Moses:* Against great odds, Moses led the Jews out of Egypt to the Promised Land.

have no slaves, can see the cruelty of his oppression; but who of you appreciate the galling humiliation, the refinements of degradation, to which women (the mothers, wives, sisters, and daughters of freemen) are subject, in this the last half of the nineteenth century? How many of you have ever read even the laws concerning them that now disgrace your statute-books? In cruelty and tyranny, they are not surpassed by any slaveholding code in the Southern states; in fact they are worse, by just so far as woman, from her social position, refinement, and education, is on a more equal ground with the oppressor.

Allow me just here to call the attention of that party now so much interested in the slave of the Carolinas, to the similarity in his condition and that of the mothers, wives, and daughters of the Empire State.[2] The Negro has no name. He is Cuffy Douglas or Cuffy Brooks, just whose Cuffy he may chance to be. The woman has no name. She is Mrs. Richard Roe or Mrs. John Doe, just whose Mrs. she may chance to be. Cuffy has no right to his earnings; he can not buy or sell, or lay up anything that he can call his own. Mrs. Roe has no right to her earnings; she can neither buy nor sell, make contracts, nor lay up anything that she can call her own. Cuffy has no right to his children; they can be sold from him at any time. Mrs. Roe has no right to her children; they may be bound out to cancel a father's debts of honor. The unborn child, even, by the last will of the father, may be placed under the guardianship of a stranger and a foreigner. Cuffy has no legal existence; he is subject to restraint and moderate chastisement. Mrs. Roe has no legal existence; she has not the best right to her own person. The husband has the power to restrain, and administer moderate chastisement.

Blackstone[3] declares that the husband and wife are one, and learned commentators have decided that that one is the husband. In all civil codes, you will find them classified as one. Certain rights and immunities, such and such privileges are to be secured to white male citizens. What have women and Negroes to do with rights? What know they of government, war, or glory?

The prejudice against color, of which we hear so much, is no stronger than that against sex. It is produced by the same cause, and manifested very much in the same way. The Negro's skin and the woman's sex are both *prima facie*[4] evidence that they were intended to

---

[2] *Empire State*: New York.

[3] *Blackstone*: Sir William Blackstone, who wrote the cornerstone legal treatise *Commentaries on the Laws of England* (1765–69), enforced the patriarchal principle that the property of a married woman belonged to her husband.

[4] *Prima facie*: On the first view.

be in subjection to the white Saxon man. The few social privileges which the man gives the woman, he makes up to the Negro in civil rights. The woman may sit at the same table and eat with the white man; the free Negro may hold property and vote. The woman may sit in the same pew with the white man in church; the free Negro may enter the pulpit and preach. Now, with the black man's right to suffrage, the right unquestioned, even by Paul,[5] to minister at the altar, it is evident that the prejudice against sex is more deeply rooted and more unreasonably maintained than that against color.

Just imagine an inhabitant of another planet entertaining himself, some pleasant evening in searching over our great national compact, our Declaration of Independence, our Constitutions, or some of our statute-books; what would he think of those "women and Negroes" that must be so fenced in, so guarded against? Why, he would certainly suppose we were monsters, like those fabulous giants or Brobdingnagians[6] of olden times, so dangerous to civilized man, from our size, ferocity, and power. Then let him take up our poets, from Pope down to Dana,[7] let him listen to our Fourth of July toasts, and some of the sentimental adulations of social life, and no logic could convince him that this creature of the law, and this angel of the family altar, could be one and the same being. Man is in such a labyrinth of contradictions with his marital and property rights; he is so befogged on the whole question of maidens, wives, and mothers, that from pure benevolence we should relieve him from this troublesome branch of legislation. We should vote, and make laws for ourselves. Do not be alarmed, dear ladies! You need spend no time reading Grotius, Coke, Puffendorf, Blackstone, Bentham, Kent, and Story[8] to find out what you need. We may safely trust the shrewd selfishness of the white man, and consent to live under the same broad code where he has so comfortably ensconced himself. Any legislation that will do for man, we may abide by most cheerfully.

Now do not think, gentlemen, we wish you to do a great many troublesome things for us. We do not ask our legislators to spend a

---

[5] *Paul:* St. Paul silences women as he deprives them of the right to teach (1 Timothy 2.12).

[6] *Brobdingnagians:* In Jonathan Swift's *Gulliver's Travels,* the Brobdingnagians are giants.

[7] *Pope down to Dana:* British male authors who satirized or romanticized women characters: Alexander Pope (1688–1744); Richard Henry Dana Sr. (1787–1879).

[8] *Grotius, . . . Story:* Paralleling Blackstone, all were influential men who wrote on legal matters.

whole session in fixing up a code of laws to satisfy a class of most unreasonable women. We ask no more than the poor devils in the Scripture asked, "Let us alone." In mercy, let us take care of ourselves, our property, our children, and our homes. True, we are not so strong, so wise, so crafty as you are, but if any kind friend leaves us a little money, or we can by great industry earn fifty cents a day, we would rather buy bread and clothes for our children than cigars and champagne for our legal protectors. There has been a great deal written and said about protection. We, as a class, are tired of one kind of protection, that which leaves us everything to do, to dare, and to suffer, and strips us of all means for its accomplishment. We would not tax man to take care of us. No, the Great Father has endowed all his creatures with the necessary powers for self-support, self-defense, and protection. We do not ask man to represent us; it is hard enough in times like these for man to carry backbone enough to represent himself. So long as the mass of men spend most of their time on the fence, not knowing which way to jump, they are surely in no condition to tell us where we had better stand. In pity for man, we would no longer hang like a millstone round his neck. Undo what man did for us in the Dark Ages, and strike out all special legislation for us; strike the words "white male" from all your constitutions, and then, with fair sailing, let us sink or swim, live or die, survive or perish together.

FANNY FERN (SARA PAYSON [WILLIS] PARTON)

"The Working-Girls of New York," from Folly As It Flies

As the first female U.S. newspaper columnist, Fanny Fern (1811–1872) began writing after two marriages left her, and her daughters, impoverished. After the death of her first husband, she married again for economic security, but was so unhappy that she left the union and supported her family by both sewing and writing.

From her woman-centered columns in Boston and, then, New York papers, Fanny Fern mixed the sentimental with the acerbic. In 1853, her book, a collection of her columns, *Fern Leaves from Fanny's Portfolio*, was a best-seller. In 1854, her novel *Ruth Hall*, the story of a woman both determined and modest, was also successful. When *New York Ledger* offered her $100 per column for an exclusive "right to publish" (her work could not appear in any other newspaper), Fern became the highest paid newswriter in America. Those columns appeared from 1856 to 1872.

As this column — "The Working-Girls of New York" — shows, Fern's skills as an empathetic observer made her writing both poignant and accurate, and her comparatively direct style made that writing readable. Known more for the topics of her columns than for their style, Fern's work brought to the attention of the literate a number of problems, characters, and issues about which they would have remained ignorant had they been left insulated within their own social circles. Gertrude Stein, too, would have been sheltered from knowledge about such lives.

This column appeared originally in Fanny Fern's best-selling collection *Folly As It Flies* (New York: G. W. Carleton & Co., 1868).

Nowhere more than in New York does the contest between squalor and splendor so sharply present itself. This is the first reflection of the observing stranger who walks its streets. Particularly is this noticeable with regard to its women. Jostling on the same pavement with the dainty fashionist is the care-worn working-girl. Looking at both these women, the question arises, which lives the more miserable life — she whom the world styles "fortunate," whose husband belongs to three clubs, and whose only meal with his family is an occasional breakfast, from year's end to year's end; who is as much a stranger to his own children as to the reader; whose young son of seventeen has already a detective on his track employed by his father to ascertain where and how he spends his nights and his father's money; swift retribution for that father who finds food, raiment, shelter, equipages for his household: but love, sympathy, companionship — never? Or she — this other woman — with a heart quite as hungry and unappeased, who also faces day by day the same appalling question: *Is this all life has for me?*

A great book is yet unwritten about women. Michelet[1] has aired his wax-doll theories regarding them. The defender of "woman's rights" has given us her views. Authors and authoresses of little, and big repute, have expressed themselves on this subject, and none of them as yet have begun to grasp it: men — because they lack spirituality,

---

[1] *Michelet:* Much of the underpinning of patriarchal law existed in works of history and literature germane to the topic. Jules Michelet (1798–1894), a French historian, assumed his views were favorable to women, but in reality he placed women firmly into the separate sphere that made them objects in men's care.

rightly and justly to interpret women; women — because they dare not, or will not tell us that which most interests us to know. Who shall write this bold, frank, truthful book remains to be seen. Meanwhile woman's millennium is yet a great way off; and while it slowly progresses, conservatism and indifference gaze through their spectacles at the seething elements of to-day, and wonder "what ails all our women?"

Let me tell you what ails the working-girls. While yet your breakfast is progressing, and your toilet unmade, comes forth through Chatham Street and the Bowery, a long procession of them by twos and threes to their daily labor. Their breakfast, so called, has been hastily swallowed in a tenement house, where two of them share, in a small room, the same miserable bed. Of its quality you may better judge, when you know that each of these girls pays but three dollars a week for board, to the working man and his wife where they lodge.

The room they occupy is close and unventilated, with no accommodations for personal cleanliness, and so near to the little Flinegans that their Celtic night-cries are distinctly heard. They have risen unrefreshed, as a matter of course, and their ill-cooked breakfast does not mend the matter. They emerge from the doorway where their passage is obstructed by "nanny goats" and ragged children rooting together in the dirt, and pass out into the street. They shiver as the sharp wind of early morning strikes their temples. There is no look of youth on their faces; hard lines appear there. Their brows are knit; their eyes are sunken; their dress is flimsy, and foolish, and tawdry; always a hat, and feather or soiled artificial flower upon it; the hair dressed with an abortive attempt at style; a soiled petticoat; a greasy dress, a well-worn sacque or shawl, and a gilt breast-pin and earrings.

Now follow them to the large, black-looking building, where several hundred of them are manufacturing hoop-skirts. If you are a woman you have worn plenty; but you little thought what passed in the heads of these girls as their busy fingers glazed the wire, or prepared the spools for covering them, or secured the tapes which held them in their places. You could not stay five minutes in that room, where the noise of the machinery used is so deafening, that only by the motion of the lips could you comprehend a person speaking.

Five minutes! Why, these young creatures bear it, from seven in the morning till six in the evening; week after week, month after month, with only an hour at midday to eat their dinner of a slice of bread and butter or an apple, which they usually eat in the building, some of them having come a long distance. As I said, the roar of machinery in that room is like the roar of Niagara. Observe them as you enter. Not one lifts her head. They might as well be machines, for any interest or curiosity they show, save always to know *what o'clock it is*. Pitiful! pitiful, you almost sob to yourself, as you look at these young girls. *Young?* Alas! it is only in years that they are young.

ANNA JULIA HAYWOOD COOPER

From *The Higher Education of Women*

One of the strongest voices for the promotion of education for women is that of North Carolina and Washington educator Anna Julia Haywood Cooper (1858?–1964). Born a slave, Cooper felt compelled to speak for the generations of mute, black women of her culture. A "pupil teacher" at age nine at St. Augustine's Normal and Collegiate Institute, she received B.A. and M.A. degrees from Oberlin College, one of the few excellent institutions then open to black women. She taught languages there and at Wilberforce University, also in Ohio, before returning in 1885 to St. Augustine's where she taught Latin, Greek, and mathematics. In 1901 she was the first woman principal of a Washington, D.C., public school, where she waged battles to offer classes for adult education. She also prevented a proposed "colored curriculum" of only vocational courses from being implemented.

In 1925 she received a Ph.D. from the Sorbonne in Paris. A founder of the Colored Women's League in Washington, she was an early activist for women's education and civil rights. This essay, taken from her 1892 book *A Voice from the South — By a Black Woman from the South,* shows that long-term commitment.

As the sister of college-educated men, and as a prospective college student herself, Gertrude Stein would have been interested in any discourse about education, particularly about women's education.

Long out of print, Cooper's important book has been republished in the Schomburg Library of Nineteenth-Century Black Women Writers (New York: Oxford University Press, 1988).

In the very first year of our century, the year 1801, there appeared in Paris a book by Silvain Maréchal, entitled "Shall Woman Learn the Alphabet." The book proposes a law prohibiting the alphabet to women, and quotes authorities weighty and various, to prove that the

woman who knows the alphabet has already lost part of her womanli-
ness. . . . Please remember this book was published at the *beginning* of
the Nineteenth Century. At the end of its first third, (in the year 1833)
one solitary college[1] in America decided to admit women within its
sacred precincts, and organized what was called a "Ladies' Course" as
well as the regular B.A. or Gentlemen's course.

It was felt to be an experiment — a rather dangerous experiment —
and was adopted with fear and trembling by the good fathers, who
looked as if they had been caught secretly mixing explosive compounds and were guiltily expecting every moment to see the foundations under them shaken and rent and their fair superstructure
shattered into fragments.

But the girls came, and there was no upheaval. They performed
their tasks modestly and intelligently. Once in a while one or two were
found choosing the gentlemen's course. Still no collapse; and the dear,
careful, scrupulous, frightened old professors were just getting their
hearts out of their throats and preparing to draw one good free breath,
when they found they would have to change the names of those
courses; for there were as many ladies in the gentlemen's course as in
the ladies', and a distinctively Ladies' Course, inferior in scope and
aim to the regular classical course, did not and could not exist.

Other colleges gradually fell into line, and to-day there are one hundred and ninety-eight colleges for women, and two hundred and seven
coeducational colleges and universities in the United States alone
offering the degree of B.A. to women, and sending out yearly into the
arteries of this nation a warm, rich flood of strong, brave, active, ener-
getic, well-equipped, thoughtful women — women quick to see and
eager to help the needs of this needy world — women who can think
as well as feel, and who feel none the less because they think —
women who are none the less tender and true for the parchment scroll
they bear in their hands — women who have given a deeper, richer,
nobler and grander meaning to the word "womanly" than any one-
sided masculine definition could ever have suggested or inspired —
women whom the world has long waited for in pain and anguish till
there should be at last added to its forces and allowed to permeate its
thought the complement of that masculine influence which has dominated it for fourteen centuries.

---

[1] *one solitary college:* Haverford College, located in Haverford, Pennsylvania, was
the first American women's institution of higher education. It was founded by Quakers
and opened its doors in 1833.

Since the idea of order and subordination succumbed to barbarian brawn and brutality in the fifth century, the civilized world has been like a child brought up by his father. It has needed the great mother heart to teach it to be pitiful, to love mercy, to succor the weak and care for the lowly. . . . As individuals, we are constantly and inevitably, whether we are conscious of it or not, giving out our real selves into our several little worlds, inexorably adding our own true ray to the flood of starlight, quite independently of our professions and our masquerading; and so in the world of thought, the influence of thinking woman far transcends her feeble declamation and may seem at times even opposed to it.

A visitor in Oberlin once said to the lady principal, "Have you no rabble in Oberlin? How is it I see no police here, and yet the streets are as quiet and orderly as if there were an officer of the law standing on every corner."

Mrs. Johnston replied, "Oh, yes; there are vicious persons in Oberlin just as in other towns — *but our girls are our police.*"

With from five to ten hundred pure-minded young women threading the streets of the village every evening unattended, vice must slink away, like frost before the rising sun: and yet I venture to say there was not one in a hundred of those girls who would not have run from a street brawl as she would from a mouse, and who would not have declared she could never stand the sight of blood and pistols.

There is, then, a real and special influence of woman. An influence subtle and often involuntary, an influence so intimately interwoven in, so intricately interpenetrated by the masculine influence of the time that it is often difficult to extricate the delicate meshes and analyze and identify the closely clinging fibers. And yet, without this influence — so long as woman sat with bandaged eyes and manacled hands, fast bound in the clamps of ignorance and inaction, the world of thought moved in its orbit like the revolutions of the moon; with one face (the man's face) always out, so that the spectator could not distinguish whether it was disc or sphere.

Now I claim that it is the prevalence of the Higher Education among women, the making it a common everyday affair for women to reason and think and express their thought, the training and stimulus which enable and encourage women to administer to the world the bread it needs as well as the sugar it cries for; in short it is the transmitting the potential forces of her soul into dynamic factors that has given symmetry and completeness to the world's agencies. So only could it be consummated that Mercy, the lesson she teaches, and Truth, the task man has set himself, should meet together: that righteousness, or *rightness*, man's ideal, — and *peace*, its necessary 'other half,' should kiss each other.

We must thank the general enlightenment and independence of woman (which we may now regard as a *fait accompli*) that both these forces are now at work in the world, and it is fair to demand from them for the twentieth century a higher type of civilization than any attained in the nineteenth. Religion, science, art, economics, have all needed the feminine flavor; and literature, the expression of what is permanent and best in all of these, may be gauged at any time to measure the strength of the feminine ingredient. You will not find theology consigning infants to lakes of unquenchable fire long after women have had a chance to grasp, master, and wield its dogmas. You will not find science annihilating personality from the government of the Universe and making of God an ungovernable, unintelligible, blind, often destructive physical force; you will not find jurisprudence formulating as an axiom the absurdity that man and wife are one, and that one the man — that the married woman may not hold or bequeath her own property save as subject to her husband's direction; you will not find political economists declaring that the only possible adjustment between laborers and capitalists is that of selfishness and rapacity — that each must get all he can and keep all that he gets, while the world cries *laissez faire* and the lawyers explain, "it is the beautiful working of the law of supply and demand;" in fine, you will not find the law of love shut out from the affairs of men after the feminine half of the world's truth is completed. . . .

All I claim is that there is a feminine as well as a masculine side to truth; that these are related not as inferior and superior, not as better and worse, not as weaker and stronger, but as complements — complements in one necessary and symmetric whole. That as the man is more noble in reason, so the woman is more quick in sympathy. That as he is indefatigable in pursuit of abstract truth, so is she in caring for the interests by the way — striving tenderly and lovingly that not one of the least of these "little ones" should perish. That while we not unfrequently see women who reason, we say, with the coolness and precision of a man, and men as considerate of helplessness as a woman, still there is a general consensus of mankind that the one trait is essentially masculine and the other is peculiarly feminine. That both are needed to be worked into the training of children, in order that our boys may supplement their virility by tenderness and sensibility, and our girls may round out their gentleness by strength and self-reliance. That, as both are alike necessary in giving symmetry to the individual, so a nation or a race will degenerate into mere emotionalism on the one hand, or bullyism on the other, if dominated by either exclusively; lastly, and most emphatically, that the feminine factor can have its proper effect only through woman's development and education so that she may fitly and intelligently stamp her force on the forces of her day, and add her modicum to the riches of the world's thought. . . .

One Farmer's Wife

As Virginia Woolf pointed out in her 1928 *A Room of One's Own*, more good writing has appeared over the undisclosive *Anon* (or *Anonymous*) than we realize. For reasons of social position, class, propriety, decorum, or family prejudices, many women through history have chosen to publish their writing without using their names. This unsigned essay appeared in an Illinois newspaper early in the twentieth century. Readers of the paper had been invited to submit essays. The obvious difference of opinion here between the writer and her husband — about literacy, about the importance of reading and writing — would have made her anonymity a necessity.

While readers today might shudder at this woman's workload, her life for her times and circumstances is typical. Gertrude Stein knew enough about women's hard work to give her portrait of "The Good Anna" realistic detail.

The anonymous essay appeared in *The Independent* for February 9, 1905.

I was an apt student at school and before I was eighteen I had earned a teacher's certificate of the second grade and would gladly have remained in school a few more years, but I had, unwittingly, agreed to marry the man who is now my husband, and tho I begged to be released, his will was so much the stronger that I was unable to free myself without wounding a love heart, and could not find it in my heart to do so. . . .

I always had a passion for reading; during girlhood it was along educational lines; in young womanhood it was for love stories, which

remained ungratified because my father thought it sinful to read stories of any kind, and especially love stories.

Later, when I was married, I borrowed everything I could find in the
line of novels and stories, and read them by stealth still, for my husband thought it a willful waste of time to read anything and that it
showed a lack of love for him if I would rather read than to talk to him
when I had a few moments of leisure, and, in order to avoid giving
offense and still gratify my desire, I would only read when he was not
at the house, thereby greatly curtailing my already too limited reading
hours. . . .

It is only during the last three years that I have had the news to
read, for my husband is so very penurious that he would never consent to subscribing for papers of any kind and that old habit of avoid-
ing that which would give offense was so fixed that I did not dare to
break it.

The addition of two children to our family never altered or interfered with the established order of things to any appreciable extent.
My strenuous out-door life agreed with me, and even when my children were born I was splendidly prepared for the ordeal and made
rapid recovery. . . .

Any bright morning in the latter part of May I am out of bed at four
o'clock; next, after I have dressed and combed my hair, I start a fire in
the kitchen stove, . . . sweep the floors and then cook breakfast.

While the other members of the family are eating breakfast I strain
away the morning's milk (for my husband milks the cows while I get
breakfast) and fill my husband's dinnerpail, for he will go to work on
our other farm for the day.

By this time it is half-past five o'clock, my husband is gone to his
work, and the stock loudly pleading to be turned into the pastures.
The younger cattle, a half-dozen steers, are left in the pasture at night,
and I now drive the two cows a half-quarter mile and turn them in
with the others, come back, and then there's a horse in the barn that
belongs in a field where there is no water, which I take to a spring quite
a distance from the barn; bring it back and turn it into a field with the
sheep, a dozen in number, which are housed at night.

The young calves are then turned out into the warm sunshine, and
the stock hogs, which are kept in a pen, are clamoring for feed, and I
carry a pailful of swill to them, and hasten to the house and turn out
the chickens and put out feed and water for them, and it is, perhaps,
6:30 a.m.

I have not eaten breakfast yet, but that can wait; I make the beds next and straighten things up in the living room, for I dislike to have the early morning caller find my house topsy-turvy. When this is done I go to the kitchen, which also serves as a dining room, and uncover the table, and take a mouthful of food occasionally as I pass to and fro at my work until my appetite is appeased.

By the time the work is done in the kitchen it is about 7:15 a.m., and the cool morning hours have flown, and no hoeing done in the garden yet, and the children's toilet has to be attended to and churning has to be done.

Finally the children are washed and churning done, and it is eight o'clock, and the sun getting hot, but no matter, weeds die quickly when cut down in the heat of the day, and I use the hoe to a good advantage until the dinner hour, which is 11:30 a.m. We come in, and I comb my hair, and put fresh flowers in it, and eat a cold dinner, put out feed and water for the chickens; set a hen, perhaps, sweep the floors again; sit down and rest and read a few moments, and it is nearly one o'clock, and I sweep the door yard while I am waiting for the clock to strike the hour.

I make and sow a flower bed, dig around some shrubbery, and go back to the garden to hoe until time to do the chores at night. . . .

I hoe in the garden till four o'clock; then I go into the house and get supper . . . when supper is all ready it is set aside, and I pull a few hundred plants of tomato, sweet potato or cabbage for transplanting . . . I then go after the horse, water him, and put him in the barn; call the sheep and house them, and go after the cows and milk them, feed the hogs, put down hay for three horses, and put oats and corn in their troughs, and set those plants and come in and fasten up the chickens. . . . It is 8 o'clock p.m.; my husband has come home, and we are eating supper; when we are through eating I make the beds ready, and the children and their father go to bed, and I wash the dishes and get things in shape to get breakfast quickly next morning. . . .

All the time that I have been going about this work I have been thinking of things I have read . . . and of other things which I have a desire to read, but cannot hope to while the present condition exists.

As a natural consequence there are daily, numerous instances of absent-mindedness on my part; many things left undone. . . . My husband never fails to remind me that it is caused by my reading so much; that I would get along much better if I should never see a book or paper. . . .

I use an old fashioned churn, and the process of churning occupies
from thirty minutes to three hours, according to the condition of the
cream, and I always read something while churning. . . .

CHARLOTTE PERKINS STETSON GILMAN

From *Women and Economics*

Charlotte Perkins Stetson Gilman (1860–1935) was the leading intellectual of the late-nineteenth-, and early-twentieth-century women's
movement. Born in Hartford, Connecticut, this cousin of Harriet Beecher
Stowe grew up in poverty after her father abandoned the family. They
moved each year of Charlotte's young life. Finally, after some courses at
Rhode Island School of Design, she began a career teaching art. Her marriage to Charles Stetson and the birth of a daughter in 1885 led to her nervous collapse, for which she was treated with rest and inactivity by the
internationally known physician S. Weir Mitchell. Her look into madness
became the point of origin for her classic story "The Yellow Wallpaper"
(1892).

Her life experiences made Gilman see how crucial a woman's ability to
support herself was, and she studied economics and developed the principles that her 1898 book, *Women and Economics,* presented. After its publication, she began a career as a lecturer. Gilman was subjected to harsh
treatment by the media because she had, first, divorced Stetson and then,
later, had allowed him and his second wife, who was a close friend of hers,
to raise her daughter. While such choices may seem reasonable to us
today, a hundred years ago they were considered unnatural. Motherhood
was the highest calling for any woman, and Gilman's willingness to allow
someone else to raise her child seemed very strange.

She later married a cousin, George Gilman, but continued to lead an
active feminist life, publishing widely. From 1911 to 1916 she almost single-
handedly wrote and published a monthly magazine, *The Forerunner*; her
utopian feminist novel, *Herland,* was serialized in its pages.

As these excerpts illustrate, Gilman drew on both clear-headed thought
and personal experience to try to state general principles for women determining ways to shape satisfying lives. When Gertrude Stein was a medical
student at Johns Hopkins Medical School, she was asked by a Baltimore
physician — Claribel Cone, the sister of her good friend Etta Cone — to
speak to a women's group in Baltimore about the advantages of college

education for women. In the unpublished typescript of Stein's talk, she makes several references to *Women and Economics*. It is hard today to appreciate how widely read and discussed Charlotte Perkins Stetson Gilman's work was. The book was first published by Small, Maynard and Co. in Boston, 1898.

Preface

This book is written to offer a simple and natural explanation of one of the most common and most perplexing problems of human life, — a problem which presents itself to almost every individual for practical solution, and which demands the most serious attention of the moralist, the physician, and the sociologist —

To show how some of the worst evils under which we suffer, evils long supposed to be inherent and ineradicable in our natures, are but the result of certain arbitrary conditions of our own adoption, and how, by removing those conditions, we may remove the evils resultant —

To point out how far we have already gone in the path of improvement, and how irresistibly the social forces of to-day are compelling us further, even without our knowledge and against our violent opposition, — an advance which may be greatly quickened by our recognition and assistance —

To reach in especial the thinking women of to-day, and urge upon them a new sense, not only of their social responsibility as individuals, but of their measureless racial importance as makers of men.

It is hoped also that the theory advanced will prove sufficiently suggestive to give rise to such further study and discussion as shall prove its error or establish its truth.

[Sex Distinctions]

In establishing the claim of excessive sex-distinction in the human race, much needs to be said to make clear to the general reader what is meant by the term. To the popular mind, both the coarsely familiar and the over-refined, "sexual" is thought to mean "sensual"; and the charge of excessive sex-distinction seems to be a reproach. This should be at once dismissed, as merely showing ignorance of the terms used. A man does not object to being called "masculine," nor a woman to being called "feminine." Yet whatever is masculine or feminine is sexual. To be distinguished by femininity is to be distinguished by sex. To be over-feminine is to be over-sexed. To manifest in excess any of the distinctions of sex, primary or secondary, is to be over-sexed. Our hypothetical peacock, with his too large and splendid tail, would be over-sexed, and no offence to his moral character!

The primary sex-distinctions in our race as in others consist merely in the essential organs and functions of reproduction. The secondary distinctions, and this is where we are to look for our largest excess — consist in all those differences in organ and function, in look and action, in habit, manner, method, occupation, behavior, which distinguish men from women. In a troop of horses, seen at a distance, the sexes are indistinguishable. In a herd of deer the males are distinguishable because of their antlers. The male lion is distinguished by his mane, the male cat only by a somewhat heavier build. In certain species of insects the male and female differ so widely in appearance that even naturalists have supposed them to belong to separate species. Beyond these distinctions lies that of conduct. Certain psychic attributes are manifested by either sex. The intensity of the maternal passion is a sex-distinction as much as the lion's mane or the stag's horns. The belligerence and dominance of the male is a sex-distinction: the modesty and timidity of the female is a sex-distinction. The tendency to "sit" is a sex-distinction of the hen: the tendency to strut is a sex-distinction of the cock. The tendency to fight is a sex-distinction of males in general: the tendency to protect and provide for, is a sex-distinction of females in general.

With the human race, whose chief activities are social, the initial tendency to sex-distinction is carried out in many varied functions. We have differentiated our industries, our responsibilities, our very virtues, along sex lines. It will therefore be clear that the claim of excessive sex-distinction in humanity, and especially in woman, does not carry with it any specific "moral" reproach, though it does in the larger sense prove a decided evil in its effect on human progress.

In primary distinctions our excess is not so marked as in the farther and subtler development; yet, even here, we have plain proof of it. Sex-energy in its primal manifestation is exhibited in the male of the human species to a degree far greater than is necessary for the processes of reproduction, — enough, indeed, to subvert and injure those processes. The direct injury to reproduction from the excessive indulgence of the male, and the indirect injury through its debilitating effect upon the female, together with the enormous evil to society produced by extra-marital indulgence, — these are facts quite generally known. We have recognized them for centuries, and sought to check the evil action by law, civil, social, moral. But we have treated it always as a field of voluntary action, not as a condition of morbid development. We have held it as right that man should be so, but wrong that man should do so. Nature does not work in that way. What it is right to be, it is right to do. What it is wrong to do, it is wrong to be. This inordinate demand in the human male is an excessive sex-distinction. In this, in a certain over-coarseness and hardness, a too great belligerence and pride, a too great subservience to the power of sex-attraction, we find the main marks of excessive sex-distinction in men. It has been always checked and offset in them by the healthful activities of racial life. Their energies have been called out and their faculties developed along all the lines of human progress. In the growth of industry, commerce, science, manufacture, government, art, religion, the male of our species has become human, far more than male. Strong as this passion is in him, inordinate as is his indulgence, he is a far more normal animal than the female of his species, — far less oversexed. To him this field of special activity is but part of life, — an incident. The whole world remains besides. To her it is the world. This has been well stated in the familiar epigram of Madame de Staël, — "Love with man is an episode, with woman a history."[1] It is in woman that we find most fully expressed the excessive sex-distinction of the human species, — physical, psychical, social. See first the physical manifestation.

To make clear by an instance the difference between normal and abnormal sex-distinction, look at the relative condition of a wild cow and a "milch cow," such as we have made. The wild cow is a female. She has healthy calves, and milk enough for them; and that is all the femininity she needs. Otherwise than that she is bovine rather than feminine. She is a light, strong, swift, sinewy creature, able to run, jump, and fight, if necessary. We, for economic uses, have artificially developed the cow's capacity for producing milk. She has become a walking milk-machine, bred and tended to that express end, her value measured in quarts. The secretion of milk is a maternal function, — a sex-function. The cow is over-sexed. Turn her loose in natural conditions, and, if she survive the change, she would revert in a very few

---

[1] "*Love with . . . history*": Epigram taken from *De L'Influence des Passions* (1796) by Madame de Staël (1766–1817), French novelist and intellectual, famous for her salons.

generations to the plain cow, with her energies used in the general activities of her race, and not all running to milk.

Physically, woman belongs to a tall, vigorous, beautiful animal species, capable of great and varied exertion. In every race and time when she has opportunity for racial activity, she developes accordingly, and is no less a woman for being a healthy human creature. In every race and time where she is denied this opportunity, — and few, indeed, have been her years of freedom, — she has developed in the lines of action to which she was confined; and those were always lines of sex-activity. In consequence the body of woman, speaking in the largest generalization, manifests sex-distinction predominantly.

Woman's femininity — and "the eternal feminine" means simply the eternal sexual — is more apparent in proportion to her humanity than the femininity of other animals in proportion to their caninity or felinity or equinity. "A feminine hand" or "a feminine foot" is distinguishable anywhere. We do not hear of "a feminine paw" or "a feminine hoof." A hand is an organ of prehension, a foot an organ of locomotion: they are not secondary sexual characteristics. The comparative smallness and feebleness of woman is a sex-distinction. We have carried it to such an excess that women are commonly known as "the weaker sex." There is no such glaring difference between male and female in other advanced species. In the long migrations of birds, in the ceaseless motion of the grazing herds that used to swing up and down over the continent each year, in the wild, steep journeys of the breeding salmon, nothing is heard of the weaker sex. And among the higher carnivora, where longer maintenance of the young brings their condition nearer ours, the hunter dreads the attack of the female more than that of the male. The disproportionate weakness is an excessive sex-distinction. Its injurious effect may be broadly shown in the Oriental nations, where the female in curtained harems is confined most exclusively to sex-functions and denied most fully the exercise of race-functions. In such peoples the weakness, the tendency to small bones, and adipose tissue of the over-sexed female, is transmitted to the male, with a retarding effect on the development of the race. Conversely, in early Germanic tribes the comparatively free and humanly developed women — tall, strong, and brave — transmitted to their sons a greater proportion of human power and much less of morbid sex-tendency.

The degree of feebleness and clumsiness common to women, the comparative inability to stand, walk, run, jump, climb, and perform other race-functions common to both sexes, is an excessive sex-distinction; and the ensuing transmission of this relative feebleness to their children, boys and girls alike, retards human development. Strong, free, active women, the sturdy, field-working peasant, the burden-bearing savage, are no less good mothers for their human strength. But our civilized "feminine delicacy," which appears somewhat less delicate when recognized as an expression of sexuality in excess, — makes us no better mothers, but worse. The relative weakness of women is a sex-distinction. It is apparent in her to a degree that injures motherhood that injures wifehood, that injures the individual. The sex-usefulness and the human usefulness of women, their general duty to their kind, are greatly injured by this degree of distinction. In every way the over-sexed condition of the human female reacts unfavorably upon herself, her husband, her children, and the race.

In its psychic manifestation this intense sex-distinction is equally apparent. The primal instinct of sex-attraction has developed under social forces into a conscious passion of enormous power, a deep and lifelong devotion, overwhelming in its force. This is excessive in both sexes, but more so in women than in men, — not so commonly in its simple physical form, but in the unreasoning intensity of emotion that refuses all guidance, and drives those possessed by it to risk every other good for this one end. It is not at first sight easy, and it may seem an irreverent and thankless task, to discriminate here between what is good in the "master passion" and what is evil, and especially to claim for one sex more of this feeling than for the other; but such discrimination can be made.

It is good for the individual and for the race to have developed such a degree of passionate and permanent love as shall best promote the happiness of individuals and the reproduction of species. It is not good for the race or for the individual that this feeling should have become so intense as to override all other human faculties, to make a mock of the accumulated wisdom of the ages, the stored power of the will; to drive the individual — against his own plain conviction — into a union sure to result in evil, or to hold the individual helpless in such an evil union, when made.

Such is the condition of humanity, involving most evil results to its offspring and to its own happiness. And, while in men the immediate dominating force of the passion may be more conspicuous, it is in women that it holds more universal sway. For the man has other powers and faculties in full use, whereby to break loose from the force of this; and the woman, specially modified to sex and denied racial activity, pours her whole life into her love, and, if injured here, she is injured irretrievably. With him it is frequently light and transient, and, when most intense, often most transient. With her it is a deep, all-absorbing force, under the action of which she will renounce all that life offers, take any risk, face any hardships, bear any pain. It is maintained in her in the face of a lifetime of neglect and abuse. The common instance of the police court trials — the woman cruelly abused who will not testify against her husband — shows this. This devotion, carried to such a degree as to lead to the mismating of individuals with its personal and social injury, is an excessive sex-distinction.

But it is in our common social relations that the predominance of sex-distinction in women is made most manifest. The fact that, speaking broadly, women have, from the very beginning, been spoken of expressively enough as "the sex," demonstrates clearly that this is the main impression which they have made upon observers and recorders. Here one need attempt no farther proof than to turn the mind of the reader to an unbroken record of facts and feelings perfectly patent to every one, but not hitherto looked at as other than perfectly natural and right. So utterly has the status of woman been accepted as a sexual one that it has remained for the woman's movement of the nineteenth century to devote much contention to the claim that women are persons! That women are persons as well as females, — an unheard of proposition!

In a "Handbook of Proverbs of All Nations," a collection comprising many thousands, these facts are to be observed: first, that the proverbs concerning women are an insignificant minority compared to those concerning men; second, that the proverbs concerning women almost invariably apply to them in general, — to the sex. Those concerning men qualify, limit, describe, specialize. It is "a lazy man," "a violent man," "a man in his cups." Qualities and actions are predicated of man individually, and not as a sex, unless he is flatly contrasted with woman, as in "A man of straw is worth a woman of gold," "Men are deeds, women are words," or "Man, woman, and the devil are the three degrees of comparison." But of woman it is always and only "a woman," meaning simply a female, and recognizing no personal distinction: "As much pity to see a woman weep as to see a goose go barefoot." "He that hath an eel by the tail and a woman by her word hath a slippery handle." "A woman, a spaniel, and a walnut-tree, — the more you beat 'em, the better they be." Occasionally a distinction is made between "a fair woman" and "a black woman"; and Solomon's "virtuous woman," who commanded such a high price, is familiar to us all. But in common thought it is simply "a woman" always. The boast of the profligate that he knows "the sex,"

so recently expressed by a new poet, — "The things you will learn
from the Yellow and Brown, they'll 'elp you an' 'eap with the White";
the complaint of the angry rejected that "all women are just alike!"—
the consensus of public opinion of all time goes to show that the characteristics common to the sex have predominated over the characteristics distinctive of the individual, — a marked excess in sex-distinction.

From the time our children are born, we use every means known to
accentuate sex-distinction in both boy and girl; and the reason that the
boy is not so hopelessly marked by it as the girl is that he has the whole
field of human expression open to him besides. In our steady insistence
on proclaiming sex-distinction we have grown to consider most
human attributes as masculine attributes, for the simple reason that
they were allowed to men and forbidden to women.

A clear and definite understanding of the difference between race-
attributes and sex-attributes should be established. Life consists of
action. The action of a living thing is along two main lines, — self-
preservation and race-preservation. The processes that keep the individual alive, from the involuntary action of his internal organs to the
voluntary action of his external organs, — every act, from breathing
to hunting his food, which contributes to the maintenance of the individual life, — these are the processes of self-preservation. Whatever
activities tend to keep the race alive, to reproduce the individual, from
the involuntary action of the internal organs to the voluntary action
of the external organs; every act from the development of germ-cells
to the taking care of children, which contributes to the maintenance of
the racial life, — these are the processes of race-preservation. In race-
preservation, male and female have distinctive organs, distinctive
functions, distinctive lines of action. In self-preservation, male and
female have the same organs, the same functions, the same lines of
action. In the human species our processes of race-preservation have
reached a certain degree of elaboration; but our processes of self-
preservation have gone farther, much farther.

All the varied activities of economic production and distribution,
all our arts and industries, crafts and trades, all our growth in science,
discovery, government, religion, — these are along the line of self-
preservation: these are, or should be, common to both sexes. To teach, to
rule, to make, to decorate, to distribute, — these are not sex-functions:
they are race-functions. Yet so inordinate is the sex-distinction of the
human race that the whole field of human progress has been considered a masculine prerogative. What could more absolutely prove the
excessive sex-distinction of the human race? That this difference

should surge over all its natural boundaries and blazon itself across every act of life, so that every step of the human creature is marked "male" or "female," — surely, this is enough to show our over-sexed condition.

Little by little, very slowly, and with most unjust and cruel opposition, at cost of all life holds most dear, it is being gradually established by many martyrdoms that human work is woman's as well as man's. Harriet Martineau[2] must conceal her writing under her sewing when callers came, because "to sew" was a feminine verb, and "to write" a masculine one. Mary Somerville[3] must struggle to hide her work from even relatives, because mathematics was a "masculine" pursuit. Sex has been made to dominate the whole human world, — all the main avenues of life marked "male," and the female left to be a female, and nothing else.

But while with the male the things he fondly imagined to be "masculine" were merely human, and very good for him, with the female the few things marked "feminine" were feminine, indeed; and her ceaseless reiterance of one short song, however sweet, has given it a conspicuous monotony. In garments whose main purpose is unmistakably to announce her sex; with a tendency to ornament which marks exuberance of sex-energy, with a body so modified to sex as to be grievously deprived of its natural activities; with a manner and behavior wholly attuned to sex-advantage, and frequently most disadvantageous to any human gain; with a field of action most rigidly confined to sex-relations; with her overcharged sensibility, her prominent modesty, her "eternal femininity," — the female of genus homo is undeniably over-sexed.

This excessive distinction shows itself again in a marked precocity of development. Our little children, our very babies, show signs of it when the young of other creatures are serenely asexual in general appearance and habit. We eagerly note this precocity. We are proud of it. We carefully encourage it by precept and example, taking pains to develope the sex-instinct in little children, and think no harm. One of the first things we force upon the child's dawning consciousness is the fact that he is a boy or that she is a girl, and that, therefore, each must regard everything from a different point of view. They must be dressed differently, not on account of their personal needs, which are exactly

---

[2] *Harriet Martineau:* Martineau (1801–1876), feminist English writer.
[3] *Mary Somerville:* Somerville (1780–1872), English writer best known for her book *The Connexion of the Physical Sciences* (1834).

similar at this period, but so that neither they, nor any one beholding them, may for a moment forget the distinction of sex.

Our peculiar inversion of the usual habit of species, in which the male carries ornament and the female is dark and plain, is not so much a proof of excess indeed, as a proof of the peculiar reversal of our position in the matter of sex-selection. With the other species the males compete in ornament, and the females select. With us the females compete in ornament, and the males select. If this theory of sex-ornament is disregarded, and we prefer rather to see in masculine decoration merely a form of exuberant sex-energy, expending itself in nonproductive excess, then, indeed, the fact that with us the females manifest such a display of gorgeous adornment is another sign of excessive sex-distinction. In either case the forcing upon girl-children of an elaborate ornamentation which interferes with their physical activity and unconscious freedom, and fosters a premature sex-consciousness, is as clear and menacing a proof of our condition as could be mentioned. That the girl-child should be so dressed as to require a difference in care and behavior, resting wholly on the fact that she is a girl, — a fact not otherwise present to her thought at that age, — is a precocious insistence upon sex-distinction, most unwholesome in its results. Boys and girls are expected, also, to behave differently to each other, and to people in general, — a behavior to be briefly described in two words. To the boy we say, "Do"; to the girl, "Don't." The little boy must "take care" of the little girl, even if she is larger than he is. "Why?" he asks. Because he is a boy. Because of sex. Surely, if she is the stronger, she ought to take care of him, especially as the protective instinct is purely feminine in a normal race. It is not long before the boy learns his lesson. He is a boy, going to be a man; and that means all. "I thank the Lord that I was not born a woman, runs the Hebrew prayer. She is a girl, "only a girl," "nothing but a girl," and going to be a woman, — only a woman. Boys are encouraged from the beginning to show the feelings supposed to be proper to their sex. When our infant son bangs about, roars, and smashes things, we say proudly that he is "a regular boy!" When our infant daughter coquettes with visitors, or wails in maternal agony because her brother has broken her doll, whose sawdust remains she nurses with piteous care, we say proudly that "she is a perfect little mother already!" What business has a little girl with the instincts of maternity? No more than the little boy should have with the instincts of paternity. They are sex-instincts, and should not appear till the period of adolescence. The most normal girl is the "tom-boy," — whose numbers increase among us in these, wiser days, — a healthy young creature, who is human through and through, not feminine till it is time to be. The most normal boy has calmness and gentleness as well as vigor and courage. He is a human creature as well as a male creature, and not aggressively masculine till it is time to be. Childhood is not the period for these marked manifestations of sex. That we exhibit them, that we admire and encourage them, shows our over-sexed condition.

GERTRUDE STEIN

From *Letter to Henri Pierre Roche* (June 12, 1912)

... construction is a man's business. Being beautiful is a woman's business but there are a great many women who are not beautiful and they act stupidly the ones that are not beautiful if they act as if they were beautiful and [e]xpect to achieve the results of one being beautiful. Now to take the instance of yourself and myself. You are a man and I am a woman but I have a much more constructive mind than you have. I am a genuinely creative artist and being such my personality determines my art just as Matisse's or Picasso's or Wagner's or anyone else. Now you if I were a man would not write me such a letter because you would respect the *inevitable* character of my art. It would be very much (if I were a man) as if Bruce were to advise Matisse. You would not do it however you might wish my art other than it is. But being a man and believing that a man's business is to be constructive you forget the much greater constructive power of my mind and the absolute nature of my art which if I were a man you would respect.

Do you see why I think your letter unimportant and stupid. I have often felt much pleasure and much encouragement ... from your appreciation but that is as far as any outsider can be of assistance to any one genuinely creative.

Sincerely yours,
Gertrude Stein

The interest in women's health meant not only power for medical doctors, but the creation of new products that would aid women in obtaining better physical conditions. "Dr. Scott's Electric Corset" promised great benefits to purchasers, even though at least some of the health problems in the later nineteenth century (respiratory and circulatory problems, deformities of the rib cage and the undue constriction of internal organs) stemmed from such unhealthful clothing as the corset. Reproduced by permission of the Bakken Library, Minneapolis, Minnesota.

The harnessing of electricity coupled with the increased focus on health care led to the development of such products as the "electropoise," a machine devised to run a healing, low-grade, electric current through the body of many disease-sufferers. Notice that the marketing illustration shows a woman — presumably a sleepless neurotic — using the device. Reproduced by permission of the Bakken Library, Minneapolis, Minnesota.

S. WEIR MITCHELL

From *Fat and Blood and How to Make Them*

The physician who influenced women's medicine the most during the late nineteenth century, S. Weir Mitchell (1829–1914) believed in the symbiosis between literal health and whole body conditioning. Unfortunately, given the state of medical understanding about women's bodies and health, he believed with much of the rest of the medical profession that the focus of treatment for women should be the womb. A woman's reproductive organs were her primary difference from men; therefore, the symptoms that a woman experienced must be rooted in those areas of her body. Medicine the world over, from about 1870 through the early years of the twentieth century, emphasized the fragility of women, their dependence on men (fathers/husbands/sons), and their tendency to neurasthenia or hysteria.

While Mitchell tempered his treatment a bit (because he had treated a number of Civil War soldiers, and knew that nervous illness was not the province of women alone), he believed firmly that rest provided a sure means of helping overwrought patients recover. His rest cure was accompanied by overfeeding, isolation from family and friends, and the absence of any intellectual effort. Mitchell treated not only Gilman, but Jane Addams and many other prominent women at the turn of the century. Immortalized in Charlotte Perkins Gilman's short story "The Yellow Wallpaper" (1892), Mitchell had an immense influence on the practice of women's medicine, and on psychology and psychiatry. While Gertrude Stein did not study with Mitchell, she did have courses with William Osler, one of her Johns Hopkins Medical School professors, who was a near-disciple of his. She therefore had to contend with an approach to women's medical treatment that she thought was often inappropriate — if not useless.

These selections are from Mitchell's early book *Fat and Blood and How to Make Them* (Philadelphia: J. B. Lippincott, 1877).

Chapter Three
Seclusion

It is rare to find any of the class of patients I have described so free from the influence of their habitual surroundings as to make it easy to treat them in their own homes. It is needful to disentangle them from the contact with those who have been the willing slaves of their caprices. I have often made the effort to treat them in their own homes and isolate them there, but I have rarely done so without promising myself that I would not again complicate my treatment by any such embarrassments. Once separate the patient from the moral and physical surroundings which have become part of her life of sickness, and you will have made a change which will be in itself beneficial, and will enormously aid in the treatment which is to follow. Of course this step is not essential in such cases as are merely anaemic and feeble and thin, owing to distinct causes, like the exhaustion of overwork and of long dyspepsia; but I am now speaking chiefly of the large and troublesome class of thin-blooded emotional women, for whom a state of weak health has become a long and almost, I might say, a cherished habit. For them there is often no success possible until we have broken up the whole daily drama of the sick-room, with its little selfishnesses and its craving for sympathy and indulgence. Nor should we hesitate to insist upon this change, for not only shall we then act in the true interests of the patient, but we shall confer on those near to her an inestimable benefit. A hysterical girl is, as Wendell Holmes has said in his decisive phrase, a vampire who sucks the blood of the healthy people about her; and I may add that pretty surely where there is one hysterical girl there will be soon or late two sick women.

Chapter Four
Rest

Sometimes the question is easy to settle. If you find a woman who is in good state as to color and flesh, and who is always able to do what it pleases her to do, and who is tired by what does not please her, that is a woman to order out of bed and to control with a firm and steady will. That is a woman who is made to be made to walk, with no regard to her aches, and to be made to persist until exertion ceases to give rise to the mimicry of fatigue. In some cases the man who can insure belief in his opinions and obedience to his decrees secures very often most brilliant and sometimes easy success. . . . There are still other cases in which the same mischievous tendencies to repose, to endless tire, to hysterical symptoms, and to emotional displays have grown out of defects of nutrition so distinct that no man ought to think for them of mere exertion as a sole means of cure. . . . Nothing upsets these cases like over-exertion, and the attempt to make them walk usually ends in some mischievous emotional display. . . . As to the two sets of cases just sketched, no one need hesitate; the one must walk, the other should not until we have bettered her nutritive state. . . . But between these two classes lies the large number of such cases, giving us every kind of real or imagined symptom, and dreadfully well fitted to puzzle the most competent physician. As a rule, no harm is done by rest. . . . the rest I like for them is not at all their notion of rest. To lie abed half the day, and sew a little and read a little, and be interesting and excite sympathy, is all very well, but when they are bidden to stay in bed a month, and neither to read, write, nor sew, and to have one nurse, — who is not a relative, — then rest becomes for some women a rather bitter medicine, and they are glad enough to accept the order to rise and go about when the doctor issues a mandate. . . .

WILLIAM OSLER

Medicine in the Nineteenth Century
Excerpts from a lecture given for the
Johns Hopkins Historical Club, January, 1901
(printed in the New York *Sun*)

One of the leading physicians and scientists of human medicine in the modern United States, William Osler (1849–1919) was instrumental in combining the scientific aspects of medicine with the humanistic. Born and educated on the Canadian frontier, he was trained at Trinity College and McGill Medical School. He studied as well in London, Berlin, and Vienna. After coming to Philadelphia in 1884 to teach at the University of Pennsylvania, he quickly gained prominence within international medical circles. He left Penn in 1889 to assume a post at Johns Hopkins Medical School, and, in 1892, published the influential text *The Principles and Practice of Medicine*. In 1905 he left Johns Hopkins — after helping to make it the premier United States medical program — to accept a position at Oxford. In England he continued his rise in international circles, becoming Sir William Osler (see the Harvey Cushing biography, *Life of Sir William Osler*, 1940).

A follower of S. Weir Mitchell, Osler began his career as a pathologist, and then became a clinician. He too believed in motivating patients to improve their health, and felt that a physician's tenacity was central to the health of the patient. Throughout his life, Osler wrote and spoke eloquently, and was much in demand for addresses. His influence on American medicine remains inestimable. As the following selection shows, however, the state of information at the turn of the century now seems lamentable.

Osler was Gertrude Stein's professor for gynecology classes at Johns Hopkins, and her failure to attend his classes is one of the reasons he made the motion during a meeting of the medical school faculty that Miss Stein not be graduated with her class. Even though, as Stein recounts in *The Autobiography of Alice B. Toklas*, she was assured that she would graduate after taking courses in summer school, she decided not to finish the degree.

Osler's remarks in 1901 show the paternalism inherent in some physicians' attitudes at the turn of the century. This lecture appears in Osler's book *Aequanimitas: With Other Addresses to Medical Students, Nurses and Practitioners of Medicine* (Philadelphia: P. Blakiston's Sons, 1932).

Venereal Diseases. — These continue to embarrass the social economist and to perplex and distress the profession. The misery and ill-health which they cause are incalculable, and the pity of it is that the cross is not always borne by the offender, but innocent women and children share the penalties. The gonorrhœal infection, so common, and often so little heeded, is a cause of much disease in parts other than those first affected. Syphilis claims its victims in every rank of life, at every age, and in all countries. We now treat it more thoroughly, but all attempts to check its ravages have been fruitless. Physicians have two important duties: the incessant preaching of continence to young men, and scrupulous care, in every case, that the disease may not be a source of infection to others, and that by thorough treatment the patient may be saved from the serious late nervous manifestations. We can also urge that in the interests of public health venereal diseases, like other infections, shall be subject to supervision by the State. The opposition to measures tending to the restriction of these diseases is most natural: on the one hand, from women, who feel that it is an aggravation of a shocking injustice and wrong to their sex; on the other, from those who feel the moral guilt in a legal recognition of the evil. It is appalling to contemplate the frightful train of miseries which a single diseased woman may entail, not alone on her associates, but on scores of the innocent — whose bitter cry should make the opponents of legislation feel that any measures of restriction, any measures of registration, would be preferable to the present disgraceful condition, which makes of some Christian cities open brothels and allows the purest homes to be invaded by the most loathsome of all diseases. . . .

Leprosy. — Since the discovery of the germ of this terrible disease systematic efforts have been made to improve the state of its victims and to promote the study of the conditions under which the disease prevails. The English Leprosy Commission has done good work in calling attention to the widespread prevalence of the disease in India and in the East. In this country leprosy has been introduced into San Francisco by the Chinese, and into the North-western States by the Norwegians, and there are foci of the disease in the Southern States, particularly Louisiana, and in the province of New Brunswick. The problem has an additional interest since the annexation of Hawaii and the Philippine Islands, in both of which places leprosy prevails extensively. By systematic measures of inspection and the segregation of affected individuals the disease can readily be held in check. It is not likely ever to increase among native Americans, or again gain such a foothold as it had in the Middle Ages.

Puerperal Fever. — Perhaps one of the most striking of all victories of preventive medicine has been the almost total abolition of so-called child-bed fever from the maternity hospitals and from private practice. In many institutions the mortality after child-birth was five or six per cent., indeed sometimes as high as ten per cent., whereas to-day, owing entirely to proper antiseptic precautions, the mortality has fallen to three-tenths to four-tenths per cent. The recognition of the contagiousness of puerperal fever was the most valuable contribution to medical science made by Oliver Wendell Holmes.[1] There had been previous suggestions by several writers, but his essay on the "Contagiousness of Puerperal Fever," published in 1843, was the first strong, clear, logical statement of the case. Semmelweis,[2] a few years later, added the weight of a large practical experience to the side of the contagiousness, but the full recognition of the causes of the disease was not reached until the recent antiseptic views had been put into practical effect. . . .

---

[1] *Oliver Wendell Holmes:* American writer and physician (1809–1894) educated at Harvard. He taught at Dartmouth and then became dean of the Harvard Medical School. Better known for his poetry and essays, he also made important contributions to the study of medicine.

[2] *Semmelweis:* Ignaz Philipp Semmelweis, Hungarian physician (1818–1865) who recognized that puerperal fever was infectious, knowledge that would eventually reduce the mortality rate from infection in childbirth. His discovery, while a physician in Vienna, led to ridicule and his termination, and, probably, to his suicide. By the 1890s, the value of his findings had been recognized.

The New School of Medicine

The nineteenth century has witnessed a revolution in the treatment of disease, and the growth of a new school of medicine. The old schools — regular and homœopathic — put their trust in drugs, to give which was the alpha and the omega of their practice. For every symptom there were a score or more of medicines — vile, nauseous compounds in one case; bland, harmless dilutions in the other. The characteristic of the New School is firm faith in a few good, well-tried drugs, little or none in the great mass of medicines still in general use. Imperative drugging — the ordering of medicine in any and every malady — is no longer regarded as the chief function of the doctor. Naturally, when the entire conception of the disease was changed, there came a corresponding change in our therapeutics. In no respect is this more strikingly shown than in our present treatment of fever — say, of the common typhoid fever. During the first quarter of the century the patients were bled, blistered, purged and vomited, and dosed with mercury, antimony, and other compounds to meet special symptoms. During the second quarter the same, with variations in different countries. After 1850 bleeding became less frequent, and the experiments of the Paris and Vienna schools began to shake the belief in the control of fever by drugs. During the last quarter sensible doctors have reached the conclusion that typhoid fever is not a disease to be treated with medicines, but that in a large proportion of all cases diet, nursing and bathing meet the indications. There is active, systematic, careful, watchful treatment, but not with drugs. The public has not yet been fully educated to this point, and medicines have sometimes to be ordered for the sake of their friends, and it must be confessed that there are still in the ranks *antiques* who would insist on a dose of some kind every few hours.

The battle against poly-pharmacy, or the use of a large number of drugs (of the action of which we know little, yet we put them into bodies of the action of which we know less), has not been fought to a finish: There have been two contributing factors on the side of progress — the remarkable growth of the sceptical spirit fostered by Paris, Vienna and Boston physicians, and, above all, the valuable lesson of homœopathy, the infinitesimals of which certainly could not do harm, and quite as certainly could not do good; yet nobody has ever claimed that the mortality among homœopathic practitioners was greater than among those of the regular school. A new school of practitioners has arisen which cares nothing for homœopathy and less for so-called allopathy. It seeks to study, rationally and scientifically, the action of drugs, old and new. It is more concerned that a physician shall know how to apply the few great medicines which all have to use, such as quinine, iron, mercury, iodide of potassium, opium and digitalis, than that he should employ a multiplicity of remedies the action of which is extremely doubtful.

The growth of scientific pharmacology, by which we now have many active principles instead of crude drugs, and the discovery of the art of making medicines palatable, have been of enormous aid in rational practice. There is no limit to the possibility of help from the scientific investigation of the properties and action of drugs. At any day the new chemistry may give to us remedies of extraordinary potency and of as much usefulness as cocaine. There is no reason why we should not even in the vegetable world find for certain diseases specifics of virtue fully equal to that of quinine in the malarial fevers.

One of the most striking characteristics of the modern treatment of disease is the return to what used to be called the natural methods — diet, exercise, bathing and massage. There probably never has been a period in the history of the profession when the value of *diet* in the prevention and the cure of disease was more fully recognized. Dyspepsia, the besetting malady of this country, is largely due to improper diet, imperfectly prepared and too hastily eaten. One of the great lessons to be learned is that the preservation of health depends in great part upon food well cooked and carefully eaten. A common cause of ruined digestion, particularly in young girls, is the eating of sweets between meals and the drinking of the abominations dispensed in the chemists' shops in the form of ice-cream sodas, etc. Another frequent cause of ruined digestion in business men is the hurried meal at the lunch-counter. And a third factor, most important of all, illustrates the old maxim, that more people are killed by over eating and drinking than by the sword. Sensible people have begun to realize that alcoholic excesses lead inevitably to impaired health. A man may take four or five drinks of whiskey a day, or even more, and think perhaps that he transacts his business better with that amount of stimulant; but it only too frequently happens that early in the fifth decade, just as business or political success is assured, Bacchus[3] hands in heavy bills for payment,

[3] *Bacchus:* Greek and Roman god of wine and revelry, vegetation and fertility, often associated with Dionysus.

in the form of serious disease of the arteries or of the liver, or there is a general breakdown. With the introduction of light beer there has been not only less intemperance, but a reduction in the number of cases of organic disease of the heart, liver and stomach caused by alcohol. While temperance in the matter of alcoholic drinks is becoming a characteristic of Americans, intemperance in the quantity of food taken is almost the rule. Adults eat far too much, and physicians are beginning to recognize that the early degenerations, particularly of the arteries and of the kidneys, leading to Bright's disease, which were formerly attributed to alcohol are due in large part to too much food. . . .

A third noteworthy feature in modern treatment has been a return to psychical methods of cure, in which *faith in something is suggested* to the patient. After all, faith is the great lever of life. Without it, man can do nothing; with it, even with a fragment, as a grain of mustard-seed, all things are possible to him. Faith in us, faith in our drugs and methods, is the great stock in trade of the profession. In one pan of the balance, put the pharmacopœias of the world, all the editions from Dioscorides to the last issue of the United States Dispensatory; heap them on the scales as did Euripides[4] his books in the celebrated contest in the "Frogs"; in the other put the simple faith with which from the days of the Pharaohs until now the children of men have swallowed the mixtures these works describe, and the bulky tomes will kick the beam. It is the *aurum potabile,* the touchstone of success in medicine. . . . The cures in the temples of Æsculapius,[5] the miracles of the saints, the remarkable cures of those noble men, the Jesuit missionaries, in this country, the modern miracles at Lourdes and at St. Anne de Beaupré in Quebec, and the wonder-workings of the so-called Christian Scientists, are often genuine, and must be considered in discussing the foundations of therapeutics. We physicians use the same power every day. If a poor lass, paralyzed apparently, helpless, bed-ridden for years, comes to me, having worn out in mind, body and estate a devoted family; if she in a few weeks or less by faith in me, and faith alone, takes up her bed and walks, the saints of old could not have

---

[4] *Euripides:* Greek playwright, born 480 or 485 B.C.E.; d. 406 B.C.E., who ranked with Aeschylus and Sophocles, though his writings were more realistic. He wrote over ninety plays.

[5] *Æsculapius:* Sometimes spelled Asclepius. Legendary Greek physician, the son of Apollo and Coronis, who became so skillful in healing that he could revive the dead. At that point, Zeus killed him. Æsculapius became the god of medicine, and temples were built to honor him.

done more, St. Anne and many others can scarcely to-day do less. We enjoy, I say, no monopoly in the faith business. The faith with which we work, the faith, indeed, which is available to-day in everyday life, has its limitations. It will not raise the dead; it will not put in a new eye in place of a bad one (as it did to an Iroquois Indian boy for one of the Jesuit fathers), nor will it cure cancer or pneumonia, or knit a bone; but, in spite of these nineteenth-century restrictions, such as we find it, faith is a most precious commodity, without which we should be very badly off. . . .

J. WHITRIDGE WILLIAMS

From *Obstetrics*

As a very young physician, John Whitridge Williams (1866–1931) earned the nickname "The Bull" while teaching in the Johns Hopkins medical program, where he was chair of the department of obstetrics and a professor in that department for thirty-two years. A brilliant student, he had graduated with an undergraduate degree from Johns Hopkins in only two years, had taken his medical degree at twenty-one from the University of Maryland, and had done further study in Vienna and Berlin. A specialist in pelvic pathology, he was committed to the belief that obstetrics and gynecology should be combined into a single field, a belief he eventually saw implemented.

Popular for his anecdotal style of recounting medical experience, Williams achieved a great deal during his lifetime. His now-classic text, *Obstetrics*, has been published in countless editions. Unfortunately for Gertrude Stein, the young professor was not swayed by her objections to his narrative methods or to his beliefs about women's bodies. Williams's course was one of those she boycotted during her years at medical school.

This excerpt is taken from Williams's *Obstetrics* in the 1923 edition, published in New York by D. Appleton.

Racial Differences in Pelves. — Considerable variations may be observed in the form of the pelvis in various races, and especially upon comparing those obtained from aboriginal and civilized peoples. But in spite of the researches of Weber, Stein, Verneau, Topinard, Turner, and others, our knowledge of the subject is still fragmentary. Stein distinguished four groups of pelves:

1. Blunt heart-shaped.
2. Elliptical, with the greatest diameter transverse.
3. Round.
4. Elliptical, with the greatest diameter anteroposterior.

Topinard attempted to classify pelves according to their "general index" — that is, the relation between their height and width, as represented by the distance between the iliac crests. His careful measurements showed that the pelves becomes increasingly lower and broader the more civilized the race from which it is obtained.

Turner based his classification upon the relation between the transverse and anteroposterior diameters of the superior strait, and divided pelves into three great groups: dolichopellic, in which the conjugata vera is greater than the transverse diameter; mesatipellic, in which the conjugata vera and transverse diameters are of equal length; and platypellic, in which the conjugata vera is shorter than the transverse diameter. He stated that the first variety had not been observed in women, though it is not infrequent in men; but the investigations of Scharlau show that Turner was in error, as it is frequently noted in the aboriginal women of Australia. The mesatipellic variety is observed in the women of the lower races, notably among the Bushmen, Hottentots,[1] and the lower classes of negroes; while the platypellic forms are found in all the higher races. But even among civilized whites considerable racial differences are frequently noted, and it is generally stated that the pelves of the English and Holstein[2] women are broader than those of other nationalities; while the Jewesses living in the vicinity of Dorpat[3] have extremely small pelves. Gache states that the pelvis is usually normal in the Argentine Republic, while it is imperfectly developed and frequently funnel-shaped in Mexico.

While the study of the racial differences in the pelvis presents a marked anthropological interest, it is, as yet, of little practical obstetrical value, as no extended studies have been made concerning the form and size of the heads of children which are born through them. The

---

[1] *Hottentots:* The word Hottentot refers to a South African tribe, also known as the Khoi.

[2] *Holstein:* Schleswig-Holstein is the northernmost state in Germany, bordering Denmark. People from this vicinity are referred to as Holsteins.

[3] *Dorpat:* Dorpat is a city in Germany.

careful work of my former assistant, T. F. Riggs, has shown that con-
tracted pelves occur several times more frequently among black than
white women in Baltimore, while operative delivery is more frequently
required among the latter. This is due to the fact that the negro children are somewhat smaller and have more compressible heads, and
thus compensate for the smaller size of the pelvis. Acosta-Sison makes
a similar statement concerning the Philippine women, and Kinoshita
of Tokio informs me that in Japan both the pelvic and cephalic measurements fall below the average observed in European and American
white women.

Pelvis of the New-born Child. — The pelvis of the child at birth is
partly bony and partly cartilaginous. The innominate bone does not
exist as such, its place being taken by the ilium, ischium, and pubis,
which are united by a large Y-shaped cartilage, the three bones meeting in the acetabulum. The iliac crests and the acetabula, as well as
the greater part of the ischiopubic rami, are entirely cartilaginous in
structure.

The cartilaginous portions of the pelvis gradually give place to
bone, but complete union in the neighborhood of the acetabulum does
not occur until about the age of puberty, and occasionally even at a
later period. Indeed, we may say that the innominate bones do not
become completely ossified and fully developed until between the
twentieth and twenty-fifth years. . . .

2

Questions of Otherness:
Sexuality ("Inversion"), Race, Gender

Stein writes in her journal about being very afraid when, as a student at Radcliffe, she read Oscar Wilde's poignant "The Ballad of Reading Gaol" (1898); (see pp. 263–71). The risks of being different were no longer abstract. It was dangerous enough to be a woman, a student at the Harvard Annex, as Radcliffe College was originally known, but to have that danger compounded by racial or religious difference was certain hardship, as black or Jewish students early recognized. (Both Leo and Gertrude were conscious of being among the few Jewish students in Cambridge.) But perhaps the most dangerous difference at the turn of the century was sexual. Here the aura of misinformation about sexual preference, or about any sexuality at all, meant that life could be ruined by the private being made public.

The case of Oscar Wilde shadowed the literate world. The brilliant writer's 1895 trial and two-year imprisonment for sodomy illustrated the dangers of crossing — or even thinking of crossing — the heterosexual norms of circumspect society. What the outcome of Wilde's trial meant to the imagination of the English-speaking world was that sexual practices that deviated from social norms could be defined as criminal, and could mean both imprisonment and death. The devastation of Wilde's literary career was punishment enough for his homosexuality; the public shame, his family's reaction, and society's abandonment of him — despite his class standing — destroyed any sense of Wilde as a recoverable person. For a college-age woman who suspected that she could love other women, despite her friendships with men at Harvard, Stein knew she had to be discreet — even secretive. Vocabulary fed into the shame of the homosexual; *inversion* and *perversion* were the colloquial terms for nonheterosexual practices. The secrecy of sexuality, implicit in polite and upperclass society, led to a strangely uninformed population. The medicalization of those aspects of human life that affected women and their lives complicated easy transmission of information about the sexual; during the 1880s and 1890s, writings about sexuality and its variants were published for the first time. Partly in reaction against the 1873 Comstock Law, which forebade the mailing of sexually explicit materials about birth control, new sexual information took on the quality of the forbidden, the titillating. The 1897 publication of Havelock Ellis's six-volume *Studies in the Psychology of Sex* shocked the world by defending male homosexuality; in 1899 Ellis's second volume, *Auto-Eroticism,* described and explained both female and male masturbation. That sex could be pleasurable — and that it could exist in something other than heterosexual pairings — contradicted the usual medical and religious views. Ellis's work also introduced readers to the notion that women's sexuality did not need to be passive (see pp. 285–88).

Excerpts from Ellis's work give the reader some insight into how flawed even supposedly "scientific" information could be; Freud's essay "The Sexual Aberrations," while somewhat more accurate, reinforced the notion that bisexual or homosexual practices were so rare as to be inherently exotic (see pp. 271–81). Judging from the print materials at the turn of the century, women homosexuals — seldom referred to as lesbians — received even more stereotypical treatment than did gay men. "Deviants" or "inverts," women who preferred same-sex relationships were scorned and ridiculed, as the writing of Otto Weininger shows. The scathing tone of his belittlement of women mirrors the sentiments expressed in his treatise on the Jew (see pp. 293–302). Weininger, a late convert to Protestantism from Judaism, was horrified by the idea of marginalization, his own in particular. Author of the popular *Sex and Character,* Weininger theorized that each human exhibited some ratio of either male or female traits. Female traits, seen as inferior, were paralleled with the characteristics of Jewish people — in Weininger's eyes the two groups were equally inferior. That Stein was fascinated with this book, buying copies in France and sending them to her women friends in the States, raises questions about her own hostility toward women. Stein all too readily agreed that most geniuses were male, and that the woman of genius might well have a high component of masculinity. After years of protesting the medical profession's infantilizing of women as patients, misreading their symptoms and their psychologies, was Stein, too, infected with that aura of intellectualized — if faulty — reasoning?

Demeaning as Weininger was on the subject of women, his venomous tirade against Judaism — which he links with the condition of being negro — defies rational explanation. Like women, he wrote, Jews have no honor, no ability, no trustworthiness, and no humanity. The travesty of Weininger's claiming that his "research" was done with the best scientific methodology removes any value from his book; a few years after its publication, the author committed suicide. Perhaps his act lent credibility to his point that "the bitterest Antisemites are to be found amongst the Jews themselves. . . . whoever detests the Jewish disposition detests it first of all in himself; that he should persecute it in others is merely his endeavour to separate himself in this way from Jewishness" (see p. 294).

Gertrude Stein's own tentative but thoughtful essay, written while she was a Radcliffe student, makes clear that she felt personally committed to her Jewish identity (see pp. 291–92). In the full text of this essay (which is here briefly excerpted), Stein discusses the various social aspects of this racial identity, concluding that a Jew must marry a Jew. The line must be continued, and preserved. Set against her self-identification as a Jewish person runs the inherently subversive identification of herself as a lesbian. Like oil and water, the two personae would never merge. One would have to give way to the other.

Margaret C. Anderson's posthumously published memoir about her lesbian existence in France gives a coded interpretation of the passion that needed to be suppressed. In the 1996 appearance of *Forbidden Fires*, a first-person account of lesbian life, the realities of the dangers of that life, even in such a nonjudgmental culture as that of Paris, are made clear (see pp. 281–85). As editor of *Little Review* and a fellow expatriate Parisian, Anderson was an acquaintance of Gertrude Stein's, and their lives often ran in parallel.

History proves as well that prejudice against blacks continued long after their legal emancipation. Ida B. Wells-Barnett's excerpted essay, "The Case Stated," describes the increase of persecution because of the black man's enfranchisement during the later years of the nineteenth century — lynchings, burning-outs by the Ku Klux Klan, and other styles of murder (see pp. 302–07). While these murders were often not publicized, well-informed people knew they existed. Because

Gertrude Stein, as a college and then a medical student, felt increasingly marginalized by her Jewishness, she developed a bond with others who were outside the cultural mainstream, and whose very lives were threatened by that marginalization. Also, because of Stein's urban experiences — in Baltimore, San Francisco, and Boston, as well as in European cities — she had a great appreciation for racial difference, and seldom felt caught in the webs of nineteenth-century prejudice that trapped so many privileged Americans.

OSCAR WILDE

From *The Ballad of Reading Gaol*

Born in Dublin, Oscar Wilde (1854–1900) studied at Trinity College and at Magdalen College, Oxford, where he was satirized in *Punch* for his art-for-art's sake philosophy. A disciple of Walter Pater and John Ruskin, Wilde published collections of poetry and his 1891 novel, *Picture of Dorian Gray*. Best known as a playwright for his 1890s comedies (*Lady Windermere's Fan, A Woman of No Importance, The Importance of Being Earnest*), Wilde's professional life ended abruptly in 1895 when he brought action for libel against the marquess of Queensberry, who objected to Wilde's friendship with his son, Lord Alfred Douglas. Wilde was then charged with sexual offenses under the Criminal Law Amendment and sentenced to prison for two years. "The Ballad of Reading Gaol" was published in 1898.

After his release from prison in 1897, Wilde lived in poverty and ill health in France until his death in 1900. His writings and accounts of his celebrated — and admonitory — trial and imprisonment lived long past that date, however, continuing into the twenty-first century. Vivid testimony to the homophobia that was then seldom expressed, Wilde's life and death became a warning to all bisexuals or homosexuals that punishment not only existed, but would be doled out, for nonconventional sexual practices.

"The Ballad of Reading Gaol," published in 1898, was the first expression of Wilde's painful situation. By writing about his incarceration in the gaol (jail) located in the town of Reading, Wilde took on the issue squarely. Gertrude Stein remembered the fearful effect of the poem when she read it as a student, and became even less disclosive about her romantic feelings for women than she had previously been.

Oscar Wilde, the then-romantic figure, at the beginning of his popularity as a British/Irish author and dramatist, educated at Trinity College, Dublin, and at Magdalen College, Oxford. Once he was imprisoned a decade later, although his literary fame grew, his stature as a humanist and wit of a superior class evaporated.

Excerpts from the poem are reprinted here from *The Works of Oscar Wilde, 1856–1900*, ed. G. F. Maine (London: Collins, 1948).

I

He did not wear his scarlet coat,
    For blood and wine are red,
And blood and wine were on his hands
    When they found him with the dead,
The poor dead woman whom he loved,   5
    And murdered in her bed.

He walked amongst the Trial Men
    In a suit of shabby gray;
A cricket cap was on his head,
    And his step seemed light and gay;   10
But I never saw a man who looked
    So wistfully at the day.

I never saw a man who looked
    With such a wistful eye
Upon that little tent of blue   15
    Which prisoners call the sky,
And at every drifting cloud that went
    With sails of silver by.

I walked, with other souls in pain,
    Within another ring,   20
And was wondering if the man had done
    A great or little thing,
When a voice behind me whispered low,
    "That fellow's got to swing."

Dear Christ! the very prison walls   25
    Suddenly seemed to reel,
And the sky above my head became
    Like a casque° of scorching steel;
And, though I was a soul in pain,
    My pain I could not feel.   30

28. *casque:* An open conical helmet, commonly used in medieval times.

I only knew what haunted thought
　　Quickened his step, and why
He looked upon the garish day
　　With such a wistful eye;
The man had killed the thing he loved,　　　　　35
　　And so he had to die.

Yet each man kills the thing he loves,
　　By each let this be heard,
Some do it with a bitter look,
　　Some with a flattering word,　　　　　　　40
The coward does it with a kiss,
　　The brave man with a sword!

Some kill their love when they are young,
　　And some when they are old;
Some strangle with the hands of Lust,　　　　45
　　Some with the hands of Gold:
The kindest use a knife, because
　　The dead so soon grow cold.

Some love too little, some too long,
　　Some sell, and others buy;　　　　　　　　50
Some do the deed with many tears,
　　And some without a sigh:
For each man kills the thing he loves,
　　Yet each man does not die.

He does not die a death of shame　　　　　　55
　　On a day of dark disgrace,
Nor have a noose about his neck,
　　Nor a cloth upon his face,
Nor drop feet foremost through the floor
　　Into an empty space.　　　　　　　　　　60

He does not sit with silent men
　　Who watch him night and day;
Who watch him when he tries to weep,
　　And when he tries to pray;
Who watch him lest himself should rob　　　　65
　　The prison of its prey.

He does not wake at dawn to see
　　Dread figures throng his room,
The shivering Chaplain robed in white,

The Sheriff stern with gloom, 70
And the Governor all in shiny black,
With the yellow face of Doom.

He does not rise in piteous haste
To put on convict-clothes,
While some coarse-mouthed Doctor gloats, and notes 75
Each new and nerve-twitched pose,
Fingering a watch whose little ticks
Are like horrible hammer-blows.

He does not feel that sickening thirst
That sands one's throat, before 80
The hangman with his gardener's gloves
Comes through the padded door,
And binds one with three leathern thongs,
That the throat may thirst no more,

He does not bend his head to hear 85
The Burial Office° read,
Nor, while the anguish of his soul
Tells him he is not dead,
Cross his own coffin, as he moves
Into the hideous shed. 90

He does not stare upon the air
Through a little roof of glass:
He does not pray with lips of clay
For his agony to pass;
Nor feel upon his shuddering cheek 95
The kiss of Caiaphas.°

. . . .

V

I know not whether Laws be right, 535
Or whether Laws be wrong;
All that we know who lie in gaol
Is that the wall is strong;

---

86. *Burial Office:* The service for burying a person.
96. *Caiaphas:* High priest of the Jews who presided at the Sanhedrin (court of law)
that condemned Jesus to death.

And that each day is like a year,
 A year whose days are long.   540

But this I know, that every Law
 That men have made for Man,
Since first Man took his brother's life,
 And the sad world began,
But straws the wheat and saves the chaff°   545
 With a most evil fan.

This too I know — and wise it were
 If each could know the same —
That every prison that men build
 Is built with bricks of shame,   550
And bound with bars lest Christ should see
 How men their brothers maim.

With bars they blur the gracious moon,
 And blind the goodly sun:
And they do well to hide their Hell,   555
 For in it things are done
That Son of God nor son of Man
 Ever should look upon!

The vilest deeds like poison weeds
 Bloom well in prison-air: 560
It is only what is good in Man
 That wastes and withers there:
Pale Anguish keeps the heavy gate,
 And the warder is Despair.

For they starve the little frightened child   565
 Till it weeps both night and day:
And they scourge the weak, and flog° the fool,
 And gibe° the old and gray,
And some grow mad, and all grow bad,
 And none a word may say.   570

Each narrow cell in which we dwell
 Is a foul and dark latrine,
And the fetid breath of living Death
 Chokes up each grated screen,

545. *chaff*: Husks of grain, disposed of after threshing.
567. *flog*: Beat, thrash.
568. *gibe*: Jeer at, taunt.

And all, but Lust, is turned to dust   575
  In Humanity's machine.
The brackish° water that we drink
  Creeps with a loathsome slime,
And the bitter bread they weigh in scales
  Is full of chalk and lime,   580
And Sleep will not lie down, but walks
  Wild-eyed, and cries to Time.

But though lean Hunger and green Thirst
  Like asp° with adder° fight,
We have little care of prison fare,   585
  For what chills and kills outright
Is that every stone one lifts by day
  Becomes one's heart by night.

With midnight always in one's heart,
  And twilight in one's cell,   590
We turn the crank, or tear the rope,
  Each in his separate Hell,
And the silence is more awful far
  Than the sound of a brazen bell.

And never a human voice comes near   595
  To speak a gentle word:
And the eye that watches through the door
  Is pitiless and hard:
And by all forgot, we rot and rot,
  With soul and body marred.   600

And thus we rust Life's iron chain
  Degraded and alone:
And some men curse, and some men weep,
  And some men make no moan:
But God's eternal Laws are kind   605
  And break the heart of stone.

And every human heart that breaks,
  In prison-cell or yard,
Is as that broken box that gave
  Its treasure to the Lord,   610

---

577. *brackish:* Displeasing, salty, brinelike.
584. *asp:* Venomous snake, as the Egyptian cobra.
584. *adder:* Common European viper, may be poisonous or nonpoisonous.

And filled the unclean leper's house
  With the scent of costliest nard.°

Ah! happy they whose hearts can break
  And peace of pardon win!
How else may man make straight his plan   615
  And cleanse his soul from Sin?
How else but through a broken heart
  May Lord Christ enter in?

And he of the swollen purple throat,
  And the stark and staring eyes,   620
Waits for the holy hands that took
  The Thief to Paradise;
And a broken and a contrite heart
  The Lord will not despise.

The man in red who reads the Law   625
  Gave him three weeks of life,
Three little weeks in which to heal
  His soul of his soul's strife,
And cleanse from every blot of blood
  The hand that held the knife.   630

And with tears of blood he cleansed the hand,
  The hand that held the steel:
For only blood can wipe out blood,
  And only tears can heal:
And the crimson stain that was of Cain°   635
  Became Christ's snow-white seal.

VI

In Reading gaol by Reading town
  There is a pit of shame,
And in it lies a wretched man
  Eaten by teeth of flame,   640
In a burning winding-sheet he lies,
  And his grave has got no name.

And there, till Christ call forth the dead,
  In silence let him lie:

612. *costliest nard:* Most valuable perfume.
635. *Cain:* First son of Adam and Eve; murdered his brother Abel.

No need to waste the foolish tear,   645
Or heave the windy sigh:
The man had killed the thing he loved,
And so he had to die.

And all men kill the thing they love,
By all let this be heard,   650
Some do it with a bitter look,
Some with a flattering word,
The coward does it with a kiss,
The brave man with a sword!

SIGMUND FREUD

From *The Sexual Aberrations*
Trans. by A. A. Brill

Sigmund Freud (1856–1939) was the Austrian neurologist who founded psychoanalysis. His various theories shaped much of twentieth-century understanding of the human mind, sexuality, and family relationships. A student of J. M. Charcot in Paris, his first published work, *On the Psychical Mechanism of Hysterical Phenomena* (1893), was coauthored with Josef Breuer, but their partnership foundered as Freud became convinced that sexuality underlay many personality problems. A virtual outcast from the medical profession at the turn of the century, Freud published the groundbreaking *The Interpretation of Dreams* (1900), *The Psychopathology of Everyday Life* (1904), and *Three Contributions to the Sexual Theory* (1905). The excerpt included here, from the latter influential book, shows how even the most enlightened theorist codified "normal" and "abnormal." When we understand that public understanding of sexuality (and deviancy from its normal definitions) was based on very primitive notions of human behavior, with Freud's beliefs being at the outer edge of conventional knowledge, it is possible to see why homosexual inclinations had to be shrouded in secrecy.

Freud's later collaborations with Swiss psychiatrists Eugen Bleuler and C. G. Jung, the Austrian Alfred Adler, and others led to the creation of whole schools of innovative psychiatric thought. Other of Freud's key books are *A General Introduction to Psychoanalysis* (1910); *The Ego and the Id* (1923); *Totem and Tabu* (1913), a study of the origin of religious beliefs; and *Moses and Monotheism* (1939), the application of psychoanalytic theory to wider cultural problems. Not only would Gertrude Stein have read Freud for her own interests, she would have studied his work when she was a student at Johns Hopkins Medical School, which was modeled on German medical programs.

1. DEVIATION IN REFERENCE TO THE SEXUAL OBJECT

The popular theory of the sexual impulse corresponds closely to the poetic fable of dividing the person into two halves — man and woman — who strive to become reunited through love. It is therefore very surprising to hear that there are men for whom the sexual object is not woman but man and that there are women for whom it is not man but woman. Such *persons* are called contrary sexuals, or better, inverts; the *condition,* that of inversion. The number of such individuals is considerable though difficult of accurate determination.[1]

A. Inversion

THE BEHAVIOR OF INVERTS. The above mentioned persons behave in many ways quite differently.

(a) They are absolutely inverted; that is, their sexual object must be always of the same sex, while the opposite sex can never be to them an object of sexual longing, but leaves them indifferent or may even evoke sexual repugnance. As men they are unable, on account of this repugnance, to perform the normal sexual act or miss all pleasure in its performance.

(b) They are amphigenously inverted (psychosexually hermaphroditic); that is, their sexual object may belong indifferently to either the same or to the other sex. The inversion lacks the character of exclusiveness.

(c) They are occasionally inverted; that is, under certain external conditions, chief among which are the inaccessibility of the normal sexual objection and imitation, they are able to take as the sexual object a person of the same sex and thus find sexual gratification.

Note: the footnotes in this selection are original to the selection.

[1] For the difficulties entailed in the attempt to ascertain the proportional number of inverts, compare the word of M. Hirschfeld in the *Jahrbuch für sexuelle Zwischenstufen,* 1904. (Cf. also Brill, "The Conception of Homosexuality," *Journal of the A.M.A.,* August 2, 1913.)

The inverted also manifest a manifold behavior in their judgment about the peculiarities of their sexual impulse. Some take the inversion as a matter of course, just as the normal person does regarding his libido, firmly demanding the same rights as the normal. Others, however, strive against the fact of their inversion and perceive in it a morbid compulsion.[2]

Other variations concern the relations of time. The characteristics of the inversion in any individual may date back as far as his memory goes, or they may become manifest to him at a definite period before or after puberty.[3] The character is either retained throughout life, or it occasionally recedes or represents an episode on the road to normal development. A periodical fluctuation between the normal and the inverted sexual object has also been observed. Of special interest are those cases in which the libido changes, taking on the character of inversion after a painful experience with the normal sexual object.

These different categories of variation generally exist independently of one another. In the most extreme cases it can regularly be assumed that the inversion has existed at all times and that the person feels contented with his peculiar state.

Many authors will hesitate to gather into a unit all the cases enumerated here and will prefer to emphasize the differences rather than the common characters of these groups, a view which corresponds with their preferred judgment of inversions. But no matter what divisions may be set up, it cannot be overlooked that all transitions are abundantly met with, so that the formation of a series would seem to impose itself.

CONCEPTION OF INVERSION. The first attention bestowed upon inversion gave rise to the conception that it was a congenital sign of nervous degeneration. This harmonized with the fact that doctors first met it among the nervous, or among persons giving such an impression. There are two elements which should be considered independently in this conception: the congenitality and the degeneration.

DEGENERATION. This term "degeneration" is open to the objections which may be urged against the promiscuous use of this word in

---

[2] Such a striving against the compulsion to inversion favors influence by suggestion or psychoanalysis.

[3] Many have justly emphasized the fact that the autobiographic statements of inverts, as to the time of the appearance of their tendency to inversion, are untrustworthy, as they may have repressed from memory any evidences of heterosexual feelings. Psychoanalysis has confirmed this suspicion in all cases of inversion accessible, and has decidedly changed their anamnesis by filling up the infantile amnesias.

general. It has in fact become customary to designate all morbid manifestations not of traumatic or infectious origin as degenerative. Indeed, Magnan's classification of degenerates makes it possible that the highest general configuration of nervous accomplishment need not exclude the application of the concept of degeneration. Under the circumstances, it is a question what use and what new content the judgment of "degeneration" still possesses. It would seem more appropriate not to speak of degeneration: (1) where there are not many marked deviations from the normal; (2) where the capacity for working and living do not in general appear markedly impaired.[4]

That the inverted are not degenerates in this qualified sense can be seen from the following facts:

1. The inversion is found among persons who otherwise show no marked deviation from the normal.

2. It is found also among persons whose capabilities are not disturbed and who, on the contrary, are distinguished by especially high intellectual development and ethical culture.[5]

3. If one disregards the patients of one's own practice and strives to comprehend a wider field of experience, he will in two directions encounter facts which will prevent him from assuming inversions as a degenerative sign.

(a) It must be considered that inversion was a frequent manifestation among the ancient nations at the height of their culture. It was an institution endowed with important functions. (b) It is found to be unusually prevalent among savages and primitive races, whereas the term "degeneration" is generally limited to higher civilization (I. Bloch). Even among the most civilized nations of Europe, climate and race have a most powerful influence on the distribution of, and attitude toward, inversion.[6]

INNATENESS. Only for the first and most extreme class of inverts, as can be imagined, has innateness been claimed, and this from their own

[4] With what reserve the diagnosis of degeneration should be made and what slight practical significance can be attributed to it can be gathered from the discussions of Moebius (*Ueber Entartung: Grenzfragen des Nerven- und Seelenlebens*, No. III, 1900). He says: "If we review the wide sphere of degeneration upon which we have here turned some light, we can conclude without further ado that it is really of little value to diagnose degeneration."

[5] We must agree with the spokesman of "Uranism" (I. Bloch) that some of the most prominent men known have been inverts and perhaps absolute inverts.

[6] In the conception of inversion the pathological features have been separated from the anthropological. For this credit is due to I. Bloch (*Beiträge zur Ätiologie der Psychopathia Sexualis*, 2 Teile, 1902–1903), who has also brought into prominence the existence of inversion in the old civilized nations.

assurance that at no time in their life has their sexual impulse followed a different course. The fact of the existence of two other classes, especially of the third, is difficult to reconcile with the assumption of its being congenital. Hence, the propensity of those holding this view to separate the group of absolute inverts from the others results in the abandonment of the general conception of inversion. Accordingly, in a number of cases the inversion would be of a congenital character, while in others it might originate from other causes.

In contradistinction to this conception is that which assumes inversion to be an *acquired* character of the sexual impulse. It is based on the following facts:

(1) In many inverts (even absolute ones) an early affective sexual impression can be demonstrated, as a result of which the homosexual inclination developed.

(2) In many others outer influences of a promoting and inhibiting nature can be demonstrated, which in earlier or later life led to a fixation of the inversion — among which are exclusive relations with the same sex, companionship in war, detention in prison, dangers of heterosexual intercourse, celibacy, genital weakness, and so on.

(3) Hypnotic suggestion may remove the inversion, which would be surprising if it were of a congenital character.

In view of all this, the existence of congenital inversion can certainly be questioned. The objection may be made to it that a more accurate examination of those claimed to be congenitally inverted will probably show that the direction of the libido was determined by a definite experience of early childhood, which has not been retained in the conscious memory of the person but which can be brought back to memory by proper influences (Havelock Ellis). According to this author inversion can be designated only as a frequent variation of the sexual impulse which may be determined by a number of external circumstances of life.

The apparent certainty thus reached is, however, overthrown by the retort that manifestly there are many persons who have experienced even in their early youth those very sexual influences such as seduction, mutual onanism, without becoming inverts, or without constantly remaining so. Hence, one is forced to assume that the alternatives congenital and acquired are either incomplete or do not cover the circumstances present in inversions.

EXPLANATION OF INVERSION. The nature of inversion is explained neither by the assumption that it is congenital nor that it is acquired. In the first case, we need to be told what there is in it of the congenital, unless we are satisfied with the roughest explanation, namely, that a person brings along a congenital sexual impulse connected with a definite sexual object. In the second case it is a question whether the manifold accidental influences suffice to explain the acquisition unless there is something in the individual to meet them halfway. The negation of this last factor is inadmissible according to our former conclusions.

THE APPROACH TO INVERSION. Since the time of Frank Lydston, Kiernan, and Chevalier, a new series of ideas has been introduced for the explanation of the possibility of sexual inversion. These contain a new contradiction to the popular belief which assumes that a human being is either a man or a woman. Science shows cases, however, in which the sexual characteristics appear blurred and thus the sexual distinction is made difficult, especially on an anatomical basis. The genitals of such persons unite the male and female characteristics (hermaphroditism). In rare cases both parts of the sexual apparatus are well developed (true hermaphroditism), but usually both are stunted.[7]

The importance of these abnormalities lies in the fact that they unexpectedly facilitate the understanding of the normal formation. A certain degree of anatomical hermaphroditism really belongs to the normal. In no normally formed male or female are traces of the apparatus of the other sex lacking; these either continue functionless as rudimentary organs, or they are transformed for the purpose of assuming other functions.

The conception which we gather from this long-known anatomical fact is the original predisposition to bisexuality, which in the course of development has changed to monosexuality, leaving slight remnants of the stunted sex.

It was natural to transfer this conception to the psychic sphere and to conceive the inversion in its aberrations as an expression of psychic hermaphroditism. In order to bring the question to a decision, it was necessary to have only one other circumstance, namely, a regular concurrence of the inversion with the psychic and somatic signs of hermaphroditism.

But this second expectation was not realized. The relations between the assumed psychical and the demonstrable anatomical androgyny should never be conceived as being so close. There is frequently found

---

[7] Compare the last detailed discussion of somatic hermaphroditism (Taruffi, *Hermaphroditismus und Zeugungsunfähigkeit*, German edit. by R. Teuscher, 1903), and the works of Neugebauer in many volumes of the *Jahrbuch für sexuelle Zwischenstufen*.

in the inverted a diminution of the sexual impulse (H. Ellis) and a slight anatomical stunting of the organs. This, however, is found frequently but by no means regularly or preponderantly. Thus we must recognize that inversion and somatic hermaphroditism are totally independent of each other.

Great importance has also been attached to the so-called secondary and tertiary sex characters, and their aggregate occurrence in the inverted has been emphasized (H. Ellis). There is much truth in this, but it should not be forgotten that the secondary and tertiary sex characteristics very frequently manifest themselves in the other sex, thus indicating androgyny without, however, involving changes in the sexual object in the sense of an inversion.

Psychic hermaphroditism would gain in substantiality if parallel with the inversion of the sexual object there should be at least a change in the other psychic qualities, such as in the impulses and distinguishing traits characteristic of the other sex. But such inversion of character can be expected with some regularity only in inverted women; in men the most perfect psychic manliness may be united with the inversion. If one firmly adheres to the hypothesis of a psychic hermaphroditism, one must add that in certain spheres its manifestations allow the recognition of only a very slight contrary determination. The same also holds true in the somatic androgyny. According to Halban, the appearance of individual stunted organs and secondary sex characters are quite independent of each other.[8]

A spokesman of the masculine inverts stated the bisexual theory in its crudest form in the following words: "It is a female brain in a male body." But we do not know the characteristics of a "female brain." The substitution of the anatomical for the psychological is as frivolous as it is unjustified. The tentative explanation by v. Krafft-Ebing seems to be more precisely formulated than that of Ulrich but does not essentially differ from it. Krafft-Ebing thinks that the bisexual predisposition supplies the individual with male and female brain centers as well as with the somatic sexual organs. These centers develop first toward puberty mostly under the influence of the independent sex glands. We can, however, say the same of the male and female "centers" as of the male and female brains; and, moreover, we do not even know whether we can assume for the sexual functions separate brain locations ("centers") such as we may assume for language.

---

[8] J. Halban, "Die Entstehung der Geschlechtscharaktere," *Arch. für Gynäkologie,* Bd. 70, 1903. See also there the literature on the subject.

At all events, after this discussion two notions persist; first, that a bisexual predisposition is to be presumed for the inversion also; only, we do not know of what it consists beyond the anatomical formations; and, second, that we are dealing with disturbances which are experienced by the sexual impulse during its development.[9]

THE SEXUAL OBJECT OF THE INVERT. The theory of psychic hermaphroditism presupposed that the sexual object of the inverted is the reverse of the normal. The inverted man, like the woman, succumbs to the charms emanating from manly qualities of body and mind; he feels like a woman and seeks a man.

But however true this may be for a great number of inverts, it by no means indicates the general character of inversion. There is no doubt that a great part of the male inverted have retained the psychic character of virility, that proportionately they show but little of the secondary characters of the other sex, and that they really look for real feminine psychic features in their sexual object. If that were not so it would be incomprehensible why masculine prostitution, in offering itself to inverts, copies in all its exterior, today as in antiquity, the dress and attitudes of woman. This imitation would otherwise be an insult to the ideal of the inverts. Among the Greeks, where the most manly men were found among inverts, it is quite obvious that it was not the masculine character of the boy which kindled the love of man; it was

---

[9] According to a report in Vol. VI of the *Jahrbuch für sexuelle Zwischenstufen*, E. Gley is supposed to have been the first to mention bisexuality as an explanation of inversion. He published a paper ("Les abérrations de l'instinct sexuel") in the *Revue Philosophique* as early as January, 1884. It is, moreover, noteworthy that the majority of authors who trace inversion back to bisexuality assume this factor not only for the inverts but also for those who have developed normally, and justly interpret the inversion as a result of disturbance in development. Among these authors are Chevalier (*Inversion Sexuelle*, 1893) and v. Krafft-Ebing ("Zur Erklärung der konträren Sexualempfindung," *Jahrbücher für Psychiatrie u. Nervenheilkunde*, XIII), who states that there are a number of observations "from which at least the virtual and continued existence of this second center (of the underlying sex) results." A Dr. Arduin (*Die Frauenfrage und die sexuellen Zwischenstufen*, Vol. II of the *Jahrbuch für sexuelle Zwischenstufen*, 1900) states that "in every man there exist male and female elements." See also the same *Jahrbuch*, Bd. I, 1899 ("Die objektive Diagnose der Homosexualität," by Mr. Hirschfeld, pp. 8–9). In the determination of sex, as far as heterosexual persons are concerned, some are disproportionately more strongly developed than others. G. Herman is firm in his belief "that in every woman there are male and in every man there are female germs and qualities" ("Genesis, das Gesetz der Zeugung," Bd. IX, *Libido und Manie*, 1903). As recently as 1906 W. Fliess *(Der Ablauf des Lebens)* has claimed ownership of the idea of bisexuality (in the sense of double sex). In noninformed circles the assertion is made that the recently deceased philosopher O. Weininger is the authority for the human bisexuality conception, since this notion is made the foundation of his rather rash work (*Geschlecht und Charakter*, 1903 [translated into English]). The citations here made show how unfounded is such a claim.

his physical resemblance to woman as well as his feminine psychic qualities, such as shyness, demureness, and the need of instruction and help. As soon as the boy himself became a man he ceased to be a sexual object for men and in turn became a lover of boys. The sexual object in this case as in many others is therefore not of the like sex, but unites both sex characters, a compromise between the impulses striving for the man and for the woman, but firmly conditioned by the masculinity of body (the genitals).[10]

[10] Although psychoanalysis has not yet given us a full explanation for the origin of inversion, it has revealed the psychic mechanism of its genesis and has essentially enriched the problems in question. In all the cases examined we have ascertained that the later inverts go through in their childhood a phase of very intense but short-lived fixation on the woman (usually on the mother), and after overcoming it they identify themselves with the woman and take themselves as the sexual object; that is, proceeding on a narcissistic basis, they look for young men resembling themselves in persons whom they wish to love as their mother has loved them. We have, moreover, frequently found that alleged inverts are by no means indifferent to the charms of women, but the excitation evoked by the woman is always transferred to a male object. They thus repeat through life the mechanism which gave origin to their inversion. Their obsessive striving for the man proves to be determined by their restless flight from the woman. Psychoanalytic research very strongly opposes the attempt to separate homosexuals from other persons as a group of a special nature. By also studying sexual excitations other than the manifestly open ones, it discovers that all men are capable of homosexual object selection and actually accomplish this in the unconscious. Indeed the attachments of libidinous feelings to persons of the same sex play no small role as factors in normal psychic life, and as causative factors of disease they play a greater role than those belonging to the opposite sex. According to psychoanalysis, it rather seems that it is the independence of the object, selection of the sex of the object, the same free disposal over male and female objects, as observed in childhood, in primitive states and in prehistoric times, which forms the origin from which the normal as well as the inversion types developed, following restrictions in this or that direction. In the psychoanalytic sense the exclusive sexual interest of the man for the woman is also a problem requiring an explanation, and is not something that is self-evident and explainable on the basis of chemical attraction. The determination as to the definite sexual behavior does not occur until after puberty, and is the result of a series of as yet not observable factors, some of which are of a constitutional, while some are of an accidental, nature. Certainly some of these factors can turn out to be so enormous that by their character they influence the result. In general, however, the multiplicity of the determining factors is reflected by the manifoldness of the outcomes in the manifest sexual behavior of the person. In the inversion types it can be ascertained that they are altogether controlled by an archaic constitution and by primitive psychic mechanisms. The importance of the *narcissistic object selection* and the *clinging* to the erotic significance of the *anal* zone seem to be their most essential characteristics. But one gains nothing by separating the most extreme inversion types from the others on the basis of such constitutional peculiarities. What is found in the latter as seemingly an adequate determinant can also be demonstrated only in lesser force in the constitution of transitional types and in manifestly normal persons. The differences in the results may be of a qualitative nature, but analysis shows that the differences in the determinants are only quantitative. As a remarkable factor among the accidental influences of the object selection, we found the sexual rejection or the early sexual intimidation, and our attention was also called to the fact that the existence of both parents plays

The conditions in the woman are more definite; here the active inverts, with special frequency, show the somatic and psychic characters of man and desire femininity in their sexual object; though even here greater variation will be found on more intimate investigation. THE SEXUAL AIM OF THE INVERT. The important fact to bear in mind is that no uniformity of the sexual aim can be attributed to inversion. Intercourse per anum in men by no means goes with inversion; masturbation is just as frequently the exclusive aim; and the limitation

---

an important role in the child's life. The disappearance of a strong father in childhood not infrequently favors the inversion. Finally, one can put forth the claim that the inversion of the sexual object should notionally be strictly separated from the mixing of the sex characteristics in the subject. A certain degree of independence is unmistakable also in this relation. A series of important points of view concerning the question of inversion have been brought forward by Ferenczi (in a contribution, "Zur Nosologie der männlichen Homosexualität" [Homoerotic], in *Int. Zeit. f. Psa.,* II, 1914). Published in English in *Contribution to Psychoanalysis,* I (Bodger, Boston, 1916). Ferenczi correctly criticizes the fact that under the term Homosexuality (which term we would replace by the better one Homoerotic) a number of different conditions are grouped which are of quite variable significance both from an organic as well as from a psychical viewpoint because the one symptom of inversion is present. He shows that there are but four very marked differences at least between two types of subject-homoerotics, who feel and act like women, and the object-homoerotic who is masculine throughout and has only (mistakenly) exchanged a female object against one of the same sex. The first he recognizes as a true "intermediary sexual stage" in the sense of Magnus Hirschfeld; the second he terms — less fortunately — a compulsion neurotic. The striving against the tendency to inversion as well as the possibility of psychical influence is possible only with the object homoerotic. It may also be added that after the recognition of these two types in many individuals a certain amount of subject homoeroticism is found mixed with a portion of object homoeroticism.

Of recent years biological workers, especially Eugen Steinach, have thrown a clear light upon the organic conditionings of homoerotism as well as upon sexual characters. Through the experimental procedure of castration followed by implanting the gonads of the opposite sex, he was able in different mammals to change males into females and vice versa. The change concerns more or less completely the somatic sexual characters and the psychosexual behavior (as subject- and object-erotic). The carriers of this sex-determining power are not that portion of the sexual glands which builds up the sexual cells but the so-called interstitial cells of the organs (the puberty glands).

In one case the sexual alteration took place in a man whose testicles had been damaged by tuberculosis. In his sexual life he had behaved as a passive homosexual woman, and showed very clearly marked secondary female sexual characters (hair distribution, nature of facial hair, fatty mammæ, and female hips). Following the implantation of a cryptorchids testicle, this man began to behave as a man and directed his libido toward the female in the normal manner. At the same time the somatic female sex character disappeared (A. Lipschütz, *Die Pubersätsdrüse und ihre Wirkungen* [Bern, 1919]).

It would be unjust to maintain that the knowledge of inversion is placed on a new basis, and premature to expect directly a way to the cure of homosexuality. W. Fliess has correctly accented the fact that this experimental experience does not solve the problem of the general bisexual anlage of the higher animals. It seems to me much more probable that a direct confirmation of the accepted bisexuality will come from such and further investigations.

of the sexual aim to mere effusion of feelings is here even more frequent than in heterosexual love. In women, too, the sexual aims of the inverted are manifest, among which contact with the mucous membrane of the mouth seems to be preferred. CONCLUSION. Though from the material on hand we are by no means in a position to explain satisfactorily the origin of inversion, we can say that through this investigation we have obtained an insight which can become of greater significance to us than the solution of the above problem. Our attention is called to the fact that we have assumed a too close connection between the sexual impulse and the sexual object. The experience gained from the so-called abnormal cases teaches us that a connection exists between the sexual impulse and the sexual object which we are in danger of overlooking in the uniformity of normal states where the impulse seems to bring with it the object. We are thus instructed to separate this connection between the impulse and the object. The sexual impulse is probably entirely independent of its object and is not originated by the stimuli proceeding from the object. . . .

MARGARET C. ANDERSON

From *Forbidden Fires*

Born into a well-educated Chicago family, Margaret C. Anderson (1886–1973) is best known for creating and editing the *Little Review* in 1914. The avant-garde magazine, with its maxim "Life for Art's sake," was begun in Chicago and grew notorious once it was based in Greenwich Village, New York City. It moved away from some of its anarchist and feminist connections, and the influence of Anderson's co-editor, Jane Heap, began to change the journal into a more experimental publication. Ezra Pound, too, played a role in its development, and secured funding for it from the lawyer John Quinn. Publishing chapters from James Joyce's *Ulysses* between 1918 and 1921 kept the magazine under the fire of the Comstock laws, and in 1921 at a Special Session Court in New York City, Anderson and Heap were found guilty of publishing obscene material (*Ulysses*) and fined $100. The Society for the Suppression of Vice had won. By 1923, however, the *Little Review* and its editors were established in Paris, where the magazine flourished until 1929. Anderson recounts the story in her memoir, *My Thirty Years War.*

As Anderson, Heap, Solita Solano, Janet Flanner, and Georgette LeBlanc formed various lesbian liaisons and partnerships, their households became salons that welcomed lesbians and gays. Much of Anderson's *Forbidden Fires* is the story of Georgette LeBlanc's leaving her relationship with playwright Maurice Maeterlinck in order to establish her love with Anderson. Given the secrecy of the affair and the fact that this written account remained unpublished until 1996, the reader gains insight into the kind of repression that was considered necessary, even in Paris during the liberal 1920s.

Similarly, even though many of the writers, composers and painters who frequented the Paris salon of Gertrude Stein and Alice B. Toklas were themselves lesbian or homosexual, none of them wrote about the Stein-Toklas relationship until years after Stein's death in 1946.

The Anderson memoir was edited by Mathilda M. Hills and published by Naiad Press, Tallahassee, Florida, in 1996.

Chapter III

The next blessing that a kind fate held in store for me was a new love — the love of a country, an old and mythical country in which, as it turned out, I was to live for the rest of my life.

For years, whenever I heard people talking about France, I had felt a thrill of nostalgia for this luminous country I had never seen. It was as if I knew in advance that Paris would be for me, as it had been for George Moore, "that white city to which we all come as beggars."

Kaye knew Paris well, had even lived there, and regarded it as the world's magical city. Therefore she agreed entirely with my idea that we should one day adopt it as our own. We began making plans to go — the kind of plans that are so wild, so intricate, and so difficult that one can't remember them afterward, or even remember how one had the courage to attempt them. But after a year's struggle — and just four years after Kaye and I had met — we were ready to leave New York. At last, one wonderful day, we stood on the deck of a ship — two beggars on their way to Le Havre . . . and the white city.

And Paris was not far from London.

The first time I saw Paris! . . . This, I thought, is the city where I belong, where I should have been born. I was at last to know patriotism, which had always been only a word to me.

The sanctuary of France . . . what was the protection it offered? What was its promise? "Rescued" was the word I repeated most often as Kaye and I walked the streets of Paris and drank in the sights and sounds of the unbelievable city. The sudden beauty of a balcony's grill-work, a faultless building, the courtyard gardens, the winding Seine, the sense of light and freedom . . .

Rescued from *what*, rescued *for* what? Rescued from sameness, from the regulated, the dull, the alien, the arid, the incomplete, the inarticulate, from what Henry James called "the whimsical retention of speech which is such a common form of American sociability." And for what? First of all for leisured living, for the cult of beauty, for the civilized, for give and take, for the culture that could always be taken for granted, for native brilliance and wit, for the gaiety that rests on seriousness, for the Gallic value that is placed on grace of manner.

The gods who were still directing my destiny produced the next miracle. I was taken by some English friends to a small left-bank theater where a famous *diseuse* (singer, actress, writer) was talking about the new French poets. Afterward we went to her flat in the rue Barbet de Jouy, near the Musée Rodin, and it was on this night of poetry that the great friendship of my life had its beginning.

My first impression of Claire Lescaut was that she possessed some special human inspiration, some illumination, that was a mystery which only time and knowledge would help me to understand. This quality was so evident that one almost stopped thinking of how beautiful her face was; I found that I was responding above all to the touching charm in which her personality was enveloped — a personality, I felt, of such strength and sweetness that it could only be founded on some great purity of nature.

When the *soirée* ended and we left her flat, I searched for words to express the emotions I had been feeling. The only ones I found were these: "She is like a stronghold." Trust in the purity of a nature . . . this is the quality which was to lead to a deathless love.

She spoke no English, and my college French was pitiable; I could scarcely remember a word of the language I was supposed to have learned. However, I was determined to communicate, so I invented a language that existed somewhere above the rational world but which Claire Lescaut could understand. Though she was as well-known for the grace of her prose as for her exquisite diction, she was no pedant, and she was so entertained by my non-grammar (I always spoke in the present tense) and my incredible accent that she refused to correct me.

It wasn't long before I was able to talk about Freud (then in great vogue) in a French that was primitive but clear.

"What I like about Americans," Claire told me, "is their instant ability to show their enthusiasm, and their disarming frankness in presenting their personalities. We are old and formal, we are always polite, we are *plutôt* reserved, so we respond to the way Americans are so immediately able to reveal their inner lives." And she added, "The Russians have the same talent."

As one always does in France, we talked a great deal about beauty. One of my English friends was irritated by the French mania for speaking of beauty before other qualities. "Don't they ever think of anything else?" she said. "I think it's silly to put so much emphasis on how people look, rather than on what they are."

"Tell her," Claire said, "that we speak of it first because it is the first observation that can be honestly made — the visual. It would be pretentious to speculate, at once, on someone's inner life. That becomes visible soon afterward, through a hundred little revelations. But beauty has such importance, physically as well as psychically. How can anyone be insensitive to it, or untrained to recognize it? Ah, the English!"

It wasn't long before I told her about my arrested adolescent experience of romantic love. How sympathetic she was, how amused, how informed. "But you must see her again," she said. "Perhaps she will be kinder to you next time."

Ah, I thought, *this* is civilization: life as it is, simply and really, no matter in what realm; no inhibitions, no need for Freud, no self-consciousness, no stoppages. "You will achieve what you are able to inspire," Claire went on. "But you must know how to proceed, you must learn the value of *prélude*. Naiveté won't help you, or candor, or simplicity — except with people who are highly organized. In this case you must learn to understand nineteenth-century England."

I discovered that in France nothing that was genuine in human experience was outlawed; nothing was discounted — at least in the artist milieu — except the uneasy, the strained, the unknowing, the bourgeois. And the French attitude toward sex eased my emotional tensions. No one was shocked by sex classifications — they spoke of *pédérastes* without any of the American leer at "fairies" or "pansies"; they looked upon lesbians as a race of charming people. These tendencies were not regarded as aberrations but as non-conformities; the unusual, the rare — perhaps the unaccountable; singular, but not abnormal; anomalous, perhaps, but original; in any case, exceptional and interesting.

"Androgynous," Claire said. "Your Audrey should have understood."

"But the word 'androgyne' isn't in common usage in America," I said. "Americans don't discuss these differentiations; they talk only of 'normal' sex; for sex variations they have only contempt, pity, hatred or ribaldry."

France had André Gide and others who wrote frankly about their natures. But though the French bourgeois tended to loathe such types, you could always talk with him easily about them as you couldn't with bourgeois Americans. Claire even told me that the subject was sometimes discussed in public debate: a well-known American ex-patriate, seconded by a titled Frenchwoman, had defended lesbianism on the lecture platform without causing the slightest scandal.

I was intrigued by the story about Verlaine who, always in financial difficulties, was urged by friends to ask for governmental assistance. He wrote to the Président du Conseil: "*Je suis pauvre, je suis poète, je suis pédéraste*"[1] . . . considering that the last qualification was sure to draw sympathetic attention to his difficulties. And I became quite used to Frenchwomen who, when they wanted to advertise their love affairs, said to you proudly, "*Oui, je suis Vénusienne.*"[2]

"No one talks like this in America," was all I could say. "It's a rest for the mind and its superfluous conflicts."

[1] "*Je suis . . . pédéraste*": "I am poor, I am a poet, I am a pederast." (Translation from the French.)
[2] *Vénusienne:* A Vénusienne is a lover of women, a lesbian.

HAVELOCK ELLIS

From *The Mechanism of Detumescence*

British psychologist Havelock Ellis (1859–1939) was one of the first sexologists, marking that field of study with his six-volume *Studies in the Psychology of Sex*, published between 1896 and 1928. Less theoretical than works by Richard von Krafft-Ebing or other nineteenth-century experts in this emerging area, Ellis's books often provided precise information about arousal, intercourse, masturbation, contraception, and various sexual pairings; they also correlated sexual imagery with its uses in literature and art. More important, they defined homosexuality as a custom resulting from societal and cultural pressures, and gave it the term *inversion*. Underlying Ellis's work was the notion that homosexuality was pathological.

His use of the case study method made his immense volumes somewhat readable, and it is his work — rather than that of Sigmund Freud — that reached a comparatively large general readership. Although trained as a physician, Ellis also wrote poetry and essays, and moved in international literary circles. His cachet as writer removed some of the censure from public reaction to his major work, the first volume of which was banned on charges of obscenity. Other books include *A Study of British Genius* (1904), *The Dance of Life* (1923), *Man and Woman* (revised 1934), and *My Life* (1940).

There is at least one mention of Havelock Ellis having visited the Stein salon early in the twentieth century, when Leo and Gertrude's late-evening salon followed the earlier reception in Mike and Sally Stein's house a few blocks away. Both households showed the art that made Paris a collector's paradise — pieces by Juan Gris, Pablo Picasso, Henri Matisse, Paul Cézanne, and countless other painters.

This text is excerpted from *Studies in the Psychology of Sex,* Vol. II (New York: Random House, 1906).

. . . In the woman the specifically sexual muscular process is less visible, more obscure, more complex, and uncertain. Before detumescence actually begins there are at intervals involuntary rhythmic contractions of the walls of the vagina, seeming to have the object of at once stimulating and harmonizing with those that are about to begin in the male organ. It would appear that these rhythmic contractions are the exaggeration of phenomenon which is normal, just as slight contraction is normal and constant in the bladder. Jastreboff has shown, in the rabbit, that the vagina is in constant spontaneous rhythmic contraction from above downward, not peristaltic, but in segments, the intensity of the contractions increasing with age and especially with sexual development. This vaginal contraction which in women only becomes well marked just before detumescence, and is due mainly to the action of the sphincter cunni (analogous to the bulbo-cavernosus in the male), is only a part of the localized muscular process. At first there would appear to be a reflex peristaltic movement of the Fallopian tubes and uterus. Dembo observed that in animals stimulation of the upper anterior wall of the vagina caused gradual contraction of the uterus, which is erected by powerful contraction of its muscular fiber and round ligaments while at the same time it descends toward the vagina, its cavity becoming more and more diminished and mucus being forced out. In relaxing, Aristotle long ago remarked, it aspirates the seminal fluid.

Although the active participation of the sexual organs in woman, to the end of directing the semen into the womb at the moment of detumescence, is thus a very ancient belief, and harmonizes with the Greek view of the womb as an animal in the body endowed with a considerable amount of activity,[1] precise observation in modern times has offered but little confirmation of the reality of this participation. Such observations as have been made have usually been the accidental result of sexual excitement and orgasm occurring during a gynæcological examination. As, however, such a result is liable to occur in erotic subjects, a certain number of precise observations have accumulated during the past century. So far as the evidence goes, it would seem that in women, as in mares, bitches, and other animals, the uterus becomes shorter, broader, and softer during the orgasm, at the same time descending lower into the pelvis, with its mouth open intermittently, so that, as one writer remarks, spontaneously recurring to the simile which commended itself to the Greeks, "the uterus might be likened to an animal gasping for breath."[2] This sensitive, responsive mobility of the uterus is, indeed, not confined to the moment of detumescence, but may occur at other times under the influence of sexual emotion.

It would seem probable that in this erection, contraction, and descent of the uterus, and its simultaneous expulsion of mucus, we have the decisive moment in the completion of detumescence in woman, and it is probable that the thick mucus, unlike the earlier more limpid secretion, which women are sometimes aware of after orgasm, is emitted from the womb at this time. This is, however, not absolutely certain. Some authorities regard detumescence in women as accomplished in the pouring out of secretions, others in the rhythmic genital contractions; the sexual parts may, however, be copiously bathed in mucus for an indefinitely long period before the final stage of detumescence is achieved, and the rhythmic contractions are also taking place at a somewhat early period; in neither respect is there any obvious increase at the final moment of orgasm. In women this would seem to be more

Note: the footnotes in this selection are original to the selection.

[1] *Cf.* the discussion of hysteria in "Auto-Erotism," vol. i of these *Studies.*

[2] Hirst, *Text-Book of Obstetrics,* 1899, p. 67.

conspicuously a nervous manifestation than in men. On the subjective side it is very pronounced, with its feeling of relieved tension and agreeable repose — a moment when, as one woman expresses it, together with intense pleasure, there is, as it were, a floating up into a higher sphere, like the beginning of chloroform narcosis — but on the objective side this culminating moment is less easy to define. . . .

While the active part played by the womb in detumescence can no longer be questioned, it need not too hastily be assumed that the belief in the active movements of the spermatozoa must therefore be denied. The vigorous motility of the tadpole-like organisms is obvious to anyone who has ever seen fresh semen under the microscope; and if it is correct, as Clifton Edgard states, that the spermatozoa may retain their full activity in the female organs for at least seventeen days, they have ample time to exert their energies. The fact that impregnation sometimes occurs without rupture of the hymen is not decisive evidence that there has been no penetration, as the hymen may dilate without rupturing; but there seems no reason to doubt that conception has sometimes taken place when ejaculation has occurred without penetration; this is indicated in a fairly objective manner when, as has been occasionally observed, conception has occurred in women whose vaginas were so narrow as scarcely to admit the entrance of a goosequill; such was the condition in the case of a pregnant woman brought forward by Roubaud. The stories, repeated in various books, of women who have conceived after homosexual relations with partners who had just left their husbands' beds are not therefore inherently impossible.[3] Janke quotes numerous cases in which there has been impregnation in virgins who have merely allowed the penis to be placed in contact with the vulva, the hymen remaining unruptured until delivery.[4]

It must be added, however, that even if the semen is effused merely at the mouth of the vagina, without actual penetration, the spermatozoa are still not entirely without any resource save their own motility in the task of reaching the ovum. As we have seen, it is not only the uterus which takes an active part in detumescence; the vagina also is in active movement, and it seems highly probable that, at all events in some women and under some circumstances, such movement favoring aspiration toward the womb may be communicated to the external mouth of the vagina. . . .

[3] The earliest story of the kind with which I am acquainted, that of a widow who was thus impregnated by a married friend, is quoted in Schurig's *Spermatologia* (p. 224) from Amatus Lusitanus, *Curationum Centuriæ Septem*, 1620.
[4] Janke, *Die Willkürliche Hervorbringen des Geschlechts*, p. 238.

Representative of the anti-Semitic caricatures published throughout Europe and the States in the late nineteenth century, Hermann-Paul's *Jeune Vierge* (Young Virgin) portrays physical characteristics, clothing, and hairstyle that would have been recognized as Jewish. Reproduced by permission of the Jane Voorhees Zimmerli Art Museum (Rutgers).

A manuscript page from Gertrude Stein's Radcliffe essay "The modern Jew who has given up the faith of his fathers can reasonably and consistently believe in isolation," which she wrote for English C, George Pierce Baker's forensics course (in which she received an A–). Reproduced by permission of The Yale Collection of American Literature, Beinecke Rare Book and Manuscript Library, Yale University.

GERTRUDE STEIN

From *The Modern Jew Who Has Given Up the Faith of His Fathers Can Reasonably and Consistently Believe in Isolation*

Before trying to substantiate the proposition that we are going to discuss in this paper, let us make an attempt to come to some kind of an agreement as to what we mean by our terms. First what do we mean by the Jewish Faith.

If we ask a number of Jews to tell him what the Jewish Faith really means, he gets answers as diverse as are the number of the questioned. On the one hand a man laying stress only on form will reply, that only a man who keeps the law believes in Judaism. If he rides on Saturday, he is no Jew, if he eat of meat not Kosher (unclean, not killed according to law) he is no Jew. Another will tell you that the observance of Yom Kipper (the day of atonement) is the last bulwark of religion, another will cite still some other custom as being the final mark of faith and so on. The only unanimity in this class is that Judaism depends on law, custom and observances. On the other hand we have the extreme liberal, the so-called Reform Jew who tells you that as long as you are born a Jew and believe in a God all is well, another tells you that even a belief in a personal God is not essential to the Jewish Faith, if you only believe in a Matthew Arnold sort of force that makes for righteousness, it will suffice. . . .

The modern Jew has to-day in great measure departed from the faith of his fathers. This is due to several causes. Firstly, because the general skeptical spirit prevalent at this period of the world's history among all classes has powerfully affected the Jewish community for, owing to the high average of brain power and the extreme formalism of their religion, they naturally have a strong tendency to embrace revolutionary ideas and skepticism in all its forms. The more religiously inclined temperamentally have also from a different stand-point begun to depart from their old faith. They are beginning to change not because of skepticism but because the strenuous religiousness of their souls make them wish to escape from a religion which has become a hard-shell of formalism with all the soul fled and the living substance gone.

Added to these spiritual causes for desertion is the practical impossibility in this bustling nineteenth-century life to keep up with the observance of the Saturday Sabbath, the holidays, the fasts, the long prayers, the particular preparation of foods and the care of the household utensils according to the law, in short, all those family ceremonials that have done so much to hold the race together.

These tendencies are today rapidly bringing the condition and future of the race to a crucial point. The Jew has become now, through his financial ability and his great clannishness, a great power, and in consequence, while, on the one hand, the doors of Christendom have been a little opened to him and the Gentile has become glad to have him for his friend, on the other, the great influence of the race is raising against them a new wave of the old prejudices and we seem to be on the eve of a worse anti-semitic crisis than ever before. To illustrate this one need but to draw the attention of the reader to the spirit prevalent in Germany, the recent anti-semitic riot in Paris and the not so very distant exodus in Russia.

In the first place the prejudice against the race will never die down as long as they keep separate. A wealthy race holding itself apart from the rest of the nation they belong to is too convenient a political scapegoat to ever hope to be left alone. So long as the Jews keep themselves isolated so long are they bound to be subject to persecution to a greater or less extent. We must ask whether they gain enough by this exclusion to make it worth while to be in this attitude of separateness and persecution.

Yes. I think they do. To be a Jew is of great practical benefit. Wherever a Jew goes no matter into what strange lands and he meets another Jew, he has found a friend. Appeal to a Jew in behalf of another Jew and he will never say you nay. Ask any Israelite no matter how liberal, no matter how numerous and intimate are his Christian friends; ask him to tell you to whom he would rather appeal if he were in any need either spiritual or material, whether he would rather go to a perfect stranger a Jew or to his most intimate Christian friend and without hesitation he will reply, "To the Jew every time.". . .

OTTO WEININGER

"Judaism" and *"Woman and Her Significance,"*
from Sex and Character

A gifted student, adept at languages, Weininger (1880–1903) studied at the Royal and Imperial State Gymnasium at the Piarist Cloister in Vienna. A top student of Wilhelm Jerusalem, he took the Ph.D. degree in philosophy at the University of Vienna, graduating in 1902. (The day of his graduation he converted to Protestantism, after being reared a practicing Jew.) He then toured Europe before publishing his book, *Geschlecht und Charakter* (Sex and character), from which these excerpts are taken.

In his work, Weininger identified traits as either masculine (m) or feminine (f), and insisted that each human being comprised a mix of both types of characteristics. Unfortunately, to prove his contentions, he thought it necessary to prove women's "moral inferiority." Paralleling that inferiority was that of the Jewish people, according to Weininger, though the force of his rejection of both women and Jews seemed to go beyond his customary dogmatism.

So influential was his study of the sexual base of human character that his ideas continued long after his suicide the following year. Ford Madox Ford reported that everyone in Europe was talking about *Sex and Character,* and evidence of their regard for the work appears in writing by Antonin Artaud, Sigmund Freud, Karl Kraus, Franz Kafka, Ludwig Wittgenstein, August Strindberg, James Joyce, D. H. Lawrence, and Gertrude Stein. Stein was said to have sent copies of *Sex and Character* to her friends in the States, but extant letters show that they were less impressed with Weininger than European readers were.

These selections are taken from the United States publication of the book, translated from the German and published by G. P. Putnam's Sons in 1908.

Judaism

The Jewish race, which has been chosen by me as a subject of discussion, because, as will be shown, it presents the gravest and most formidable difficulties for my views, appears to possess a certain anthropological relationship with both negroes and Mongolians. The readily curling hair points to the negro; admixture of Mongolian blood is suggested by the perfectly Chinese or Malay formation of face and skull which is so often to be met with amongst the Jews and which is associated with a yellowish complexion. This is nothing more than the result of everyday experience, and these remarks must not be otherwise understood; the anthropological question of the origin of the Jewish race is apparently insoluble, and even such an interesting answer to it as that given by H. S. Chamberlain has recently met with much opposition. The author does not possess the knowledge necessary to treat of this; what will be here briefly, but as far as pos≠sible profoundly analysed, is the psychical peculiarity of the Jewish race.

This is an obligatory task imposed by psychological observation and analysis. It is undertaken independently of past history, the details of which must be uncertain. The Jewish race offers a problem of the deepest significance for the study of all races, and in itself it is intimately bound up with many of the most troublesome problems of the day.

I must, however, make clear what I mean by Judaism; I mean neither a race nor a people nor a recognised creed. I think of it as a tendency of the mind, as a psychological constitution which is a possibility for all mankind, but which has become actual in the most conspicuous fashion only amongst the Jews. Antisemitism itself will confirm my point of view. . . .

Thus the fact is explained that the bitterest Antisemites are to be found amongst the Jews themselves. For only the quite Jewish Jews, like the completely Aryan Aryans, are not at all Antisemitically disposed; among the remainder only the commoner natures are actively Antisemitic and pass sentence on others without having once sat in judgment on themselves in these matters; and very few exercise their Antisemitism first on themselves. This one thing, however, remains none the less certain: whoever detests the Jewish disposition detests it first of all in himself; that he should persecute it in others is merely his endeavour to separate himself in this way from Jewishness; he strives to shake it off and to localise it in his fellow-creatures, and so for a moment to dream himself free of it. Hatred, like love, is a projected phenomenon; that person alone is hated who reminds one unpleasantly of oneself.

The Antisemitism of the Jews bears testimony to the fact that no one who has had experience of them considers them loveable — not even the Jew himself; the Antisemitism of the Aryans grants us an insight no less full of significance: it is that the Jew and the Jewish race must not be confounded. There are Aryans who are more Jewish than Jews, and real Jews who are more Aryan than certain Aryans. I need not enumerate those non-semites who had much Jewishness in them, the lesser (like the well-known Frederick Nicolai of the eighteenth century) nor those of moderate greatness (here Frederick Schiller can scarcely be omitted), nor will I analyse their Jewishness. Above all Richard Wagner — the bitterest Antisemite — cannot be held free from an accretion of Jewishness even in his art, however little one be misled by the feeling which sees in him the greatest artist enshrined in historical humanity; and this, though indubitably his Siegfried is the most un-Jewish type imaginable. . . .

As there is no real dignity in women, so what is meant by the word "gentleman" does not exist amongst the Jews. The genuine Jew fails in this innate good breeding by which alone individuals honour their own individuality and respect that of others. There is no Jewish nobility, and this is the more surprising as Jewish pedigrees can be traced back for thousands of years.

The familiar Jewish arrogance has a similar explanation; it springs from want of true knowledge of himself and the consequent overpowering need he feels to enhance his own personality by depreciating that of his fellow-creatures. And so, although his descent is incomparably longer than that of the members of Aryan aristocracies, he has an inordinate love for titles. The Aryan respect for his ancestors is rooted in the conception that they were *his* ancestors; it depends on his valuation of his own personality, and, in spite of the communistic strength and antiquity of the Jewish traditions, this individual sense of ancestry is lacking.

The faults of the Jewish race have often been attributed to the repression of that race by Aryans, and many Christians are still disposed to blame themselves in this respect. But the self-reproach is not justified. Outward circumstances do not mould a race in one direction, unless there is in the race the innate tendency to respond to the moulding forces; the total result comes at least as much from the natural disposition as from the modifying circumstances. We know now that the proof of the inheritance of acquired characters has broken down, and, in the human race still more than the lower forms of life, it is certain that individual and racial characters persist in spite of all adaptive moulding. When men change, it is from within, outwards, unless the change, as in the case of women, is a mere superficial imitation of real change, and is not rooted in their natures. And how can we reconcile the idea that the Jewish character is a modern modification with the history of the foundation of the race, given in the Old Testament without any disapprobation of how the patriarch Jacob deceived his dying father, cheated his brother Esau and over-reached his father-in-law, Laban?

The defenders of the Jew have rightly acquitted him of any tendency to heinous crimes, and the legal statistics of different countries confirm this. The Jew is not really anti-moral. But, none the less, he does not represent the highest ethical type. He is rather non-moral, neither very good nor very bad, with nothing in him of either the angel or the devil. Notwithstanding the Book of Job and the story of Eden, it is plain that the conceptions of a Supreme Good and a Supreme Evil are not truly Jewish; I have no wish to enter upon the lengthy and controversial topics of Biblical criticism, but at the least I shall be on sure ground when I say that these conceptions play the least significant part in modern Jewish life. Orthodox or unorthodox, the modern Jew does not concern himself with God and the Devil, with Heaven and Hell. If he does not reach the heights of the Aryan, he is also less inclined to commit murder or other crimes of violence.

So also in the case of the woman; it is easier for her defenders to point to the infrequency of her commission of serious crimes than to prove her intrinsic morality. The homology of Jew and woman becomes closer the further examination goes. There is no female devil, and no female angel; only love, with its blind aversion from actuality, sees in woman a heavenly nature, and only hate sees in her a prodigy of wickedness. Greatness is absent from the nature of the woman and the Jew, the greatness of morality, or the greatness of evil. In the Aryan man, the good and bad principles of Kant's religious philosophy are ever present, ever in strife. In the Jew and the woman, good and evil are not distinct from one another. . . .

Woman and Her Significance

Woman's incapacity for truth — which I hold to be consequent on her lack of free will with regard to the truth, in accordance with Kant's "Indeterminism" — conditions her falsity. Any one who has had anything to do with women knows how often they give offhand quite patently untrue reasons for what they have said or done, under the momentary necessity of answering a question. It is, however, hysterical subjects who are most careful to avoid unveracity (in a most marked and premeditated way before strangers); but however paradoxical it may sound it is exactly in this that their untruthfulness lies! They do not know that this desire for truth has come to them from outside and is no part of their real nature.

Gertrude Stein found Harvard and Radcliffe much less prejudiced than Johns Hopkins, but this cover of *The Harvard Lampoon* from January 11, 1900, suggests that being Jewish was seen as a comic situation even in Cambridge, Massachusetts. Reproduced by permission of The Harvard University Archives.

Hermann-Paul's *Le Rire: La belle juive va aux provisions* (A Chuckle: The Beautiful Jewess Goes Shopping), 1896, deepens the satire by having the Jewish woman intent on succeeding in her shopping endeavors. Reproduced by permission of the Jane Voorhees Zimmerli Art Museum (Rutgers).

They have slavishly accepted the postulate of morality, and, therefore, wish to show at every opportunity, like a good servant, how faithfully they follow instructions. . . .

The untruthfulness of the hysterical is proportional to their belief in their own accuracy. From their complete inability to attain personal truth, to be honest about themselves — the hysterical never think for themselves, they want other people to think about them, they want to arouse the interest of others — it follows that the hysterical are the best mediums for hypnotic purposes. But any one who allows him or herself to be hypnotised is doing the most immoral thing possible. It is yielding to complete slavery; it is a renunciation of the will and consciousness; it means allowing another person to do what he likes with the subject. Hypnosis shows how all possibility of truth depends upon the wish to be truthful, but it must be the real wish of the person concerned: when a hypnotised person is told to do something, he does it when he comes out of the trance, and if asked his reasons will give a plausible motive on the spot, not only before others, but he will justify his action to himself by quite fanciful reasons. In this we have, so to speak, an experimental proof of Kant's "Ethical Code."

All women can be hypnotised and like being hypnotised, but this proclivity is exaggerated in hysterical women. Even the memory of definite events in their life can be destroyed by the mere suggestion of the hypnotiser. Breuer's experiments on hypnotised patients show clearly that the consciousness of guilt in them is not deeply seated, as otherwise it could not be got rid of at the mere suggestion of the hypnotiser. But the sham conviction of responsibility, so readily exhibited by women of hysterical constitution, rapidly disappears at the moment when nature, the sexual impulse, appears to drive through the superficial restraints. In the hysterical paroxysm what happens is that the woman, while no longer believing it altogether herself, asseverates more and more loudly: "I do not want that at all, some one not really me is forcing it on me, but I do not want it at all." Every stimulation from outside will now be brought into relation with that demand, which, as she partly believes, is being forced on her, but which, in reality, corresponds with the deepest wish of her nature. That is why women in a hysterical attack are so easily seduced. The "attitudes passionelles" of the hysterical are merely passionate repudiations of sexual desire, which are loud merely because they are not real, and are more plaintive than at other times because the danger is greater. It is easy to understand why the sexual experiences of the time preceding puberty play so large a part in acute hysteria. The influence of extraneous moral views can be imposed comparatively easily on the child, as they have little to overcome in the almost unawakened state of the sexual inclinations. But, later on, the suppressed, although not wholly vanquished, nature, lays hold of these old experiences, reinterprets them in the light of the new contents of consciousness, and the crisis takes place. The different forms that the paroxysms assume and their shifting nature are due very largely to the fact that the subject does not admit the true cause, the presence of a sexual desire, any consciousness of it being attributed by her to some extraneous influence, some self that is not her "real self."

Medical observation or interpretation of hysteria is wrong; it allows itself to be deceived by the patients, who in turn deceive themselves. It is not the rejecting ego but the rejected which is the true and original nature of the hysterical patients, however much they pretend to themselves and others that it is foreign to them. . . .

But it may be asked, with reason, why all women are not hysterical, since all women are liars? This brings us to a necessary inquiry as to the hysterical constitution. If my theory has been on the right lines, it ought to be able to give an answer in accordance with facts. According to it, the hysterical woman is one who has passively accepted in entirety the masculine and conventional valuations instead of allowing her own mental character its proper play. The woman who is not to be led is the antithesis of the hysterical woman. I must not delay over this point; it really belongs to special female characterology. The hysterical woman is hysterical because she is servile; mentally she is identical with the maid-servant. Her opposite (who does not really exist) is the shrewish dame. So that women may be subdivided into the maid who serves, and the woman who commands.[1] . . .

Woman is not a free agent; she is altogether subject to her desire to be under man's influence, herself and all others: she is under the sway of the phallus, and irretrievably succumbs to her destiny, even if it leads to actively developed sexuality. At the most a woman can reach an indistinct feeling of her un-freedom, a cloudy idea of the possibility of controlling her destiny — manifestly only a flickering spark of the free, intelligible subject, the scanty remains of inherited maleness in

---

[1] We may find the analogy to this in men: there are masculine "servants" who are so by nature, and there is the masculine form of the shrew — e.g., the policeman. It is a noticeable fact that a policeman usually finds his sexual complement in the housemaid.

her, which, by contrast, gives her even this slight comprehension. It is also impossible for a woman to have a clear idea of her destiny, or of the forces within her: it is only he who is free who can discern fate, because he is not chained by necessity; part of his personality, at least, places him in the position of spectator and a combatant outside his own fate and makes him so far superior to it. One of the most conclusive proofs of human freedom is contained in the fact that man has been able to create the idea of causality. Women consider themselves most free when they are most bound; and they are not troubled by the passions, because they are simply the embodiment of them. It is only a man who can talk of the "dira necessitas" within him; it is only he could have created the idea of destiny, because it is only he who, in addition to the empirical, conditioned existence, possesses a free, intelligible ego. . . .

Woman, the normal receptive woman of whom I am speaking, is impregnated by the man not only physically (and I set down the astonishing mental alteration in women after marriage to a physical phenomenon akin to telegony[2]), but at every age of her life, by man's consciousness and by man's social arrangements. Thus it comes about that although woman lacks all the characters of the male sex, she can assume them so cleverly and so slavishly that it is possible to make mistakes such as the idea of the higher morality of women.

But this astounding receptivity of woman is not isolated, and must be brought into practical and theoretical connection with the other positive and negative characteristics of woman.

What has the match-making instinct in woman to do with her plasticity? What connection is there between her untruthfulness and her sexuality? How does it come about that there is such a strange mixture of all these things in woman? . . .

The last and absolute proof of the thoroughly negative character of woman's life, of her complete want of a higher existence, is derived from the way in which women commit suicide.

Such suicides are accompanied practically always by thoughts of other people, what they will think, how they will mourn over them, how grieved — or angry — they will be. Every woman is convinced

---

[2] *telegony:* used metaphorically here to suggest the transformation of woman through heterosexual marriage (Weininger implies intimacy with the male). As a scientific term, the hypothetical influence of a previous sire seen in the children of a later sire from the same mother.

that her unhappiness is undeserved at the time she kills herself; she pities herself exceedingly with the sort of self-compassion which is only a "weeping with others when they weep."

How is it possible for a woman to look upon her unhappiness as personal when she possesses no idea of a destiny? The most appallingly decisive proof of the emptiness and nullity of women is that they never once succeed in knowing the problem of their own lives, and death leaves them ignorant of it, because they are unable to realise the higher life of personality.

I am now ready to answer the question which I put forward as the chief object of this portion of my book, the question as to the significance of the male and female in the universe. Women have no existence and no essence; they are not, they are nothing. Mankind occurs as male or female, as something or nothing. Woman has no share in ontological reality, no relation to the thing-in-itself, which, in the deepest interpretation, is the absolute, is God. Man in his highest form, the genius, has such a relation, and for him the absolute is either the conception of the highest worth of existence, in which case he is a philosopher; or it is the wonderful fairyland of dreams, the kingdom of absolute beauty, and then he is an artist. But both views mean the same. Woman has no relation to the idea, she neither affirms nor denies it; she is neither moral nor anti-moral; mathematically speaking, she has no sign; she is purposeless, neither good nor bad, neither angel nor devil, never egoistical (and therefore has often been said to be altruistic); she is as non-moral as she is non-logical. But all existence is moral and logical existence. So woman has no existence.

IDA B. WELLS-BARNETT

From *The Case Stated*

Ida B. Wells-Barnett (1862–1931) was a newspaper editor, lecturer, and clubwoman who wrote and spoke effectively against the inhumane, late-nineteenth-century practice of lynching African Americans. As in "The Case Stated," her theme — whether she wrote for her own newspaper, *The Free Speech and Headlight* in Memphis, Tennessee, or for others' — was racially motivated violence. She campaigned against such violence in both the United States and England, and, during the 1890s, published a series of influential pamphlets on the topic.

The daughter of slaves and the oldest of eight children, Wells-Barnett was born in Mississippi. Educated at Shaw University, she taught school until 1883, supporting her siblings after her parents' deaths from yellow fever. Then she became a journalist.

In 1895 she married Ferdinand L. Barnett, a Chicago lawyer, with whom she had four children. She continued her activism until her death in 1931.

Although such essays as this were privately published, for the most part, Gertrude Stein's being a student at the Harvard Annex (later, Radcliffe College) gave her access to such marginalized publications. Boston — and, frequently, Harvard College itself — was a center for abolitionist and postwar activism; if materials were available anywhere in the United States, they would be accessible at Harvard.

These excerpts are taken from Wells-Barnett's 1892 pamphlet, *Southern Horrors: Lynch Law in all its Phases*.

The student of American sociology will find the year 1894 marked by a pronounced awakening of the public conscience to a system of anarchy and outlawry which had grown during a series of ten years to be so common, that scenes of unusual brutality failed to have any visible effect upon the humane sentiments of the people of our land.

Beginning with the emancipation of the Negro, the inevitable result of unbridled power exercised for two and a half centuries, by the white man over the Negro, began to show itself in acts of conscienceless outlawry. During the slave regime, the Southern white man owned the Negro body and soul. It was to his interest to dwarf the soul and preserve the body. Vested with unlimited power over his slave, to subject him to any and all kinds of physical punishment, the white man was still restrained from such punishment as tended to injure the slave by abating his physical powers and thereby reducing his financial worth. While slaves were scourged mercilessly, and in countless cases inhumanly treated in other respects, still the white owner rarely permitted his anger to go so far as to take a life, which would entail upon him a loss of several hundred dollars. The slave was rarely killed, he was too valuable; it was easier and quite as effective, for discipline or revenge, to sell him 'Down South.'

But Emancipation came and the vested interests of the white man in the Negro's body were lost. The white man had no right to scourge the emancipated Negro, still less has he a right to kill him. But the southern white people had been educated so long in that school of practice, in which might makes right, that they disdained to draw strict lines of action in dealing with the Negro. In slave times the Negro was kept subservient and submissive by the frequency and severity of the scourging, but, with freedom, a new system of intimidating came in vogue; the Negro was not only whipped and scourged; he was killed.

Not all nor nearly all of the murders done by white men, during the past thirty years in the South, have come to light, but the statistics as gathered and preserved by white men, and which have not been questioned, show that during these years more than ten thousand Negroes have been killed in cold blood, without the formality of judicial trial and legal execution. And yet, as evidence of the absolute impunity with which the white man dares to kill a Negro, the same record shows that during all these years, and for all these murders only three white men have been tried, convicted, and executed. As no white man has been lynched for the murder of colored people, these three executions are the only instances of the death penalty being visited upon white men for murdering Negroes.

Naturally enough the commission of these crimes began to tell upon the public conscience, and the Southern white man, as a tribute to the nineteenth century civilization, was in a manner compelled to give excuses for his barbarism. His excuses have adapted themselves to the emergency, and are aptly outlined by that greatest of all Negroes, Frederick Douglass, in an article of recent date, in which he shows that there have been three distinct eras of southern barbarism, to account for which three distinct excuses have been made.

The first excuse given to the civilized world for the murder of unoffending Negroes was the necessity of the white man to repress and stamp out alleged 'race riots.' For years immediately succeeding the war there was an appalling slaughter of colored people, and the wires usually conveyed to northern people and the world the intelligence, first, that an insurrection was being planned by Negroes, which, a few hours later, would prove to have been vigorously resisted by white men, and controlled with a resulting loss of several killed and wounded. It was always a remarkable feature in these insurrections and riots that only Negroes were killed during the rioting, and that all the white men escaped unharmed.

From 1865 to 1872, hundreds of colored men and women were mercilessly murdered and the almost invariable reason assigned was that they met their death by being alleged participants in an insurrection or riot. But this story at last wore itself out. No insurrection ever materialized; no Negro rioter was ever apprehended and proven guilty, and no dynamite ever recorded the black man's protest against oppression and wrong. It was too much to ask thoughtful people to believe this transparent story, and the southern white people at last made up their minds that some other excuse must be had.

Then came the second excuse, which had its birth during the turbulent times of reconstruction. By an amendment to the Constitution the Negro was given the right of franchise, and, theoretically at least, his ballot became his invaluable emblem of citizenship. In a government 'of the people, for the people, and by the people,' the Negro's vote became an important factor in all matters of state and national politics. But this did not last long. The southern white man would not consider that the Negro had any right which a white man was bound to respect, and the idea of a republican form of government in the southern states grew into general contempt. It was maintained that 'This is a white man's government,' and regardless of numbers, the white man should rule. 'No Negro domination' became the new legend on the sanguinary banner of the sunny South, and under it rode the Ku Klux Klan, the Regulators, and the lawless mobs, which for any cause chose to murder one man or a dozen as suited their purpose best. It was a long, gory campaign; the blood chills and the heart almost loses faith in Christianity when one thinks of . . . the countless massacres of defenceless Negroes, whose only crime was the attempt to exercise their right to vote. . . .

Brutality still continued; Negroes were whipped, scourged, exiled, shot and hung whenever and wherever it pleased the white man so to treat them, and as the civilized world with increasing persistency held the white people of the South to account for its outlawry, the murderers invented the third excuse — that Negroes had to be killed to avenge their assaults upon women. There could be framed no possible excuse more harmful to the Negro and more unanswerable if true in its sufficiency for the white man.

Humanity abhors the assailant of womanhood, and this charge upon the Negro at once placed him beyond the pale of human sympathy. With such unanimity, earnestness and apparent candor was this charge made and reiterated that the world has accepted the story that the Negro is the monster which the southern white man has painted him. And today, the Christian world feels, that while lynching is a crime, and lawlessness and anarchy the certain precursors of a nation's fall, it can not by word or deed, extend sympathy or help to a race of outlaws, who might mistake their plea for justice and deem it an excuse for their continued wrongs. . . .

It was for the assertion of this fact, in the defense of her own race, that the writer hereof [Wells-Barnett] became an exile; her property destroyed and her return to her home forbidden under penalty of death, for writing the following editorial which was printed in her paper, *The Free Speech,* in Memphis, Tenn., May 21, 1892.

'Eight Negroes lynched since the last issue of the "Free Speech" one at Little Rock, Ark., last Saturday morning where the citizens broke (?) into the penitentiary and got their man; three near Anniston, Ala., one near New Orleans; and three at Clarksville, Ga., the last three for killing a white man, and five on the same old racket — the new alarm about raping white women. The same programme of hanging, then shooting bullets into the lifeless bodies was carried out to the letter. Nobody in this section of the country believes the old thread bare lie that Negro men rape white women. If Southern white men are not careful, they will over-reach themselves and public sentiment will have a reaction; a conclusion will then be reached which will be very damaging to the moral reputation of their women.'

But threats cannot suppress the truth, and while the Negro suffers the soul deformity, resultant from two and a half centuries of slavery, he is no more guilty of this vilest of all vile charges than the white man who would blacken his name.

During all the years of slavery, no such charge was ever made, not even during the dark days of the rebellion, when the white man, following the fortunes of war went to do battle for the maintenance of slavery. While the master was away fighting to forge the fetters upon the slave, he left his wife and children with no protectors save the Negroes themselves. And yet during those years of trust and peril, no Negro proved recreant to his trust and no white man returned to a home that had been despoiled.

Likewise during the period of alleged 'insurrection,' and alarming 'race riots,' it never occurred to the white man, that his wife and children were in danger of assault. Nor in the Reconstruction era, when the hue and cry was against 'Negro Domination,' was there ever a thought that the domination would ever contaminate a fireside or strike to death the virtue of womanhood. It must appear strange indeed, to every thoughtful and candid man, that more than a quarter of a century elapsed before the Negro began to show signs of such infamous degeneration. . . .

These pages are written in no spirit of vindictiveness, for all who give the subject consideration must concede that far too serious is the condition of that civilized government in which the spirit of unrestrained outlawry constantly increases in violence, and casts its blight over a continually growing area of territory. We plead not for the colored people alone, but for all victims of the terrible injustice which puts men and women to death without form of law. During the year 1894, there were 132 persons executed in the United States by due form of law, while in the same year, 197 persons were put to death by mobs who gave the victims no opportunity to make a lawful defense. No comment need be made upon a condition of public sentiment responsible for such alarming results. . . .

Discussions of Mind and Philosophy

"One cannot come back too often to the question what is knowledge and to the answer knowledge is what one knows."

Throughout Gertrude Stein's writing career there appear echoes of her discourse from the years she was a philosophy major at Radcliffe. The intertextual relationship between her writing in fiction and poetry and the essays she wrote during the 1890s (as well as during the 1920s, 1930s, and 1940s) has seldom been investigated. While the disciplines of creative writing and philosophy today might seem clearly separate, in the categorization of turn-of-the-century educational divisions, they were fused in a number of important ways.

For instance, Stein learned a great deal about writing from the linguistic interests of philosopher and poet George Santayana, with whom she spent her freshman year in the study of the philosophy of various religious belief systems. As the sixteen women in the class studied Judaism, Catholicism, mysticism, and Eastern beliefs, Santayana forced them to create dialogues about not only subject matter but the ideas inherent in that material. He emphasized that language was the basis of discourse, that an understanding of language systems (even grammars) was essential for coherent thought (and its delivery), and that prose and poetry were inherently different models of communication.

Santayana's course gave Stein the first of her frequent *A* grades (he awarded only three to his class of sixteen). More important, it introduced her to the realities of mysticism and to the possible relationship between the cognitive levels of concentration necessary to write well and a successful written product. According to Santayana, the artist/writer made correct choices not through reason but through "contemplation," what he called "the intuition of essences." His belief that writing, like the acquisition of all knowledge, started with mysticism — and that writing was in some ways itself an acquisition of knowledge — set him at odds with most philosophers. Santayana believed that the human mind operated in two modes, one of participation ("the sense of existence") and the other of disengagement ("the intuition of pure being"). While most people lived largely in the participatory mode, the artist needed at times to be disengaged in order to create (see pp. 319–21).

It was with Hugo Münsterberg, who later called Stein his "ideal student," that she, as a freshman, began work in the Harvard Psychological Laboratory, the place she was to consider home territory during her four years at Radcliffe. Münsterberg, the brilliant young German, had been hired by William James to teach in his place during a sabbatical; James was still a presence in the lab because Münsterberg used his two-volume *Principles of Psychology* as the text for the course. Stein admired James, too, to the point of adoration. James investigated all aspects of the human mind; little was foreign or objectionable to him. Séances, drugs, spiritualism, hypnotism, mescal, faith cures, yoga, telepathy, clairvoyance — James was a founding member of the British Society for Psychical Research in 1882, and founded the American Society in 1884. With him, Stein, again, enrolled for Psychological Laboratory, and she also took his yearlong seminar in feelings and emotion; studying with James was a validation of her belief that learning was wholistic, that experience played a part in knowledge. As she later wrote, James was "the important person" in her undergraduate years. He simply "delighted her. His personality and his teaching and his way of amusing himself with himself and his students all pleased her. Keep your mind open, he used to say" (*Autobiography* 78–79).

Because so much of James's work was with states of consciousness and the problem of human attention, Stein involved herself in two projects that studied consciousness; both were eventually published in the Harvard journal. These studies led to her valuing the human ability to concentrate as a means of divining knowledge that might have remained hidden from the conscious mind. As late as the 1920s, when Stein had become a famous Parisian writer and art critic, surrounded by crowds of young American and European writers and painters, she spoke about aesthetics in language that echoed James's pronouncements about attention and what he was to call the stream of consciousness (see pp. 310–19). During the period of Stein's writing *Three Lives,* she implicitly played with the notions of the empirical, the continuous present (especially in "Melanctha"), and an almost experimental structure of demanding the reader's attention just to read her novellas.

The other excerpts included here represent the philosophical thinking of two men important to Stein. Like all modernists, she was impressed with the contributions of Henri Bergson, who helped bring the insistence on the concrete to modern writing, particularly poetry (see pp. 321–25). And, after years of admiring the work of Alfred North Whitehead, she and her companion, Alice B. Toklas, were stranded with the Whitehead family at the outbreak of World War I, when they were forbidden to cross international borders for nearly two months after war was declared. The long walks that Whitehead and Stein took during those weeks of comparative isolation cemented the friendship that Whitehead's writing in the abstract had initiated. Whitehead's essay is a concise statement of an all-inclusive thrust to "philosophy," a definition that allowed the study of psychology to become integral to its discipline (see pp. 325–28). As Stein's own innovative and imaginative consciousness proved, she had learned well from these masters of philosophy and psychology.

WILLIAM JAMES

"The Varieties of Attention" and "The Stream of Thought," from The Principles of Psychology

Few educated people were as hungry for learning during the late nineteenth century as William James (1842–1910). The eldest son of Swedenborgian theologian Henry James and elder brother of American novelist Henry James, Jr., this artist-philosopher-physician and creator of the discipline of psychology remarkably imprinted the world of intellectual thought.

William Jámes (on left) early in the twentieth century, photographed here with his Harvard colleague Josiah Royce, another well-known philosopher, at James's New Hampshire retreat. Reproduced by permission of The Harvard Archives.

From 1872 until 1907 James taught at Harvard College, first in the area of anatomy and physiology and, after 1880, in the Philosophy Department. It was here that he founded the discipline of psychology, and, after his 1890 two-volume *The Principles of Psychology* saw print, Harvard became the center for that topic of research. A fluent and approachable writer, James also wrote widely for popular journals. By way of those publications and his lectures throughout the country and Europe, much of his belief system penetrated middle-class consciousness during those same years.

Considered by some America's first philosopher, James brought the concepts of pragmatism, volunteerism, and "radical empiricism" together. Like the belief system of Charles S. Pierce, James's pragmatism insisted that experience foreground knowledge — will and interest were primary and learning, per se, only an instrument. Learning became, therefore, an active state rather than a passive one. "Radical empiricism" carried this premise further, insisting that experience underlaid all abstract philosophy.

Besides *The Principles of Psychology,* James's other influential books are *The Will to Believe* (1897), *The Varieties of Religious Experience* (1902), *Pragmatism* (1907), *A Pluralistic Universe* and *The Meaning of Truth* (1909), and *Essays in Radical Empiricism* (1912).

In addition to Gertrude Stein's being James's student at Harvard for several years, and, upon his recommendation, studying to become a physician at Johns Hopkins Medical School, she and her family kept in touch with James personally. When she published *Three Lives* in 1909, she sent him a copy, and he visited her in Paris during his last year abroad before his death in 1910. She had also studied the two-volume *The Principles of Psychology* in her first year at the Harvard Annex, when James was on leave and the psychology laboratory was under the direction of his young German colleague, Hugo Münsterberg.

These excerpts come from the 1890 edition of *The Principles of Psychology,* published by Henry Holt in New York.

The Varieties of Attention

The things to which we attend are said to *interest* us. Our interest in them is supposed to be the *cause* of our attending. What makes an object interesting we shall see presently; and later inquire in what sense interest may cause attention. Meanwhile

Attention may be divided into kinds in various ways. It is either to

a) Objects of sense (sensorial attention); or to

b) Ideal or represented objects (intellectual attention).

It is either

c) Immediate; or

d) Derived: immediate, when the topic or stimulus is interesting in itself, without relation to anything else; derived, when it owes its interest to association with some other immediately interesting thing. What I call derived attention has been named 'apperceptive' attention. Furthermore, Attention may be either

e) Passive, reflex, non-voluntary, effortless; or

f) Active and voluntary.

Voluntary attention is always derived; we never make an *effort* to attend to an object except for the sake of some *remote* interest which the effort will serve. But both sensorial and intellectual attention may be either passive or voluntary.

In *passive immediate sensorial attention* the stimulus is a sense-impression, either very intense, voluminous, or sudden, — in which case it makes no difference what its nature may be, whether sight, sound, smell, blow, or inner pain, — or else it is an *instinctive* stimulus, a perception which, by reason of its nature rather than its mere force, appeals to some one of our normal congenital impulses and has a directly exciting quality. In the chapter on Instinct we shall see how these stimuli differ from one animal to another, and what most of them are in man: strange things, moving things, wild animals, bright things, pretty things, metallic things, words, blows, blood, etc., etc., etc.

Sensitiveness to immediately exciting sensorial stimuli characterizes the attention of childhood and youth. In mature age we have generally selected those stimuli which are connected with one or more so-called permanent interests, and our attention has grown irresponsive to the rest.[1] But childhood is characterized by great active energy, and has few organized interests by which to meet new impressions and decide whether they are worthy of notice or not, and the consequence is that extreme mobility of the attention with which we are all familiar in children, and which makes their first lessons such rough affairs. Any strong sensation whatever produces accommodation of the organs which perceive it, and absolute oblivion, for the time being, of the task in hand. This reflex and passive character of the attention which, as a French writer says, makes the child seem to belong less to himself than to every object which happens to catch his notice, is the first thing which the teacher must overcome. It never is overcome in some people, whose work, to the end of life, gets done in the interstices of their mind-wandering.

The passive sensorial attention is *derived* when the impression, without being either strong or of an instinctively exciting nature, is connected by previous experience and education with things that are so. These things may be called the *motives* of the attention. The impression draws an interest from them, or perhaps it even fuses into a single complex object with them; the result is that it is brought into the focus of the mind. A faint tap *per se* is not an interesting sound; it may well escape being discriminated from the general rumor of the world. But when it is a signal, as that of a lover on the window-pane, it will hardly go unperceived. . . .

Note: The footnotes in this selection are original to the selection.

[1] Note that the permanent interests are themselves grounded in certain objects and relations in which our interest is immediate and instinctive.

Passive intellectual attention is immediate when we follow in thought a train of images exciting or interesting *per se;* derived, when the images are interesting only as means to a remote end, or merely because they are associated with something which makes them dear. Owing to the way in which immense numbers of real things become integrated into single objects of thought for us, there is no clear line to be drawn between immediate and derived attention of an intellectual sort. When absorbed in intellectual attention we may become so inattentive to outer things as to be 'absent-minded,' 'abstracted,' or *'distraits.'* All revery or concentrated meditation is apt to throw us into this state. . . .

The absorption may be so deep as not only to banish ordinary sensations, but even the severest pain. Pascal, Wesley, Robert Hall, are said to have had this capacity. Dr. Carpenter says of himself that

> he has frequently begun a lecture, whilst suffering neuralgic pain so severe as to make him apprehend that he would find it impossible to proceed; yet no sooner has he, by a determined effort, fairly launched himself into the stream of thought, than he has found himself continuously borne along without the least distraction, until the end has come, and the attention has been released; when the pain has recurred with a force that has over-mastered all resistance, making him wonder how he could have ever ceased to feel it.[2]

Dr. Carpenter speaks of launching himself by a determined *effort.* This effort characterizes what we called *active or voluntary attention.* It is a feeling which everyone knows, but which most people would call quite indescribable. We get it in the sensorial sphere whenever we seek to catch an impression of extreme *faintness,* be it of sight, hearing, taste, smell, or touch; we get it whenever we seek to *discriminate* a sensation merged in a mass of others that are similar; we get it whenever we *resist the attractions* of more potent stimuli and keep our mind occupied with some object that is naturally unimpressive. We get it in the intellectual sphere under exactly similar conditions: as when we strive to sharpen and make distinct an idea which we but vaguely seem to have; or painfully discriminate a shade of meaning from its similars; or resolutely hold fast to a thought so discordant with our impulses that, if left unaided, it would quickly yield place to images of an exciting and impassioned kind. All forms of attentive effort would be exercised

---

[2] *Mental Physiology,* § 124. The oft-cited case of soldiers not perceiving that they are wounded is of an analogous sort.

at once by one whom we might suppose at a dinner-party resolutely to listen to a neighbor giving him insipid and unwelcome advice in a low voice, whilst all around the guests were loudly laughing and talking about exciting and interesting things. *There is no such thing as voluntary attention sustained for more than a few seconds at a time.* What is called sustained voluntary attention is a repetition of successive efforts which bring back the topic to the mind.[3] The topic once brought back, if a congenial one, *develops*; and if its development is interesting it engages the attention passively for a time. Dr. Carpenter, a moment back, described the stream of thought, once entered, as 'bearing him along.' This passive interest may be short or long. As soon as it flags, the attention is diverted by some irrelevant thing, and then a voluntary effort may bring it back to the topic again; and so on, under favorable conditions, for hours together. During all this time, however, note that it is not an identical *object* in the psychological sense, but a succession of mutually related objects forming an identical *topic* only, upon which the attention is fixed. *No one can possibly attend continuously to an object that does not change.*

The Stream of Thought[4]

We now begin our study of the mind from within. Most books start with sensations, as the simplest mental facts, and proceed synthetically, constructing each higher stage from those below it. But this is abandoning the empirical method of investigation. No one ever had a simple sensation by itself. Consciousness, from our natal day, is of a teeming multiplicity of objects and relations, and what we call simple sensations are results of discriminative attention, pushed often to a very high degree. It is astonishing what havoc is wrought in psychology by admitting at the outset apparently innocent suppositions, that nevertheless contain a flaw. The bad consequences develop themselves later on, and are irremediable, being woven through the whole texture

---

[3] Prof. J. M. Cattell made experiments to which we shall refer further on, on the degree to which reaction-times might be shortened by distracting or voluntarily concentrating the attention. He says of the latter series that "the averages show that the attention can be held strained, that is, the centres kept in a state of unstable equilibrium for one second" (*Mind*, xi, 240).

[4] A good deal of this chapter is reprinted from an article "On Some Omissions of Introspective Psychology" which appeared in *Mind* for January 1884.

of the work. The notion that sensations, being the simplest things, are the first things to take up in psychology is one of these suppositions. The only thing which psychology has a right to postulate at the outset is the fact of thinking itself, and that must first be taken up and analyzed. If sensations then prove to be amongst the elements of the thinking, we shall be no worse off as respects them than if we had taken them for granted at the start.

The first fact for us, then, as psychologists is that thinking of some sort goes on. I use the word thinking, in accordance with what was said on p. 186, for every form of consciousness indiscriminately. If we could say in English 'it thinks,' as we say 'it rains' or 'it blows,' we should be stating the fact most simply and with the minimum of assumption. As we cannot, we must simply say that *thought goes on.*

FIVE CHARACTERS IN THOUGHT

How does it go on? We notice immediately five important characters in the process, of which it shall be the duty of the present chapter to treat in a general way:

1) Every thought tends to be part of a personal consciousness.

2) Within each personal consciousness thought is always changing.

3) Within each personal consciousness thought is sensibly continuous.

4) It always appears to deal with objects independent of itself.

5) It is interested in some parts of these objects to the exclusion of others, and welcomes or rejects — *chooses* from among them, in a word — all the while.

In considering these five points successively, we shall have to plunge *in medias res* as regards our vocabulary, and use psychological terms which can only be adequately defined in later chapters of the book. But everyone knows what the terms mean in a rough way; and it is only in a rough way that we are now to take them. This chapter is like a painter's first charcoal sketch upon his canvas, in which no niceties appear.

1) *Thought tends to Personal Form*

When I say *every 'thought' is part of a personal consciousness,* 'personal consciousness' is one of the terms in question. Its meaning we know so long as no one asks us to define it, but to give an accurate account of it is the most difficult of philosophic tasks. This task we must confront in the next chapter; here a preliminary word will suffice.

In this room — this lecture-room, say — there are a multitude of thoughts, yours and mine, some of which cohere mutually, and some not. They are as little each-for-itself and reciprocally independent as they are all-belonging-together. They are neither: no one of them is separate, but each belongs with certain others and with none beside. My thought belongs with my other thoughts, and your thought with your other thoughts. Whether anywhere in the room there be a mere thought, which is nobody's thought, we have no means of ascertaining, for we have no experience of its like. The only states of consciousness that we naturally deal with are found in personal consciousnesses, minds, selves, concrete particular I's and you's.

Each of these minds keeps its own thoughts to itself. There is no giving or bartering between them. No thought even comes into direct *sight* of a thought in another personal consciousness than its own. Absolute insulation, irreducible pluralism, is the law. It seems as if the elementary psychic fact were not *thought* or *this thought* or *that thought,* but *my thought,* every thought being *owned.* Neither contemporaneity, nor proximity in space, nor similarity of quality and content are able to fuse thoughts together which are sundered by this barrier of belonging to different personal minds. The breaches between such thoughts are the most absolute breaches in nature. Everyone will recognize this to be true, so long as the existence of *something* corresponding to the term 'personal mind' is all that is insisted on, without any particular view of its nature being implied. On these terms the personal self rather than the thought might be treated as the immediate datum in psychology. . . .

The Stream of Thought

3) *Within each personal consciousness,*
   thought is sensibly continuous

I can only define 'continuous' as that which is without breach, crack, or division. I have already said that the breach from one mind to another is perhaps the greatest breach in nature. The only breaches that can well be conceived to occur within the limits of a single mind would either be *interruptions, time*-gaps during which the consciousness went out altogether to come into existence again at a later moment; or they would be breaks in the *quality,* or content, of the thought, so abrupt that the segment that followed had no connection whatever with the one that went before. The proposition that within each personal consciousness thought feels continuous, means two things:

1. That even where there is a time-gap the consciousness after it feels as if it belonged together with the consciousness before it, as another part of the same self;

2. That the changes from one moment to another in the quality of the consciousness are never absolutely abrupt.

The case of the time-gaps, as the simplest, shall be taken first. And first of all, a word about time-gaps of which the consciousness may not be itself aware. . . .

With the felt gaps the case is different. On waking from sleep, we usually know that we have been unconscious, and we often have an accurate judgment of how long. The judgment here is certainly an inference from sensible signs, and its ease is due to long practice in the particular field.[5] The result of it, however, is that the consciousness is, *for itself*, not what it was in the former case, but interrupted and discontinuous, in the mere time-sense of the words. But in the other sense of continuity, the sense of the parts being inwardly connected and belonging together because they are parts of a common whole, the consciousness remains sensibly continuous and one. What now is the common whole? The natural name for it is *myself, I,* or *me.*

When Paul and Peter wake up in the same bed, and recognize that they have been asleep, each one of them mentally reaches back and makes connection with but *one* of the two streams of thought which were broken by the sleeping hours. As the current of an electrode buried in the ground unerringly finds its way to its own similarly buried mate, across no matter how much intervening earth; so Peter's present instantly finds out Peter's past, and never by mistake knits itself on to that of Paul. Paul's thought in turn is as little liable to go astray. The past thought of Peter is appropriated by the present Peter alone. He may have a *knowledge,* and a correct one too, of what Paul's last drowsy states of mind were as he sank into sleep, but it is an entirely different sort of knowledge from that which he has of his own last states. He *remembers* his own states, whilst he only *conceives* Paul's. Remembrance is like direct feeling; its object is suffused with a warmth and intimacy to which no object of mere conception ever attains. This quality of warmth and intimacy and immediacy is what Peter's *present* thought also possesses for itself. So sure as this present is me, is mine, it says, so sure is anything else that comes with the same

---

[5] The accurate registration of the 'how long' is still a little mysterious.

warmth and intimacy and immediacy, me and mine. What the qualities called warmth and intimacy may in themselves be will have to be matter for future consideration. But whatever past feelings appear with those qualities must be admitted to receive the greeting of the present mental state, to be owned by it, and accepted as belonging together with it in a common self. This community of self is what the time-gap cannot break in twain, and is why a present thought, although not ignorant of the time-gap, can still regard itself as continuous with certain chosen portions of the past.

Consciousness, then, does not appear to itself chopped up in bits. Such words as 'chain' or 'train' do not describe it fitly as it presents itself in the first instance. It is nothing jointed; it flows. A 'river' or a 'stream' are the metaphors by which it is most naturally described. *In talking of it hereafter, let us call it the stream of thought, of consciousness, or of subjective life.*

GEORGE SANTAYANA

"Form in Words," from Persons and Places

Born in Madrid, George Santayana (1863–1952) was brought to the United States as a child and was educated at Harvard, with a year of study at Cambridge, England. He taught philosophy at Harvard from 1889 through 1912, at which time he resigned from the college and moved to Italy, where he lived in retirement in a convent. It was then that he wrote the philosophical works for which he is famous.

His major aesthetic work, *The Sense of Beauty,* appeared in 1896, while he was still at Harvard. It was followed in 1905–06 by his five-volume treatise *The Life of Reason,* an exploration of the role of reason in the history of human interests and life. It was Santayana's premise that reason was the union of impulse with ideation. Unlike many philosophers of the time, he also insisted on the role of language in the formation of thought. During these years, his conflicts with the Philosophy Department at Harvard were not disruptive, although he consistently differed from William James in finding mystical beauty in reality, rather than in the supernatural.

Publications from the 1923 *Scepticism and Animal Faith* showed his belief in the role of faith as the dominating force in human life, a clear modification of his previous emphasis on reason. Other seminal works were *The Realms of Being* (its four volumes, *The Realm of Essence,* 1927; *The Realm of Matter,* 1930; *The Realm of Truth,* 1937; and *The Realm of Spirit,* 1940). His interest in the formally creative led to publications in poetry (his 1923 sonnets); a novel, *The Last Puritan* (1935); and his three-volume memoir, *Persons and Places* (1944–53).

Because Santayana was himself so fascinated by language, it seems likely that Gertrude Stein — who received one of only several As in his yearlong Harvard philosophy course — was greatly influenced by his emphasis on the crafting of thought.

This excerpt is from Santayana's *Persons and Places* (New York: Charles Scribner's Sons, 1963).

The main effect of language consists in its meaning in the ideas which it expresses. But no expression is possible without a presentation, and this presentation must have a form. This form of the instrument of expression is itself an element of effect, although in practical life we may overlook it in our haste to attend to the meaning it conveys. It is, moreover, a condition of the kind of expression possible, and often determines the manner in which the object suggested shall be apperceived. No word has the exact value of any other in the same or in another language.[1] But the intrinsic effect of language does not stop there. The single word is but a stage in the series of formations which constitute language, and which preserve for men the fruit of their experience, distilled and concentrated into a symbol.

This formation begins with the elementary sounds themselves, which have to be discriminated and combined to make recognisable symbols. The evolution of these symbols goes on spontaneously, suggested by our tendency to utter all manner of sounds, and preserved by the ease with which the ear discriminates these sounds when made. Speech would be an absolute and unrelated art, like music, were it not controlled by utility. The sounds have indeed no resemblance to the

---

Note: The footnote in this selection is original to the selection.

[1] Not only are words untranslatable when the exact object has no name in another language, as "home" or "mon ami," but even when the object is the same, the attitude toward it, incorporated in one word, cannot be rendered by another. Thus, to my sense, "bread" is as inadequate a translation of the human intensity of the Spanish "pan" as "Dios" is of the awful mystery of the English "God." This latter word does not designate an object at all, but a sentiment, a psychosis, not to say a whole chapter of religious history. English is remarkable for the intensity and variety of the colour of its words. No language, I believe, has so many words specifically poetic.

objects they symbolise; but before the system of sounds can represent the system of objects, there has to be a correspondence in the groupings of both. The structure of language, unlike that of music, thus becomes a mirror of the structure of the world as presented to the intelligence.

Grammar, philosophically studied, is akin to the deepest metaphysics, because in revealing the constitution of speech, it reveals the constitution of thought, and the hierarchy of those categories by which we conceive the world. It is by virtue of this parallel development that language has its function of expressing experience with exactness, and the poet — to whom language is an instrument of art — has to employ it also with a constant reference to meaning and veracity; that is, he must be a master of experience before he can become a true master of words. Nevertheless, language is primarily a sort of music, and the beautiful effects which it produces are due to its own structure, giving, as it crystallises in a new fashion, an unforeseen form to experience.

Poets may be divided into two classes: the musicians and the psychologists. The first are masters of significant language as harmony; they know what notes to sound together and in succession; they can produce, by the marshalling of sounds and images, by the fugue of passion and the snap of wit, a thousand brilliant effects out of old materials. The Ciceronian orator, the epigrammatic, lyric, and elegiac poets, give examples of this art. The psychologists, on the other hand, gain their effect not by the intrinsic mastery of language, but by the closer adaptation of it to things. . . .

HENRI BERGSON

From *Brain and Thought: A Philosophical Illusion*

One of the seminal philosophers of modernism, Henri Bergson (1859–1941) posited the centrality of time (*duree*, duration) and the opposition of the human life force (*élan vital*), which includes imagination, to the resistance of the material world. His various theories moved from the abstract to the concrete in such works as *Time and Free Will* (1889), *Matter and Memory* (1896), *Laughter* (1901), and *Creative Evolution* (1907).

A professor at the College de France starting in 1900, Bergson was part of an international network that privileged the new in all fields of creativity. During and after World War I, he was particularly involved in truly international efforts and, during the postwar period, became more active in describing political thought.

The recipient of the Nobel Prize for literature in 1927, Bergson provided the foundational ideas for much of the innovation which came to dominate early-twentieth-century art and literature. This excerpt from his 1904 paper, which was read at the International Congress of Philosophy in Geneva, shows his ability to present cogently what were, at the time, radical philosophical principles. This paper was later published in the *Revue de metaphysique et de morale* under the title "Le Praalogisme psycho-physiologique."

Gertrude Stein would have known of Bergson's groundbreaking work through philosophical and literary circles, as well as artistic ones.

. . . When we speak of external objects, we have to choose, in fact, between two notation-systems. We can treat external objects, and the changes they exhibit, as a system of *things* or as a system of *ideas*. And either of these two systems will work, provided we keep strictly to the one we have chosen.

Let us, first of all, try to distinguish the two systems with precision. When realism speaks of things and idealism of ideas, it is not merely a dispute about words; realism and idealism are two different notation-systems, that is to say, two different ways of setting about the analysis of reality. For the idealist, there is nothing in reality over and above what appears to his consciousness or to consciousness in general. It would be absurd to speak of a property of matter which could not be represented in idea. There is no virtuality, or, at least, nothing definitely virtual; whatever exists is actual or could become so. Idealism is, then, a notation-system which implies that everything essential in matter is displayed or displayable in the idea which we have of it, and that the real world is articulated in the very same way as it is presented in idea. The hypothesis of realism is the exact reverse. When realism affirms that matter exists independently of the idea, the meaning is that beneath our idea of matter there is an inaccessible cause of that idea, that behind perception, which is actual, there are hidden powers and virtualities; in short, realism assumes that divisions and articulations visible in our perception are purely relative to our manner of perceiving.

I am not questioning that profounder definitions could be given of the two tendencies, realist and idealist, such as they are to be found throughout the history of philosophy. I have myself indeed used the words "realism" and "idealism" in a somewhat different meaning. This is as much as to say that I have no particular liking for the definitions I have just given. They may characterize an idealism like that of Berkeley[1] and the realism opposed to it. They may also fairly well represent our ordinary notion of the two tendencies — the tendency of idealism to include the whole reality in what can be presented to our mind, the tendency of realism to claim to pass beyond what is presented to our mind. But the argument I am about to put forward is independent of any historical conception of realism and idealism. If any one is inclined to dispute the generality of my two definitions, I simply ask him to accept the words *realism* and *idealism* as conventional terms by which I intend to indicate, in the course of this study, two notations of reality, one of which implies the possibility, the other the impossibility, of identifying things with their ideas, that is with the presentations, spread out and articulated in space, which they offer to a human consciousness. That these two postulates are mutually exclusive, that consequently it is illegitimate to apply the two notation-systems at the same time to the same object, every one will agree. Now, I require nothing more for my present purpose.

I propose to establish the three following points: (1) If we choose the idealist notation, the affirmation of parallelism (in the meaning of equivalence) between the psychic state and the cerebral state is a self-contradiction. (2) If, on the other hand, we choose the realist notation, there is the same contradiction, but transposed. (3) The thesis of parallelism appears consistent only when we employ at the same time, in the same proposition, both notation-systems together. That is to say, the thesis is intelligible only because, by an unconscious trick of intellectual conjuring, we pass instantly from realism to idealism and from idealism to realism, showing ourselves in the one at the very moment when we are going to be caught in the act of self-contradiction in the other. The trick, moreover, is quite natural; we are, in this case, born conjurors, because the problem we are concerned with, the psycho-physiological problem of the relation of brain and thought, itself suggests by its very terms the two points of view of realism and

---

[1] *Berkeley:* George Berkeley (1685–1753), British philosopher who carried Locke's concept of idealism much farther, arguing that all qualities (color, taste, weight) are known only in the mind, and that there is no existence of matter independent of perception — *esse est percipii*. Qualities, not things, are perceived; perception is subjective, relative to the perceiver.

idealism, — the term "brain" making us think of a *thing*, the term "thought" of an *idea*. By the very wording of the question is prepared the double meaning which vitiates the answer. . . .

The truth is that *the philosopher unconsciously passes from the idealist to a pseudo-realist point of view.* He began by viewing the brain as an idea or picture exactly like all other ideas or pictures, encased in the other pictures and inseparable from them: the internal motion of the brain, being then a picture in the midst of pictures, was not required to provide the other pictures, since these were given with it and around it. But insensibly he comes to changing the brain and the intra-cerebral motion into *things,* that is to say, into *causes* hidden behind a particular picture and whose power extends far beyond what is presented. Whence this sliding from idealism to realism? It is favoured by many subtle fallacies; yet it would not be so smooth and easy were there not facts that seem to point in the same direction.

For, besides perception, there is memory. When I remember objects once perceived, the objects may be gone. One only has remained, my body; and yet the other objects may become visible again in the form of memory-images. Surely, then, it seems, my body, or some part of my body, has the power of evoking these images. Let us assume it does not create them; at least it is able to arouse them. How could it do this, were it not that to definite cerebral states correspond definite memory-images, and were there not, in this precise meaning, a parallelism between cerebral work and thought?

The reply is obvious: in the idealist hypothesis it is impossible for an object to be presented as an idea in the complete absence of the object itself. If there be nothing in the object over and above what is ideally present, if the presence of the object coincide with the idea we have of it, any part of the idea of the object must be in some sort a part of its presence. The recollection is no longer the object itself, I grant. Many things are wanted before it can be that. In the first place, it is fragmentary, for usually the recollection retains only some elements of the primitive perception. Again, it exists only for the person who evokes it, whereas the object forms part of a common experience. Lastly, when the memory-image arises, the accompanying modifications of the brain-image are no longer, as in perception, movements strong enough to excite the organism-image to react immediately. The body no longer feels *uplifted* by the perceived object, and since it is in the *suggestion of activity* that the *feeling of actuality* consists, the object presented no longer appears actual: this is what we express by saying that it is no longer present. The fact is that, in the idealist hypothesis, the memory-image can only be a pellicle detached from the primitive presentation or, what amounts to the same thing, from the object. It is always present, but consciousness turns its attention away from it so long as there is no reason for consciousness to consider it. Consciousness has an interest in perceiving it only when it feels itself capable of making use of it, that is to say, when the present cerebral state already outlines some of the nascent motor reactions which the real object (that is, the complete idea) would have determined: this beginning of bodily activity confers on the idea a beginning of actuality. But, then, there is no such thing as "parallelism" or "equivalence" between the memory-image and the cerebral state. For the nascent motor reactions portray some of the possible effects of the idea which is about to reappear, but they do not portray the idea; and as the same motor reaction may follow many very different recollections, it is not a definite recollection which is evoked by a definite bodily state; on the contrary, many different recollections are equally possible, and among them consciousness exercises a choice. They are subject to only one common condition — that of entering the same motor frame: in this lies their "resemblance," a term which is vague in current association theories, but which acquires a precise meaning when we define it by the identity of motor articulations. However, I shall not press this. I am content to say that in the idealist hypothesis the perceived objects are coincident with the complete and completely acting presentation, the remembered objects with the same, but incomplete and incompletely acting, presentation, and that neither in the case of perception nor in the case of memory is the cerebral state equivalent to the presentation, for the simple reason that it is part of it. . . .

ALFRED NORTH WHITEHEAD

"The Aim of Philosophy," from Modes of Thought

A Londoner by birth, Alfred North Whitehead (1861–1947) was educated at Sherbourne and Trinity College, Cambridge, and became both a stellar mathematician and a philosopher. Elected Fellow of the Royal Society in 1903, he coauthored, with Bertrand Russell, *Principia Mathematica* (1910–39). An avowed idealist, he published more characteristic works such as *Process and Reality* (1929), *Adventures of Ideas* (1933), and *Modes of Thought* (1938). In 1945 he was appointed to the Order of Merit.

Similar to William James in that his philosophical concepts were conveyed simply and in commanding language, Whitehead attracted Gertrude Stein intellectually. When, in 1914, she met Whitehead and his wife in England, and received their invitation to spend a weekend, she accepted with alacrity. Strangely, after Stein and Alice B. Toklas arrived, they found themselves stranded with the Whiteheads and their daughter Jessie — World War I had been declared. During the women's eleven-week stay, Whitehead and Stein had many long discussions on their afternoon walks and became good friends.

As his concise essay "The Aim of Philosophy" illustrates, Whitehead believed in the interrelationships of all fields of knowledge, seeing that all were devoted, in greater or lesser degree, to the pursuit of truth. All were based on an understanding of the human will.

The text reprinted here is taken from Whitehead's *Modes of Thought* (New York: Macmillan, 1957).

The task of a University is the creation of the future, so far as rational thought, and civilized modes of appreciation, can affect the issue. The future is big with every possibility of achievement and of tragedy.

Amid this scene of creative action, What is the special function of philosophy?

In order to answer this question, we must first decide what constitutes the philosophic character of any particular doctrine. What makes a doctrine philosophical? No one truth, thoroughly understood in all the infinitude of its bearings, is more or less philosophical than any other truth. The pursuit of philosophy is the one avocation denied to omniscience.

Philosophy is an attitude of mind towards doctrines ignorantly entertained. By the phrase 'ignorantly entertained' I mean that the full meaning of the doctrine in respect to the infinitude of circumstances to which it is relevant, is not understood. The philosophic attitude is a resolute attempt to enlarge the understanding of the scope of application of every notion which enters into our current thought. The philosophic attempt takes every word, and every phrase, in the verbal expression of thought, and asks, What does it mean? It refuses to be satisfied by the conventional presupposition that every sensible person knows the answer. As soon as you rest satisfied with primitive ideas, and with primitive propositions, you have ceased to be a philosopher.

Of course you have got to start somewhere for the purposes of discourse. But the philosopher, as he argues from his premises, has already marked down every word and phrase in them as topics for future enquiry. No philosopher is satisfied with the concurrence of sensible people, whether they be his colleagues, or even his own previous self. He is always assaulting the boundaries of finitude.

The scientist is also enlarging knowledge. He starts with a group of primitive notions and of primitive relations between these notions, which defines the scope of his science. For example, Newtonian dynamics assumes Euclidean space, massive matter, motion, stresses and strains, and the more general notion of force. There are also the laws of motion, and a few other concepts added later. The science consisted in the deduction of consequences, presupposing the applicability of these ideas.

In respect to Newtonian Dynamics, the scientist and the philosopher face in opposite directions. The scientist asks for the consequences, and seeks to observe the realization of such consequences in the universe. The philosopher asks for the meaning of these ideas in terms of the welter of characterizations which infest the world.

It is evident that scientists and philosophers can help each other. For the scientist sometimes wants a new idea, and the philosopher is enlightened as to meanings by the study of the scientific consequences. Their usual mode of intercommunication is by sharing in the current habits of cultivated thought.

There is an insistent presupposition continually sterilizing philosophic thought. It is the belief, the very natural belief, that mankind has consciously entertained all the fundamental ideas which are applicable to its experience. Further it is held that human language, in single words or in phrases, explicitly expresses these ideas. I will term this presupposition, The Fallacy of the Perfect Dictionary.

It is here that the philosopher, as such, parts company with the scholar. The scholar investigates human thought and human achievement, armed with a dictionary. He is the main support of civilized thought. Apart from scholarship, you may be moral, religious, and delightful. But you are not wholly civilized. You will lack power of delicate accuracy of expression.

It is obvious that the philosopher needs scholarship, just as he needs science. But both science and scholarship are subsidiary weapons for philosophy.

The Fallacy of the Perfect Dictionary divides philosophers into two schools, namely, the 'Critical School' which repudiates speculative philosophy, and the 'Speculative School' which includes it. The critical school confines itself to verbal analysis within the limits of the dictionary. The speculative school appeals to direct insight, and endeavours to indicate its meanings by further appeal to situations which promote such specific insights. It then enlarges the dictionary. The divergence between the schools is the quarrel between safety and adventure.

The strength of the critical school lies in the fact that the doctrine of evolution never entered, in any radical sense, into ancient scholarship. Thus there arises the presupposition of a fixed specification of the human mind; and the blue print of this specification is the dictionary.

I appeal to two great moments in the history of philosophy. Socrates[1] spent his life in analysing the current presuppositions of the Athenian world. He explicitly recognized that his philosophy was an attitude in the face of ignorance. He was critical and yet constructive.

Harvard is justly proud of the great period of its philosophic department about thirty years ago. Josiah Royce, William James, Santayana, George Herbert Palmer, Münsterberg,[2] constitute a group to be proud of. Among them Palmer's achievements centre chiefly in literature and in his brilliance as a lecturer. The group is a group of men individually great. But as a group they are greater still. It is a group of adventure, of speculation, of search for new ideas. To be a philosopher is to make some humble approach to the main characteristic of this group of men.

The use of philosophy is to maintain an active novelty of fundamental ideas illuminating the social system. It reverses the slow descent of accepted thought towards the inactive commonplace. If you like to phrase it so, philosophy is mystical. For mysticism is direct insight into depths as yet unspoken. But the purpose of philosophy is to rationalize mysticism: not by explaining it away, but by the introduction of novel verbal characterizations, rationally coördinated.

Philosophy is akin to poetry, and both of them seek to express that ultimate good sense which we term civilization. In each case there is reference to form beyond the direct meanings of words. Poetry allies itself to metre, philosophy to mathematic pattern.

[1] *Socrates:* Greek philosopher of Athens, 469–399 B.C.E., famous for his view of philosophy as a necessary study for all sentient men. Admired by Plato, he was eventually tried, condemned, and executed for, supposedly, corrupting the young.

[2] *Josiah Royce, William James, George Santayana, George Herbert Palmer, Hugo Münsterberg:* Key professors of philosophy (whether that be defined as religious thought, psychology, linguistics, natural science, or the variants that occurred in the discipline between the 1870s and the 1910s) at Harvard, when that department was an international leader in human thought.

4

Aesthetic Questions: Realism and the Modern

Literary historians cannot agree on the parameters of either realism or naturalism, but they do trace much of the tendency toward those modes in American letters back to European models. As Gustave Flaubert notes in his 1874 letter to the controversial Émile Zola, the latter's writing achieved a "ferocity of passion" usually not found in elite literature (see pp. 332–34). In particular, Flaubert likes the woman character, whose "hysteric state and . . . final avowal are marvellous" (see p. 333). Clearly, Zola has conceived an innovative kind of novel, and Flaubert is envious. Using one of Flaubert's short stories as a model, thirty years later, Gertrude Stein attempts to achieve the same kind of freshness with *Three Lives*.

Like Zola, Gertrude Stein writes about women. Even as she hides their passions, she flaunts their lower-class affiliations, their almost inarticulate expressions (at least in the case of Lena), and their definite lack of sophisticated mannerisms and speech. Stein wanted to become a part of the American realistic movement: the directions of William Dean Howells, Henry James, and, more important, of Hamlin Garland, Stephen Crane, and Frank Norris satisfied what she saw as the need to eradicate the imitation of established British writers. In Crane's probing of urban poverty (see pp. 341–47), even though he was not of the class he described, Stein found a style of observation that seemed more genuine to her than the achievements of either Howells or James. The intense focus on a character unable to express his or

Paintings hung at 27, rue de Fleurus, Leo and Gertrude Stein's salon, circa 1907–08. Work by Pierre Bonnard (central painting, *Siesta*), Pablo Picasso (girl with basket of flowers), Maurice Denis (nursing mother), Jean-Édouard Vuillard, Jean Renoir, and Paul Cézanne made the salon a showplace of as yet unappreciated contemporary art. Reproduced by permission of The Baltimore Museum of Art, from the Cone Archives.

her *own* inner longings provided the writer with a different kind of canvas.

Surrounded as she was with the literal canvases of Renoir, Bonnard, Matisse, Cézanne, Denis, Vuillard, Braque, Picasso, and other painters whose work could still be purchased reasonably because it was considered laughable in its grotesque originality, Gertrude Stein took comfort in the fact that the "new" was seldom valued. She drew sustenance from her friendships with many of these painters, people who lived in great poverty in order to accomplish their art. In sitting for Picasso's portrait of her, Stein walked four miles each way to reach his very meager living quarters — she made this journey between eighty and ninety times during the winter she was writing "Melanctha." Something of the sacrifice such painters as Picasso made daily, hourly, in order to create their art seeped into her own comfortable existence; her comprehension of the artist's seriousness became a part of her physical presence. Compared with what they were risking, her risks — of a brother's disapproval and, perhaps, society's — seemed small. It was for these personal reasons that Stein found herself identifying wholeheartedly with Henry James's pronouncements in his 1884 essay "The Art of the Novel" (see pp. 335–37). In this now-famous essay, James announced that if literature were to remain imitative, it would die. Rather, "[a]rt lives upon discussion, upon experiment, upon curiosity, upon variety of attempt, upon exchange of views and the comparison of standpoints; and there is a presumption that those times when no one has anything particular to say about it . . . are times . . . a little of dullness" (see p. 336). Like many of her friends who had graduated from college, Gertrude Stein had been content to follow established parameters of good taste — after all, developing that taste had been one reason for becoming better educated. But the stale conversation about aesthetics and art that dominated social gatherings of her class could not compete with the intense appreciation among these rebellious painters for the achievement of the truly new. Reading James, Stein found the rationale — and the word — that was to become a touchstone for her own development, and her own personality. The word was *interesting*. As James wrote,

> The only obligation to which in advance we may hold a novel, without incurring the accusation of being arbitrary, is that it be interesting. That general responsibility rests upon it, but it is the only one I can think of. The ways in which it is at liberty to accomplish this result (of interesting us) strike me as innumerable, and such as can only suffer from being marked out or fenced in by prescription. They are as various as the temperament of man, and they are successful in proportion as they reveal a particular mind, different from others. (p. 337)

There was no question that Stein's *Three Lives* fulfilled Henry James's mandate, for the book showed both the particular mind, different from others, of all three women characters — as well as that of the author herself.

In recent critical discourse, Stein's willingness to assume the class and race of her very diverse women characters is testimony to what Sieglinda Lemke calls "cross-over aesthetics" (146). Because Stein wrote as if in the spoken language of Melanctha, as well as that of Anna and Lena, she accepted their cultures and thought them worthy of mimicry. Even though Michael North points out in his book *The*

Dialect of Modernism that such "transformation" cannot hide its intrinsic "duality, a tension between what is and what might be," that attempt was crucial to the experimentation of modernism (67). Stein's prescience in anticipating the interest of white culture in the African American — whether black culture was represented by African sculpture, by Jean Toomer's *Cane,* by Carl Van Vechten's *Nigger Heaven,* or by her own representation of black characters' speech — was yet another of her forward-looking innovations. For North, Stein's "Melanctha" was a written script of Picasso's *Les Demoiselles d'Avignon,* which was finished in 1907, shortly after Stein's story. It goes without saying that Picasso's portrait of Stein also shares in the reliance on a highly stylized, masklike representation of women's faces.

GUSTAVE FLAUBERT

Letter to Émile Zola (1874)

It is important to realize that, in the nineteenth century, much of the literate population of the United States read widely in European and Continental literature. The impact of the British, French, and Russian novel during the nineteenth century was, in many ways, more significant than that of American fiction. Gustave Flaubert (1821–1880), a French novelist best known for his controversial 1857 work, *Madame Bovary,* became a key figure to twentieth-century modernists. An early proponent of realism, which did not transform United States literature until much later in the nineteenth century, Flaubert also brought a new insistence of the importance of craft to narrative: his emphasis on *le mot juste* (the one exact word) led directly to the self-consciousness of such early modernists as James Joyce, Ernest Hemingway, Ford Madox Ford, and — somewhat differently in execution — Gertrude Stein.

Flaubert also saw that the authorial self need not remain remote from one's writing. The several versions of his *Éducation sentimentale* (the first written early, the second published in 1870) became instructive texts for writers who valued their own experience as a component of their creative products. He was also revered for his *Salammbo* (1863) and *Three Tales* (1877), a collection that includes "A Simple Heart," one of the models for Stein's *Three Lives.* Much of his writing dealt with a woman protagonist, a somewhat new emphasis for Continental fiction. Choosing literature over his formal study of the law, Flaubert gave the rigor of precision in choices of words and diction a new kind of primacy.

While Stein spoke English more often than French, she read French. She was, in fact, in the process of translating one of Flaubert's stories from French into English when she began writing "The Good Anna."

Flaubert's letter to Émile Zola appears in *The Complete Works of Gustave Flaubert* (New York: M. Walter Dunne, 1904).

CROISSET, near Rouen, *June 3*, 1874.

I have read it — *La Conquête de Plassans* — read it all at one breath, as one swallows a glass of good wine; then I ruminated over it, and now, my dear friend, I can talk sensibly about it. I feared, after the *Ventre de Paris,* that you would bury yourself in the "system" in your resolution. But no! You are a good fellow! And your latest book is a fine, swaggering production!

Perhaps it fails in making prominent any special place, or having a central scene (a thing that never happens in real life), and perhaps also there is a little too much dialogue among the accessory characters. There! in picking you to pieces carefully, these are the only defects I discover. But what power of observation! what depth! what a masterly hand!

That which struck me most forcibly in the general tone of the work was the ferocity of passion underlying the surface of good-fellowship. That is very strong, old friend, very strong and broad, and well sustained.

What a perfect *bourgeois* is Mouret, with his curiosity, his avarice, his resignation, and his flatness! The Abbé Faujas is sinister and great — a true director! How well he manages the woman, how ably he makes himself her master, first in taking her up through charity, and then in brutalising her!

As to her (Marthe), I cannot express to you how much I admire her, and the art displayed in developing her character, or rather her malady. Her hysteric state and her final avowal are marvellous. How well you describe the breaking-up of the household!

I forgot to mention the Tronches, who are adorable ruffians, and the Abbé Bouvelle, who is exquisite with his fears and his sensibility.

Provincial life, the little gardens, the Paloque family, the Rastoil, and the tennis-parties, — perfect, perfect!

Your treatment of details is excellent, and you use the happiest words and phrases: "The tonsure like a cicatrice;" "I should like it better if he went to see the women;" "Mouret had stuffed the stove," etc.

And the circle of youth — that was a true invention! I have noted many other things on the margins, viz.:

The physical details which Olympe gives regarding her brother; the strawberry; the mother of the abbé ready to become his pander; and her old trunk.

The harshness of the priest, who waves away the handkerchief of his poor sweetheart, because he detects thereon "an odour of woman."

The description of the sacristy, with the name of M. Delangre on the wall — the whole phrase is a jewel.

But that which surpasses everything, that which crowns the whole work, is the end! I know of nothing more powerful than that *dénouement*. Marthe's visit at her uncle's house, the return of Mouret, and his inspection of the house! One is seized by fear, as in the reading of some fantastic tale, and one arrives at this effect by the tremendous realism, the intensity of truth. The reader feels his head turned, in sympathy with Mouret.

The insensibility of the *bourgeois*, who watches the fire seated in his armchair, is charming, and you wind up with one sublime stroke: the apparition of the *soutane* of the Abbé Serge at the bedside of his dying mother, as a consolation or a chastisement!

There is one bit of chicanery, however. The reader (that has no memory) does not know by instinct what motive prompts M. Rougon and Uncle Macquart to act as they do. Two paragraphs of explanation would have been sufficient.

Never mind! it is what it is, and I thank you for the pleasure it has given me.

Sleep on both ears, now your work is done!

Lay aside for me all the stupid criticisms it draws forth. That kind of document interests me very much.

HENRY JAMES

From *The Art of the Novel*

The epitome of the cultured, privately educated American, Henry James (1843–1916) was the son of Swedenborgian theologian Henry James, Sr., and the younger brother of philosopher William James. In 1862 he studied law at Harvard, but chose to become a literary person. Encouraged by the Cambridge circle and William Dean Howells at the *Atlantic Monthly,* James published critical essays to a comparatively wide readership; then, in 1876, after a decade of traveling throughout Europe (where he knew Gustave Flaubert, Guy de Maupassant, and Ivan Turgenev), he published his first novel, *Roderick Hudson.* What was to become known as the "international novel" found its roots in *The American* (1877), *The Europeans* (1878), and *Daisy Miller* (1879). A London resident for much of his life after 1876, James was an acute observer of United States culture — and many of his American characters are sympathetic, their naiveté a kind of untouched innocence.

Influential in part because of his own powerful fiction and in part because of the intellectual circles in which he moved, James created the vocabulary for serious, craft-based, realistic fiction. "The Art of the Novel" appeared in 1884, in the midst of such novels as *Washington Square* (1880), *The Portrait of a Lady* (1881), and *The Bostonians* and *The Princess Casamassima* (both 1886). His belief that the form of a fictional work was governed by its theme and characters, its emotional tone, foregrounded the innovations of modernism to come.

Other of his novels and novellas are *What Maisie Knew* and *The Spoils of Poynton* (1897), *The Turn of the Screw* (1898), *The Awkward Age* (1899), and the last three novels, usually considered his greatest, written after his last sojourn in the United States: *The Wings of the Dove* (1902), *The Ambassadors* (1903), and *The Golden Bowl* (1904). James took British citizenship in 1915.

One of the disappointments of Gertrude Stein and Alice B. Toklas's literary lives was that they never met Henry James. Both admired his writing immensely, and often read aloud from a James novel — they owned the entire New York edition — during the evening. On several visits to England, friends arranged for them to meet James, but such plans were cancelled because of his ill health.

These excerpts are taken from James's essay "The Art of the Novel" in his *Partial Portraits* (New York: Macmillan, 1888).

. . . Art lives upon discussion, upon experiment, upon curiosity, upon variety of attempt, upon the exchange of views and the comparison of standpoints; and there is a presumption that those times when no one has anything particular to say about it, and has no reason to give for practice or preference, though they may be times of honor, are not times of development — are times, possibly even, a little of dullness. The successful application of any art is a delightful spectacle, but the theory too is interesting; and though there is a great deal of the latter without the former I suspect there has never been a genuine success that has not had a latent core of conviction. Discussion, suggestion, formulation, these things are fertilizing when they are frank and sincere. . . . Literature should be either instructive or amusing, and there is in many minds an impression that these artistic preoccupations, the search for form, contribute to neither end, interfere indeed with both. They are too frivolous to be edifying, and too serious to be diverting; and they are moreover priggish and paradoxical and superfluous. That, I think, represents the manner in which the latent thought of many people who read novels as an exercise in skipping would explain itself if it were to become articulate. They would argue, of course, that a novel ought to be "good," but they would interpret this term in a fashion of their own, which indeed would vary considerably from one critic to another. One would say that being good means representing virtuous and aspiring characters, placed in prominent positions; another would say that it depends on a "happy ending," on a distribution at the last of prizes, pensions, husbands, wives, babies, millions, appended paragraphs, and cheerful remarks. Another still would say that it means being full of incident and movement, so that we shall wish to jump ahead, to see who was the mysterious stranger, and if the stolen will was ever found, and shall not be distracted from this pleasure by any tiresome analysis or "description." But they would all agree that the "artistic" idea would spoil some of their fun. One would hold it accountable for all the description, another would see it revealed in the absence of sympathy. Its hostility to a happy ending would be evident, and it might even in some cases render any ending at all impossible. . . . It matters little that as a work of art it should really be as little or as much of its essence to supply happy endings, sympathetic characters, and an objective tone, as if it were a work of mechanics: the association of ideas, however incongruous, might easily be too much for it if an eloquent voice were not sometimes raised to call attention to the fact that it is at once as free and as serious a branch of literature as any other. . . . the good health of an art which undertakes so immediately to reproduce life must demand that it be perfectly free. It lives upon exercise, and the very meaning of exercise is freedom. The only obligation to which in advance we may hold a novel, without incurring the accusation of being arbitrary, is that it be interesting. That general responsibility rests upon it, but it is the only one I can think of. The ways in which it is at liberty to accomplish this result (of interesting us) strike me as innumerable, and such as can only suffer from being marked out or fenced in by prescription. They are as various as the temperament of man, and they are successful in proportion as they reveal a particular mind, different from others. A novel is in its broadest definition a personal, a direct impression of life: that, to begin with, constitutes its value, which is greater or less according to the intensity of the impression. But there will be no intensity at all, and therefore no value, unless there is freedom to feel and say. The tracing of a line to be followed, of a tone to be taken, of a form to be filled out, is a limitation of that freedom and a suppression of the very thing that we are most curious about. The form, it seems to me, is to be appreciated after the fact: then the author's choice has been made, his standard has been indicated; then we can follow lines and directions and compare tones and resemblances. Then in a word we can enjoy one of the most charming of pleasures, we can estimate quality, we can apply the test of execution. The execution belongs to the author alone; it is what is most personal to him, and we measure him by that. The advantage, the luxury, as well as the torment and responsibility of the novelist, is that there is no limit to what he may attempt as an executant — no limit to his possible experiments, efforts, discoveries, successes. Here it is especially that he works, step by step, like his brother of the brush, of whom we may always say that he has painted his picture in a manner best known to himself. His manner is his secret, not necessarily a jealous one. He cannot disclose it as a general thing if he would; he would be at a loss to teach it to others. . . .

HAMLIN GARLAND

From *Crumbling Idols,* *Twelve Essays on Art and Literature*

An expert observer of the lives of poor and middle-class Americans, Hamlin Garland (1860–1940) himself never lived a privileged life. Born in western Wisconsin, he moved with his family to Iowa, then to the Dakota territory, and then to Illinois and Wisconsin before he homesteaded in South Dakota in 1883. As a graduate of the Cedar Valley Seminary in Osage, Iowa, in 1884 he went east to teach at the Boston School of Oratory. There the contrast between his life experiences and those of people in the more urban and educated areas of the United States sparked his imagination, and he began writing the stories that comprised his 1891 book, *Main-Travelled Roads.* In all, he published thirty of these dramatic fictions based on the hard lives of frontier characters — works that led to the movement for more realistic fiction which William Dean Howells and others were promoting. To have a truly American literature, Howells believed, writers must face the true conditions of life in the United States.

Garland's only theoretical work was *Crumbling Idols, Twelve Essays on Art and Literature,* published in 1894 by Stone and Kimball (New York). There he professed his allegiance to Howells' realism, but also pointed out that a better name for the incipient aesthetics of realism might be *veritism,* a mode in which the writer's passion for his or her locale would inscribe the tenets of accurate observation and recording. The writer must care about his subjects. Gertrude Stein's insistence on writing about American subjects, even when she lived abroad, was a reflection of her deep allegiance to her country and its literary currents.

Garland had a successful career as a popular novelist, writing such books as *Rose of Dutcher's Coolly* (1895), *Hesper* (1903), *Money Magic* (1907), and *The Forester's Daughter* (1914), but his best works tended to the autobiographical. *Boy Life on the Prairie* (1899) led to *A Son of the Middle Border* (1917) and *A Daughter of the Middle Border* (1921). He also wrote prolifically about Native American culture and about spiritualism.

II. New Fields.

The secret of every lasting success in art or literature lies, I believe, in a powerful, sincere, emotional concept of life first, and, second, in the acquired power to convey that concept to others. This leads necessarily to individuality in authorship, and to freedom from past models.

This *theory* of the veritist is, after all, a statement of his passion for truth and for individual expression. The passion does not spring from theory; the theory rises from the love of the verities, which seems to increase day by day all over the Western world.

The veritist, therefore, must not be taken to be dogmatic, only so far as he is personally concerned. He is occupied in stating his sincere convictions, believing that only in that way is the cause of truth advanced. He addresses himself to the mind prepared to listen. He destroys by displacement, not by attacking directly.

It is a settled conviction with me that each locality must produce its own literary record, each special phase of life utter its own voice. There is no other way for a true local expression to embody itself. The sun of truth strikes each part of the earth at a little different angle; it is this angle which gives life and infinite variety to literature. It is the subtle differences which life presents in California and Oregon, for example, which will produce, and justify, a Pacific-Coast literature.

. . . Veritism, as I understand it, puts aside all models, even living writers. Whatever he may do unconsciously, the artist must consciously stand alone before nature and before life. . . .

There is no necessity of treating the same material, however. Vast changes, already in progress, invite the writer. The coming in of horticulture, the immigration of farmers from all the Eastern States; the mingling of races; the feudalistic ownership of lands; the nomadic life of the farm-hands, the growth of cities, the passing Spanish civilization, — these are a few of the subjects which occur to me as worthy the best work of novelist and dramatist. . . .

V. Local Color in Art.

Local color in fiction is demonstrably the life of fiction. It is the native element, the differentiating element. It corresponds to the endless and vital charm of individual peculiarity. It is the differences which interest us; the similarities do not please, do not forever stimulate and feed as do the differences. Literature would die of dry rot if it chronicled the similarities only, or even largely. . . .

The fatal blight upon most American art has been, and is to-day, its imitative quality, which has kept it characterless and factitious, — a forced rose-culture rather than the free flowering of native plants.

Our writers despised or feared the home market. They rested their immortality upon the "universal theme," which was a theme of no interest to the public and of small interest to themselves.

During the first century and a half, our literature had very little national color. It was quite like the utterance of corresponding classes in England. But at length Bryant[1] and Cooper[2] felt the influence of our mighty forests and prairies. Whittier[3] uttered something of New England boy-life, and Thoreau[4] prodded about among newly discovered wonders, and the American literature got its first start.

Under the influence of Cooper came the stories of wild life from Texas, from Ohio, and from Illinois. The wild, rough settlements could not produce smooth and cultured poems or stories; they only furnished forth rough-and-ready anecdotes, but in these stories there were hints of something fine and strong and native. . . .

To-day we have in America, at last, a group of writers who have no suspicion of imitation laid upon them. Whatever faults they may be supposed to have, they are at any rate, themselves. American critics can depend upon a characteristic American literature of fiction and the drama from these people.

The corn has flowered, and the cotton-boll has broken into speech.

[1] *William Cullen Bryant:* Nineteenth-century American poet, essayist, and newspaper editor (1794–1878) beloved for his lyrics about nature's affect on human beings ("To a Waterfowl," "Thanatopsis," "The Yellow Violet").

[2] *James Fenimore Cooper:* Early United States novelist (1789–1851) who seized upon American subject matter — the frontier, the Native American culture — for his many popular novels.

[3] *John Greenleaf Whittier:* American Quaker poet and abolitionist-reformer (1807–1892) whose work was inspirational, again drawing on people's relationships with nature.

[4] *Henry David Thoreau:* Poet, naturalist, and essayist (1817–1862), best known for his *Walden,* the account of his life at Walden Pond, and *A Week on the Concord and Merrimack Rivers.*

STEPHEN CRANE

From *An Experiment in Misery*

Characteristic of his memorable, and always naturalistic, writing, this column from the New York press appeared the year after Stephen Crane (1871–1900) published *Maggie: A Girl of the Streets* at his own expense. As the son of a devout Methodist minister, he was influenced by the fact that both his parents had been journalists — as were two of his brothers. While his mother was still alive, Crane often collaborated on her columns; he was well published before he was twenty. Sadly, Crane was orphaned during his twenty-first year.

First a student in Asbury Park and then at the Pennington Seminary in New Jersey, he later attended Claverack College and the Hudson River Institute; by 1890 he was enrolled at Lafayette College, and during his semester at Syracuse University in the spring of 1891, rebelling against the religious mien of his family, he wrote the first version of *Maggie* while living in his fraternity house. Ubiquitous in his energy and appetites, he believed that writing must reflect the actuality of turn-of-the-century life. The publication in 1895 and 1896 of all three of his novels — *The Red Badge of Courage,* a reissue of *Maggie,* and *George's Mother* — made him a literary celebrity, both in the United States and in England. The cosmopolitan world saw the parallel between *Maggie* and such great European novels as *Madame Bovary,* and they revered *Red Badge* as a distinctively new genre — realistic in subject, yet poetic in form. *The Black Riders,* a collection of his short and often bitter poems appeared in 1895 as well; his first book of short stories, *The Little Regiment and Other Episodes of the American Civil War,* was published in 1896; and his fourth novel, *The Third Violet,* came out in 1897.

Beset by financial problems and tuberculosis, burned out by world travels as the consummate journalist, Crane died of tuberculosis in 1900, leaving the literary world a legacy — that literature should be all-encompassing, that no subject is unsuitable for the writer's consideration, and that a unique style is an important way of impressing readers with the value of the subject.

Stein appreciated Crane's innovations in style but she particularly admired his choosing lower-class characters as the subjects of his journalism and fiction.

These excerpts are from "An Experiment in Misery," which was published in *New York Press,* April 22, 1894.

AN EXPERIMENT IN MISERY/AN EVENING, A NIGHT AND A
MORN-/ING WITH THOSE CAST OUT./THE TRAMP LIVES
LIKE A KING/BUT HIS ROYALTY, TO THE NOVITIATE, HAS
DRAWBACKS OF SMELLS AND BUGS./LODGED WITH AN
ASSASSIN/A WONDERFULLY VIVID PICTURE OF A STRANGE
PHASE OF NEW YORK LIFE,/WRITTEN FOR "THE PRESS"
BY/THE AUTHOR OF "MAGGIE."[1]

Two men stood regarding a tramp.

"I wonder how he feels," said one, reflectively. "I suppose he is homeless, friendless, and has, at the most, only a few cents in his pocket. And if this is so, I wonder how he feels."

The other being the elder, spoke with an air of authoritative wisdom. "You can tell nothing of it unless you are in that condition yourself. It is idle to speculate about it from this distance."

"I suppose so," said the younger man, and then he added as from an inspiration: "I think I'll try it. Rags and tatters, you know, a couple of dimes, and hungry, too, if possible. Perhaps I could discover his point of view or something near it."

'Well, you might," said the other, and from those words begins this veracious narrative of an experiment in misery.

The youth went to the studio of an artist friend, who, from his store, rigged him out in an aged suit and a brown derby hat that had been made long years before. And then the youth went forth to try to eat as the tramp may eat, and sleep as the wanderers sleep. It was late at night, and a fine rain was swirling softly down, covering the pavements with a bluish luster. He began a weary trudge toward the downtown places, where beds can be hired for coppers. By the time he had reached City Hall Park he was so completely plastered with yells of "bum" and "hobo," and with various unholy epithets that small boys had applied to him at intervals that he was in a state of profound dejection, and looked searchingly for an outcast of high degree that the two might share miseries. But the lights threw a quivering glare over rows and circles of deserted benches that glistened damply, show-

Note: the footnote in this selection is original to the selection.
[1] *Reprinted from New York* Press: *22 April 1894.*
This autobiographical sketch has as its *mise en scène* the City Hall Park, located a few blocks from the Bowery. Crane is the young man who encounters another outcast to share his poverty. Together they go in search of a cheap boardinghouse where cots are seven cents for the night. This meeting with the seedy outcast and the flophouse business comprise the whole sketch, but what gives it importance is Crane's tone and style: irony combined with pity. It is the same as in *Maggie.*

ing patches of wet sod behind them. It seemed that their usual freights of sorry humanity had fled on this night to better things. There were only squads of well dressed Brooklyn people, who swarmed toward the Bridge.

[He Finds His Field.]

The young man loitered about for a time, and then went shuffling off down Park row. In the sudden descent in style of the dress of the crowd he felt relief. He began to see others whose tatters matched his tatters. In Chatham square there were aimless men strewn in front of saloons and lodging houses. He aligned himself with these men, and turned slowly to occupy himself with the pageantry of the street.

The mists of the cold and damp night made an intensely blue haze, through which the gaslights in the windows of stores and saloons shone with a golden radiance. The street cars rumbled softly, as if going upon carpet stretched in the aisle made by the pillars of the elevated road. Two interminable processions of people went along the wet pavements, spattered with black mud that made each shoe leave a scar-like impression. The high buildings lurked a-back, shrouded in shadows. Down a side street there were mystic curtains of purple and black, on which lamps dully glittered like embroidered flowers.

A saloon stood with a voracious air on a corner. A sign leaning against the front of the doorpost announced: "Free hot soup tonight." The swing doors snapping to and fro like ravenous lips, made gratified smacks, as if the saloon were gorging itself with plump men.

Caught by the delectable sign, the young man allowed himself to be swallowed. A bartender placed a schooner of dark and portentous beer on the bar. Its monumental form up-reared until the froth a-top was above the crown of the young man's brown derby.

[He Finds His Supper.]

"Soup over there, gents," said the bartender, affably. A little yellow man in rags and the youth grasped their schooners and went with speed toward a lunch counter, where a man with oily but imposing whiskers ladled genially from a kettle until he had furnished his two mendicants with a soup that was steaming hot and in which there were little floating suggestions of chicken. The young man, sipping his broth, felt the cordiality expressed by the warmth of the mixture, and he beamed at the man with oily but imposing whiskers, who was presiding like a priest behind an altar. "Have some more, gents?" he inquired of the two sorry figures before him. The little yellow man accepted with a swift gesture, but the youth shook his head and went out, following a man whose wondrous seediness promised that he would have a knowledge of cheap lodging houses.

On the sidewalk he accosted the seedy man. "Say, do you know a cheap place t' sleep?"

The other hesitated for a time, gazing sideways. Finally he nodded in the direction of up the street. "I sleep up there," he said, "when I've got th' price."

"How much?"

"Ten cents."

The young man shook his head dolefully. "That's too rich for me.". . .

[A Place of Smells.]

Shortly after the beginning of this journey the young man felt his liver turn white, for from the dark and secret places of the building there suddenly came to his nostrils strange and unspeakable odors that assailed him like malignant diseases with wings. They seemed to be from human bodies closely packed in dens; the exhalations from a hundred pairs of reeking lips; the fumes from a thousand bygone debauches; the expression of a thousand present miseries.

A man, naked save for a little snuff colored undershirt, was parading sleepily along the corridor. He rubbed his eyes, and, giving vent to a prodigious yawn, demanded to be told the time.

"Half past one."

The man yawned again. He opened a door, and for a moment his form was outlined against a black, opaque interior. To this door came the three men, and as it was again opened the unholy odors rushed out like released fiends, so that the young man was obliged to struggle as against an overpowering wind.

It was some time before the youth's eyes were good in the intense gloom within, but the man with benevolent spectacles led him skillfully, pausing but a moment to deposit the limp assassin upon a cot. He took the youth to a cot that lay tranquilly by the window, and, showing him a tall locker for clothes that stood near the head with the ominous air of a tombstone, left him.

[To the Polite, Horrors.]

The youth sat on his cot and peered about him. There was a gas jet in a distant part of the room that burned a small flickering orange hued flame. It caused vast masses of tumbled shadows in all parts of the place, save where, immediately about it, there was a little gray haze. As the young man's eyes became used to the darkness he could see upon the cots that thickly littered the floor the forms of men sprawled out, lying in deathlike silence or heaving and snoring with tremendous effort, like stabbed fish.

The youth locked his derby and his shoes in the mummy case near him and then lay down with his old and familiar coat around his shoulders. A blanket he handled gingerly, drawing it over part of the coat. The cot was leather covered and cold as melting snow. The youth was obliged to shiver for some time on this affair, which was like a slab. Presently, however, his chill gave him peace, and during this period of leisure from it he turned his head to stare at his friend, the assassin, whom he could dimly discern where he lay sprawled on a cot in the abandon of a man filled with drink. He was snoring with incredible vigor. His wet hair and beard dimly glistened and his inflamed nose shone with subdued luster like a red light in a fog.

Within reach of the youth's hand was one who lay with yellow breast and shoulders bare to the cold drafts. One arm hung over the side of the cot and the fingers lay full length upon the wet cement floor of the room. Beneath the inky brows could be seen the eyes of the man exposed by the partly opened lids. To the youth it seemed that he and this corpse-like being were exchanging a prolonged stare and that the other threatened with his eyes. He drew back, watching this neighbor from the shadows of his blanket edge. The man did not move once through the night, but lay in this stillness as of death, like a body stretched out, expectant of the surgeon's knife.

[Men Lay Like the Dead.]

And all through the room could be seen the tawny hues of naked flesh, limbs thrust into the darkness, projecting beyond the cots; upreared knees; arms hanging, long and thin, over the cot edges. For the most part they were statuesque, carven, dead. With the curious lockers standing all about like tombstones there was a strange effect of a graveyard, where bodies were merely flung.

Yet occasionally could be seen limbs wildly tossing in fantastic nightmare gestures, accompanied by guttural cries, grunts, oaths. And there was one fellow off in a gloomy corner, who in his dreams was oppressed by some frightful calamity, for of a sudden he began to utter long wails that went almost like yells from a hound, echoing wailfully and weird through this chill place of tombstones, where men lay like the dead.

The sound, in its high piercing beginnings that dwindled to final melancholy moans, expressed a red and grim tragedy of the unfathomable possibilities of the man's dreams. But to the youth these were not merely the shrieks of a vision pierced man. They were an utterance of the meaning of the room and its occupants. It was to him the protest of the wretch who feels the touch of the imperturbably granite wheels and who then cries with an impersonal eloquence, with a strength not from him, giving voice to the wail of a whole section, a class, a people. This, weaving into the young man's brain and mingling with his views of these vast and somber shadows that like mighty black fingers curled around the naked bodies, made the young man so that he did not sleep, but lay carving biographies for these men from his meager experience. At times the fellow in the corner howled in a writhing agony of his imaginations.

[Then Morning Came.]

Finally a long lance point of gray light shot through the dusty panes of the window. Without, the young man could see roofs drearily white in the dawning. The point of light yellowed and grew brighter, until the golden rays of the morning sun came in bravely and strong. They touched with radiant color the form of a small, fat man, who snored in stuttering fashion. His round and shiny bald head glowed suddenly with the valor of a decoration. He sat up, blinked at the sun, swore fretfully and pulled his blanket over the ornamental splendors of his head.

The youth contentedly watched this rout of the mystic shadows before the bright spears of the sun and presently he slumbered. When he awoke he heard the voice of the assassin raised in valiant curses. Putting up his head he perceived his comrade seated on the side of the cot engaged in scratching his neck with long finger nails that rasped like files.

"Hully Jee dis is a new breed. They've got can openers on their feet," he continued in a violent tirade.

The young man hastily unlocked his closet and took out his clothes.

As he sat on the side of the cot, lacing his shoes, he glanced about and saw that daylight had made the room comparatively commonplace and uninteresting. The men, whose faces seemed stolid, serene or absent, were engaged in dressing, while a great crackle of bantering conversation arose. A few were parading in unconcerned nakedness. Here and there were men of brawn, whose skins shone clear and ruddy. They took splendid poses, standing massively, like chiefs. When they had dressed in their ungainly garments there was an extraordinary change. They then showed bumps and deficiencies of all kinds. . . .

FRANK NORRIS

"Simplicity in Art," from The Responsibilities of the Novelist and Other Literary Essays

Born in Chicago, Frank Norris (1870–1902) was taken to Europe by his father when he was only seventeen because he had revealed an interest in painting. A student at the University of California in 1900, he read Émile Zola, Rudyard Kipling, and other realists, and, in 1894, he took the first draft of *McTeague* with him to Harvard. After a year in the east, he worked reporting the Boer War for the *San Francisco Chronicle* and *Collier's*. He later joined the staff of the San Francisco *Wave*, and reported the Santiago campaign in Cuba for *McClure's Magazine*, where he became ill with conditions that would lead to his death.

Norris's novels appeared at the end of the century: *Moran of the Lady Letty* (1898), *McTeague* (1899), *Blix* (1899), and *A Man's Woman* (1900). Writing of a different kind may have made a more lasting name for him — *The Octopus* (1901) showed how useful his kind of relentless naturalism was for reporting; *The Pit*, the second volume of his naturalistic trilogy on the production and distribution of wheat, appeared posthumously in 1902, after Norris had died during an operation for appendicitis. In 1914, the lost manuscript of *Vandover and the Brute* was published.

Presented here, in the essay "Simplicity in Art," Norris entreats the writer to deal with the facts of life, no matter how disgusting they might be, and to do so in the plainest and most direct manner possible. Cutting through the facade of intellectualism, of pretension in language, was the real mission of the new artist. Norris insisted that writers avoid "shamming . . . shoddyism . . . humbug," and that they do so without additional "comment." In his ideal twentieth-century prose, "There is little more than bare outline, but in the care with which it is drawn, how much thought, what infinite pains go to the making of each stroke, so that when it is made it falls just at the right place and exactly in its right sequence" (pp. 350–51).

Norris's words sound as if they might have been directing Gertrude Stein to her spare, innovative characterizations. While her first novellas (not published until after her death in 1946) show that she is trying to break out of conventional literary molds, no clear changes occur until she begins writing the stories of *Three Lives*.

The Norris essay was first published in *The Responsibilities of the Novelist and Other Literary Essays* (New York: Doubleday, Page and Co., 1901).

Simplicity in Art

Once upon a time I had occasion to buy so uninteresting a thing as a silver soup-ladle. The salesman at the silversmith's was obliging and for my inspection brought forth quite an array of ladles. But my purse was flaccid, anemic, and I must pick and choose with all the discrimination in the world. I wanted to make a brave showing with my gift — to get a great deal for my money. I went through a world of soup-ladles — ladles with gilded bowls, with embossed handles, with chased arabesques, but there were none to my taste. "Or perhaps," says the salesman, "you would care to look at something like this," and he brought out a ladle that was as plain and as unadorned as the unclouded sky — and about as beautiful. Of all the others this was the most to my liking. But the price! ah, that anemic purse; and I must put it from me! It was nearly double the cost of any of the rest. And when I asked why, the salesman said:

"You see, in this highly ornamental ware the flaws of the material don't show, and you can cover up a blow-hole or the like by wreaths and beading. But this plain ware has got to be the very best. Every defect is apparent."

And there, if you please, is a conclusive comment upon the whole business — a final basis of comparison of all things, whether commercial or artistic; the bare dignity of the unadorned that may stand before the world all unashamed, panoplied rather than clothed in the consciousness of perfection. We of this latter day, we painters and poets and writers — artists — must labour with all the wits of us, all the strength of us, and with all that we have of ingenuity and perseverance to attain simplicity. But it has not always been so. At the very earliest, men — forgotten, ordinary men — were born with an easy, unblurred vision that to-day we would hail as marvelous genius. Suppose, for instance, the New Testament was all unwritten and one of us were called upon to tell the world that Christ was born, to tell of how we had seen Him, that this was the Messiah.

How the adjectives would marshal upon the page, how the exclamatory phrases would cry out, how we would elaborate and elaborate, and how our rhetoric would flare and blazen till — so we should imagine — the ear would ring and the very eye would be dazzled; and even then we would believe that our words were all so few and feeble. It is beyond words, we should vociferate. So it would be. That is very true — words of ours. Can you not see how we should dramatize it? We would make a point of the transcendent stillness of the hour, of the deep blue of the Judean midnight, of the lip-lapping of Galilee, the murmur of Jordan, the peacefulness of sleeping Jerusalem. Then the stars, the descent of the angel, the shepherds — all the accessories. And our narrative would be as commensurate with the subject as the flippant smartness of a "bright" reporter in the Sistine chapel. We would be striving to cover up our innate incompetence, our impotence to do justice to the mighty theme by elaborateness of design and arabesque intricacy of rhetoric.

But on the other hand — listen:

"The days were accomplished that she should be delivered, and she brought forth her first born son and wrapped him in swaddling clothes and laid him in a manger, because there was no room for them in the inn."

Simplicity could go no further. Absolutely not one word unessential, not a single adjective that is not merely descriptive. The whole matter stated with the terseness of a military report, and yet — there is the epic, the world epic, beautiful, majestic, incomparably dignified, and no ready writer, no Milton[1] nor Shakspere,[2] with all the wealth of their vocabularies, with all the resources of their genius, with all their power of simile or metaphor, their pomp of eloquence or their royal pageantry of hexameters, could produce the effect contained in these two simple declarative sentences.

[1] *Milton:* John Milton, English poet (1608–1674) famous for his early odes "L'Allegro" and "Il Penseroso," his interrogation of church law, and his magnificent study of religious belief, *Paradise Lost.*

[2] *Shakspere:* William Shakespeare, English dramatist and poet (1564–1616) considered the greatest writer in English literature.

The mistake that we little people are so prone to make is this: that the more intense the emotional quality of the scene described, the more "vivid," the more exalted, the more richly coloured we suppose should be the language.

When the crisis of the tale is reached there is where we like the author to spread himself, to show the effectiveness of his treatment. But if we would only pause to take a moment's thought we must surely see that the simplest, even the barest statement of fact is only not all-sufficient but all-appropriate.

Elaborate phrase, rhetoric, the intimacy of metaphor and allegory and simile is forgivable for the unimportant episodes where the interest of the narrative is languid; where we are willing to watch the author's ingenuity in the matter of scrolls and fretwork and mosaics-rococo work. But when the catastrophe comes, when the narrative swings clear upon its pivot and we are lifted with it from out the world of our surroundings, we want to forget the author. We want no adjectives to blur our substantives. The substantives may now speak for themselves. We want no metaphor, no simile to make clear the matter. If at this moment of drama and intensity the matter is not of itself preëminently clear no verbiage, however ingenious, will clarify it. Heighten the effect. Does exclamation and heroics on the part of the bystanders ever make the curbstone drama more poignant? Who would care to see Niagara through coloured fire and calcium lights.

The simple treatment, whether of a piece of silversmith work or of a momentous religious epic, is always the most difficult of all. It demands more of the artist. The unskilful story-teller as often as not tells the story to himself as well as to his hearers as he goes along. Not sure of exactly how he is to reach the end, not sure even of the end itself, he must feel his way from incident to incident, from page to page, fumbling, using many words, repeating himself. To hide the confusion there is one resource — elaboration, exaggerated outline, violent colour, till at last the unstable outline disappears under the accumulation, and the reader is to be so dazzled with the wit of the dialogue, the smartness of the repartee, the felicity of the diction, that he will not see the gaps and lapses in the structure itself — just as the "nobby" drummer wears a wide and showy scarf to conceal a soiled shirt-bosom.

But in the master-works of narrative there is none of this shamming, no shoddyism, no humbug. There is little more than bare outline, but in the care with which it is drawn, how much thought, what infinite pains go to the making of each stroke, so that when it is made it falls just at the right place and exactly in its right sequence. This attained, what need is there for more? Comment is superfluous. If the author make the scene appear terrible to the reader he need not say in himself or in the mouth of some protagonist, "It is terrible!" If the picture is pathetic so that he who reads must weep, how superfluous, how intrusive should the author exclaim, "It was pitiful to the point of tears." If beautiful, we do not want him to tell us so. We want him to make it beautiful and our own appreciation will supply the adjectives.

Beauty, the ultimate philosophical beauty, is not a thing of elaboration, but on the contrary of an almost barren nudity: a jewel may be an exquisite gem, a woman may have a beautiful arm, but the bracelet does not make the arm more beautiful, nor the arm the bracelet. One must admire them separately, and the moment that the jewel ceases to have a value or a reason upon the arm it is better in the case, where it may enjoy an undivided attention.

But after so many hundreds of years of art and artists, of civilization and progress, we have got so far away from the sane old homely uncomplex way of looking out at the world that the simple things no longer charm, and the simple declarative sentence, straightforward, plain, seems flat to our intellectual palate — flat and tasteless and crude.

What we would now call simple our forbears would look upon as a farrago of gimcrackery, and all our art — the art of the better-minded of us — is only a striving to get back to the unblurred, direct simplicity of those writers who could see that the Wonderful, the Counselor, the mighty God, the Prince of Peace, could be laid in a manger and yet be the Saviour of the world.

It is this same spirit, this disdaining of simplicity that has so warped and inflated The First Story, making of it a pomp, an affair of gold-embroidered vestments and costly choirs, of marbles, of jeweled windows and of incense, unable to find the thrill as formerly in the plain and humble stable, and the brown-haired, grave-eyed peasant girl, with her little baby; unable to see the beauty in the crumbling mud walls, the low-ceiled interior, where the only incense was the sweet smell of the cow's breath, the only vestments the swaddling clothes, rough, coarse-fibered, from the hand-looms of Nazareth, the only pomp the scanty gifts of three old men, and the only chanting the crooning of a young mother holding her first-born babe upon her breast.

WILLIAM EDWARD BURGHARDT DU BOIS

From *The Souls of Black Folk*

William Edward Burghardt Du Bois (1868–1963), a pioneering African American leader and twentieth-century intellectual force, exerted his influence in a wide variety of spheres, including politics, literature, education, sociology, and philosophy. Although his work traversed interdisciplinary boundaries and his philosophical thought changed continually over the course of his lifetime, his primary objective remained constant: to understand and improve race relations. Born of African, French Huguenot, and Dutch ancestry, and raised for the first seventeen years of his life in the small, predominantly white community of Great Barrington, Massachusetts, Du Bois gained exposure early on to a variety of perspectives on the "problem of the color line." His experience as an undergraduate at the all-black Fisk University in Nashville, Tennessee, facilitated his growing awareness of the widespread discrimination practiced against African Americans in the United States. Upon graduating from Fisk, Du Bois entered Harvard University as a junior. He then spent two years studying at the University of Berlin, and went on to become the first African American to receive his doctorate from Harvard.

Perhaps best known for his role as a social leader, Du Bois played an influential role in the Niagara Movement of 1905, a group of pioneering African American scholars and leaders, which ultimately led to the formation of the National Association for the Advancement of Colored People (NAACP) in 1910. His political leanings evolved over his lifetime from an early endorsement of democratic individualism to an alignment with socialist thinkers who related the problems of African Americans in terms of capitalist oppression. Having grown disillusioned with the American government by 1961, Du Bois joined the Communist Party; in that same year, he left the United States with his wife, Shirley Graham Du Bois, and emigrated to Ghana. His reconnection to his African roots, along with his recognition of the importance of a communal basis for social action, presaged his role in the founding of the Pan African Movement.

In addition to practicing social activism, Du Bois left a formidable literary legacy, producing twenty-three books on sociology, history, political science, fiction, and race relations. Of these works, some of the most influential included *The Suppression of the African Slave Trade* (1896), *The Philadelphia Negro* (1899), *The Souls of Black Folk* (1903), *John Brown* (1909), and *Black Reconstruction* (1935).

The fourteen essays included in *The Souls of Black Folk* chart the trajectory of Du Bois's inner journey and represent the range of his thought on the American racial dilemma. His provocative and often quoted concept of African American "double-consciousness" appears in the first chapter, "Of Our Spiritual Strivings." This concept draws attention to the unique subject position of African Americans, wherein individuals come to know themselves exclusively "through the eyes of others," and yet find themselves unequivocally "shut out" from membership in this community of judgment. In this sense, the "double-consciousness" functions both as a gift of "second sight" as well as an uneasy condition — characterized by continual "longing" for unity in the self (pp. 354–56). Despite the existential despair implicit in the concept, Du Bois remained optimistic about the possibility of restoring America's moral idealism and rectifying the race problem through knowledge and education. Throughout the course of his career and up until his death at the age of ninety-five, Du Bois maintained an optimistic and poetic vision of humankind, resolutely approaching the problem of racial inequality within the context of its fuller human significance.

Du Bois's link with both Gertrude and Leo Stein was their shared Harvard training. Although Leo Stein graduated from Johns Hopkins, he had spent several years as an undergraduate student at Harvard, and another year in Harvard Law. He and his sister had many friends in common (she began a year after Leo, attending Radcliffe, which was at that time called the Harvard Annex), friends that also either lived or visited them in Paris. They were heavily invested in what happened at Harvard, and what happened to alumni of the college.

This essay appeared in *The Souls of Black Folk: Essays and Sketches,* published in 1903 in Chicago by A. C. McClurg.

Chapter III
Of Mr. Booker T. Washington and Others

From birth till death enslaved; in word, in deed, unmanned!
. . . .
Hereditary bondsmen! Know ye not
Who would be free themselves must strike the blow?[1]
                                        — Byron

---

[1] *From birth . . . blow?*: Epigraph from Lord George Gordon Byron's *Childe Harold's Pilgrimage,* canto II.

Easily the most striking thing in the history of the American Negro since 1876[2] is the ascendancy of Mr. Booker T. Washington. It began at the time when war memories and ideals were rapidly passing; a day of astonishing commercial development was dawning; a sense of doubt and hesitation overtook the freedmen's sons, — . . . .

After the Egyptian and Indian, the Greek and Roman, the Teuton and Mongolian, the Negro is a sort of seventh son, born with a veil, and gifted with second-sight in this American world, — a world which yields him no true self-consciousness, but only lets him see himself through the revelation of the other world. It is a peculiar sensation, this double-consciousness, this sense of always looking at one's self through the eyes of others, of measuring one's soul by the tape of a world that looks on in amused contempt and pity. One ever feels his two-ness, — an American, a Negro; two souls, two thoughts, two unreconciled strivings; two warring ideals in one dark body, whose dogged strength alone keeps it from being torn asunder.

The history of the American Negro is the history of this strife, — this longing to attain self-conscious manhood, to merge his double self into a better and truer self. In this merging he wishes neither of the older selves to be lost. He would not Africanize America, for America has too much to teach the world and Africa. He would not bleach the Negro soul in a flood of white Americanism, for he knows that Negro blood has a message for the world. He simply wishes to make it possible for a man to be both a Negro and an American, without being cursed and spit upon by his fellows, without having the doors of Opportunity closed roughly in his face.

This, then, is the end of his striving: to be a co-worker in the kingdom of culture, to escape both death and isolation, to husband and use his best powers and his latent genius. These powers of body and mind have in the past been strangely wasted, dispersed, or forgotten. The shadow of a mighty Negro past flits through the tale of Ethiopia the Shadowy and of Egypt the Sphinx. Throughout history, the powers of single black men flash here and there like falling stars, and die sometimes before the world has rightly gauged their brightness. Here in America, in the few days since Emancipation, the black man's turning hither and thither in hesitant and doubtful striving has often made his very strength to lose effectiveness, to seem like absence of power, like weakness. And yet it is not weakness, — it is the contradiction of

---

[2] *1876:* This is the year federal troops were removed from the South and, in effect, support for African American political power ended.

double aims. The double-aimed struggle of the black artisan — on the one hand to escape white contempt for a nation of mere hewers of wood and drawers of water, and on the other hand to plough and nail and dig for a poverty-stricken horde — could only result in making him a poor craftsman, for he had but half a heart in either cause. By the poverty and ignorance of his people, the Negro minister or doctor was tempted toward quackery and demagogy; and by the criticism of the other world, toward ideals that made him ashamed of his lowly tasks. The would-be black *savant* was confronted by the paradox that the knowledge his people needed was a twice-told tale to his white neighbors, while the knowledge which would teach the white world was Greek to his own flesh and blood. The innate love of harmony and beauty that set the ruder souls of his people a-dancing and a-singing raised but confusion and doubt in the soul of the black artist; for the beauty revealed to him was the soul-beauty of a race which his larger audience despised, and he could not articulate the message of another people. This waste of double aims, this seeking to satisfy two unreconciled ideals, has wrought sad havoc with the courage and faith and deeds of ten thousand thousand people, — has sent them often wooing false gods and invoking false means of salvation, and at times has even seemed about to make them ashamed of themselves.

Away back in the days of bondage they thought to see in one divine event the end of all doubt and disappointment; few men ever worshipped Freedom with half such unquestioning faith as did the American Negro for two centuries. To him, so far as he thought and dreamed, slavery was indeed the sum of all villainies, the cause of all sorrow, the root of all prejudice; Emancipation was the key to a promised land of sweeter beauty than ever stretched before the eyes of wearied Israelites. In song and exhortation swelled one refrain — Liberty; in his tears and curses the God he implored had Freedom in his right hand. At last it came, — suddenly, fearfully, like a dream. With one wild carnival of blood and passion came the message in his own plaintive cadences: —

"Shout, O children!
Shout, you're free!
For God has bought your liberty!"

Years have passed away since then, — ten, twenty, forty; forty years of national life, forty years of renewal and development, and yet the swarthy spectre sits in its accustomed seat at the Nation's feast. In vain do we cry to this our vastest social problem: —

"Take any shape but that, and my firm nerves
Shall never tremble!"

The Nation has not yet found peace from its sins; the freedman has not
yet found in freedom his promised land. Whatever of good may have
come in these years of change, the shadow of a deep disappointment
rests upon the Negro people, — a disappointment all the more bitter
because the unattained ideal was unbounded save by the simple ignorance of a lowly people.

The first decade was merely a prolongation of the vain search for
freedom, the boon that seemed ever barely to elude their grasp, — like
a tantalizing will-of-the-wisp, maddening and misleading the headless
host. The holocaust of war, the terrors of the Ku-Klux-Klan, the lies of
carpet-baggers, the disorganization of industry, and the contradictory
advice of friends and foes, left the bewildered serf with no new watch-
word beyond the old cry for freedom. As the time flew, however, he
began to grasp a new idea. The ideal of liberty demanded for its attainment powerful means, and these the Fifteenth Amendment gave him.
The ballot, which before he had looked upon as a visible sign of freedom, he now regarded as the chief means of gaining and perfecting the
liberty with which war had partially endowed him. And why not? Had
not votes made war and emancipated millions? Had not votes enfran-
chised the freedmen? Was anything impossible to a power that had
done all this? A million black men started with renewed zeal to vote
themselves into the kingdom. So the decade flew away, the revolution
of 1876 came, and left the half-free serf weary, wondering, but still
inspired. Slowly but steadily, in the following years, a new vision
began gradually to replace the dream of political power, — a powerful
movement, the rise of another ideal to guide the unguided, another
pillar of fire by night after a clouded day. It was the ideal of "book-
learning"; the curiosity, born of compulsory ignorance, to know and
test the power of the cabalistic letters of the white man, the longing to
know. Here at last seemed to have been discovered the mountain path
to Canaan; longer than the highway of Emancipation and law, steep
and ragged, but straight, leading to heights high enough to overlook
life.

Up the new path the advance guard toiled, slowly, heavily, dog-
gedly; only those who have watched and guided the faltering feet, the
misty minds, the dull understandings, of the dark pupils of those
schools know how faithfully, how piteously, this people strove to
learn. It was weary work. The cold statistician wrote down the inches

of progress here and there, noted also where here and there a foot had slipped or some one had fallen. To the tired climbers, the horizon was ever dark, the mists were often cold, the Canaan was always dim and far away. If, however, the vistas disclosed as yet no goal, no resting-place, little but flattery and criticism, the journey at least gave leisure for reflection and self-examination; it changed the child of Emancipation to the youth with dawning self-consciousness, self-realization, self-respect. In those sombre forests of his striving his own soul rose before him, and he saw himself, — darkly as through a veil; and yet he saw in himself some faint revelation of his power, of his mission. He began to have a dim feeling that, to attain his place in the world, he must be himself, and not another. For the first time he sought to analyze the burden he bore upon his back, that dead weight of social degradation partially masked behind a half-named Negro problem. He felt his poverty; without a cent, without a home, without land, tools, or savings, he had entered into competition with rich, landed, skilled neighbors. To be a poor man is hard, but to be a poor race in a land of dollars is the very bottom of hardships. He felt the weight of his ignorance, — not simply of letters, but of life, of business, of the humanities; the accumulated sloth and shirking and awkwardness of decades and centuries shackled his hands and feet. Nor was his burden all poverty and ignorance. The red stain of bastardy, which two centuries of systematic legal defilement of Negro women had stamped upon his race, meant not only the loss of ancient African chastity, but also the hereditary weight of a mass of corruption from white adulterers, threatening almost the obliteration of the Negro home.

A people thus handicapped ought not to be asked to race with the world, but rather allowed to give all its time and thought to its own social problems. But alas! while sociologists gleefully count his bastards and his prostitutes, the very soul of the toiling, sweating black man is darkened by the shadow of a vast despair. Men call the shadow prejudice, and learnedly explain it as the natural defence of culture against barbarism, learning against ignorance, purity against crime, the "higher" against the lower" races. To which the Negro cries Amen! and swears that to so much of this strange prejudice as is founded on just homage to civilization, culture, righteousness, and progress, he humbly bows and meekly does obeisance. But before that nameless prejudice that leaps beyond all this he stands helpless, dismayed, and well-nigh speechless; before that personal disrespect and mockery, the ridicule and systematic humiliation, the distortion of fact and wanton license of fancy, the cynical ignoring of the better and the

boisterous welcoming of the worse, the all-pervading desire to incul-
cate disdain for everything black, from Toussaint[3] to the devil, —
before this there rises a sickening despair that would disarm and dis-
courage any nation save that black host to whom "discouragement" is
an unwritten word.

But the facing of so vast a prejudice could not but bring the
inevitable self-questioning, self-disparagement, and lowering of ideals
which ever accompany repression and breed in an atmosphere of contempt and hate. Whisperings and portents came borne upon the four
winds: Lo! we are diseased and dying, cried the dark hosts; we cannot
write, our voting is vain; what need of education, since we must
always cook and serve? And the Nation echoed and enforced this self-
criticism, saying: Be content to be servants, and nothing more; what
need of higher culture for half-men? Away with the black man's ballot,
by force or fraud, — and behold the suicide of a race! Nevertheless,
out of the evil came something of good, — the more careful adjustment of education to real life, the clearer perception of the Negroes'
social responsibilities, and the sobering realization of the meaning of
progress.

So dawned the time of *Sturm und Drang:* storm and stress to-day
rocks our little boat on the mad waters of the world-sea; there is
within and without the sound of conflict, the burning of body and
rending of soul; inspiration strives with doubt, and faith with vain
questionings. The bright ideals of the past, — physical freedom, political power, the training of brains and the training of hands, — all these
in turn have waxed and waned, until even the last grows dim and
overcast. Are they all wrong, — all false? No, not that, but each alone
was over-simple and incomplete, — the dreams of a credulous race-
childhood, or the fond imaginings of the other world which does not
know and does not want to know our power. To be really true, all
these ideals must be melted and welded into one. The training of the
schools we need to-day more than ever, — the training of deft hands,
quick eyes and ears, and above all the broader, deeper, higher culture
of gifted minds and pure hearts. The power of the ballot we need in
sheer self-defence, — else what shall save us from a second slavery?
Freedom, too, the long-sought we still seek, — the freedom of life and
limb, the freedom to work and think, the freedom to love and aspire.

[3] *Toussaint:* Toussaint L'Overture was a Haitian general who led peasants in Haiti
against the French army during the time of the French Revolution. He symbolizes the
unquenchable spirit of blacks in the New World.

Work, culture, liberty, — all these we need, not singly but together, not successively but together, each growing and aiding each, and all striving toward that vaster ideal that swims before the Negro people, the ideal of human brotherhood, gained through the unifying ideal of Race; the ideal of fostering and developing the traits and talents of the Negro, not in opposition to or contempt for other races, but rather in large conformity to the greater ideals of the American Republic, in order that some day on American soil two world-races may give to each those characteristics both so sadly lack. We the darker ones come even now not altogether empty-handed: there are to-day no truer exponents of the pure human spirit of the Declaration of Independence than the American Negroes; there is no true American music but the wild sweet melodies of the Negro slave; the American fairy tales and folk-lore are Indian and African; and, all in all, we black men seem the sole oasis of simple faith and reverence in a dusty desert of dollars and smartness. Will America be poorer if she replace her brutal dyspeptic blundering with light-hearted but determined Negro humility? or her coarse and cruel wit with loving jovial good-humor? or her vulgar music with the soul of the Sorrow Songs?

Merely a concrete test of the underlying principles of the great republic is the Negro Problem, and the spiritual striving of the freedmen's sons is the travail of souls whose burden is almost beyond the measure of their strength, but who bear it in the name of an historic race, in the name of this the land of their fathers' fathers, and in the name of human opportunity. . . .

5

Modernism and the Avant-Garde

When Cézanne wrote to his son Paul, with some surprise, that eight of his paintings were to be exhibited at the *Salon d'Automme* (1906) one sees the transition into modernism coming slyly — to the amazement of even its greatest artists, who were long used to being ignored. So it is with all aesthetic and literary "movements," despite the fact that in retrospect critics like to pretend that everyone was on the same page and of the same opinion. Disagreements such as Henri Matisse experienced (literal fisticuffs) were more likely to be the tenor of incipient change. And change took years; indeed, some change is never fully achieved. Think of the controversy in our own lifetimes about the Cubist paintings of Picasso — the fact that many viewers appreciate his "blue period" paintings and refuse to see the genius in the stark angles and nonhuman lines and shapes of his work of a decade later.

To Gertrude Stein during her early Paris years, however, the "new" was exciting (or, as she spelled the word, *xciting*). As Leo Stein's letter to Mabel Weeks shows, he, too, was excited by the "new"; he was truly well-educated (though, naturally, self-educated) in the aesthetics of the European art world, and his later published columns and books are standards of aesthetic appreciation. Gertrude learned much from Leo, as well as from her older brother and his wife, Michael and Sally Stein, and each household purchased as much contemporary art as it could. The Saturday salons that each came to hold — because of the demands of friends and acquaintances that they be allowed to see these paintings and sculptures — fed into each other: Mike and Sally's salon was earlier, followed by the less sedate showings at Leo and Gertrude's. The latter sometimes went until dawn.

In the essays included here by Mabel Dodge Luhan and Carl Van Vechten, the position of Stein's fiction in relation to the near-hysteria over the impressionist, modern, and cubist painters is accurately described (see pp. 372–78). But rather than let the achievement of *Three Lives* (and her later 1914 experiment in poetry, the word paintings of *Tender Buttons*) rest on its association with the graphic art the world now recognizes as monumental, presented here is the assessment of the greatness of Stein's writing in the opinion of another fine modern American writer. Wright Morris, in *About Fiction,* writes: "The triumph of Stein is that of the emerging vernacular. Joyce, Mann, and Proust mark a summing up, *Three Lives* an inexhaustible beginning. . . . [here] the voice of Gertrude Stein is both provocative and deceptively simple. The apparently simple proves to be complex. Events are subordinated to consciousness. The language is musical but without strain, as if it came as naturally as conversation. And that point is crucial." Long before our present-day interest in gender, Morris concludes, "*Three Lives* is the first fiction that seems entirely free of the obsessions that dominate the male-written novel: it's all a matter of feeling, an utterly womanly rendering of the complex simple heart" (50, 128, 142).

Less a matter of gender than Morris might admit, Stein's emphasis on creating portraits that ring true emotionally stemmed — as we have seen — from her studies in philosophy and psychology, from her intimacy with that true humanist, William James, and from her willingness to accept experiential knowledge as well as the knowledge of scholarship. As Edmund Wilson wrote about her being "a master of fiction" in the 1923 *Vanity Fair,* Stein's *Three Lives* was "realism of rather a novel kind" (60). Instead of using her portraits as Flaubert did, to make the reader conscious of the author's brilliance, Stein made the reader stay closer to the subjects of her portraits — "the projection of three actual human beings as complex and as complete as life. The style itself . . . takes on the very accents and the rhythms of the minds whose adventures it is recording" (80).

For William Carlos Williams, writing in 1930, Stein's earliest work reveals "the essence of all knowledge." He particularly admired "Melanctha," which he called "a thrilling clinical record of the life of a colored woman in the present day United States, told with directness and truth. It is without question one of the best bits of characterization produced in America. It is universally admired" (55, 57).

Pablo Picasso's portrait of Gertrude Stein, painted during the winter of 1906 when she was writing "Melanctha." Reproduced by permission of The Metropolitan Museum of Art, Bequest of Gertrude Stein, 1946 (47.106).

In the cases of both Wilson and Williams, Stein's writing was mysterious — less realistic than the realistic mentors she probably knew, more direct and factual than most descriptive writing. For each critic, a solution to the mystery could have been found in another realm of art — that of painting. According to the lecture notes of Henri Matisse, the congruence of aesthetic currents in both painting and literature occurs because in them one finds many of the same principles: an insistence on the whole expression, parts subordinate to the total effect; a demand that the center of a painting be "the human figure . . . neither still life nor landscape" (38). In Matisse's words, "It is that which best permits me to express my almost religious awe towards life. I do not insist upon all the details of the face, on setting them down one-by-one with anatomical exactitude. . . . A work of art must carry within itself its complete significance and impose that upon the beholder" (38). To conclude, Matisse admits, "What I dream of is an art of balance, of purity and serenity. . . . The simplest means are those which best enable an artist to express himself. His means of expression must derive almost of necessity from his temperament" (38–39). Or as Gertrude Stein echoed, "my only thought is a complicated simplicity. I like a thing simple, but it must be simple through complication. Everything must come into your scheme; otherwise you cannot achieve real simplicity" (Haas 34).

The simplicity of Stein's *Three Lives* proves her modest contention.

LEO STEIN

Letter to Mabel Weeks about European Painting

Leo (1872–1947) was Gertrude's immediately older brother. The two of them formed a unit, set somewhat apart from the three older children. Alike in their interests in language and literature, they were both bereaved with the death of first their mother (in their young adolescence) and a few years later, their father.

Leo attended the University of California, Berkeley, as well as Harvard and Johns Hopkins: he majored in a number of things — political science, biology, law — but always saw himself as a literary and aesthetic expert. His published works are on the aesthetics of art; various reviews of exhibits and painters can be found in *The Nation, Seven Arts, The American Scholar, Dial, American Mercury,* and *The New Republic,* and in his books, *The ABC of Aesthetics* (1927) and *Appreciation: Painting, Poetry and Prose* (1947). In some circles, Leo was as well known as his younger sister.

Their respective fame was partially to blame for the end of their household living arrangements. Early in the twentieth century, Leo decided to live abroad. After trying Italy, he settled in Paris, at what was to become the famous address, 27, rue de Fleurus. He began buying paintings long before Gertrude joined him (after her years of medical school and personal heartbreak in Baltimore). By 1907–08, they had compiled the fascinating holdings that marked their fame as collectors, and had formed friendships with both Henri Matisse and Pablo Picasso. Once Alice B. Toklas joined the household, however, Leo felt less in charge, and eventually — and stormily — left, taking with him the Renoirs and leaving the Picassos. He later married Nina Auzias, a woman Gertrude thought unsuitable, and they lived in Italy through World War II. Leo Stein died of the same intestinal cancer that took several of his siblings, and perhaps his mother.

This letter is published in a commemorative volume for Leo Stein, *Journey into the Self,* ed. Edmund Fuller (New York: Crown, 1950).

27 Rue de Fleurus
[Undated]

My dear Mabel,

There is so much to be said and since I have taken to the brush I have a greater aversion than ever to the pen so that I am doubtful if even a small portion of it will come to utterance. At all events, I will make a beginning and if this proves to be a treatise, not a letter, the responsibility will lie with the obligation that I have been under ever since the Autumn Salon, of expounding L'Art Moderne[1] (you will observe that this is not the same thing as L'Art Nouveau). The men whose pictures we have bought — Renoir, Cézanne, Gauguin, Maurice Divois — and others whose pictures we have not bought but would like to — Manet, Degas, Vuillard, Bonnard, Van Gogh[2] for example — all belong. To make the subject clear requires a discussion of the qualities of the men of '70 of whom the Big Four and Puvis de Chavannes[3] are the great men and the inspirers in the main of the vital art of today. The Big Four are Manet, Renoir, Degas and Cézanne.

---

[1] *L'Art Moderne:* The European painters who were attempting something entirely different from the great mid-nineteenth-century artists; Leo Stein was sure enough of his judgment to go beyond existing art criticism of the day.

[2] *Manet, . . . Van Gogh:* Those who comprise what Leo Stein considered "the moderns." See below.

[3] *Puvis de Chavannes:* French mural painter (1824–1898) who studied in Paris with Delacroix and Couture. His work is found in the Hotel de Ville, the Sorbonne, and the Pantheon, Paris, and in the Boston Public Library.

Of these Manet[4] is the painter par excellence. He is not the great colorist that is Renoir but in sheer power of handling he has perhaps not had his equal in modern times. He had a great conception of art but few great conceptions. Almost all his work is largely haphazard because with all his splendid qualities he had not a great intellect. He rarely realized a thing completely yet everything he did was superb. There is a limpid purity in his color and feeling for effect, a realization of form and vitality in rendering, all of which are most admirable, but that power of conception which could fuse all this and cast it into the form of stable equilibrium, that perfect poise of the completely achieved — all that was beyond his range of mental grasp.

Renoir[5] was the colorist of the group. He again was a man of limited intellectuality but he had the gift of color as no one perhaps since Rubens[6] except perhaps Fragonard[7] has had it — what you might call the feeling for absolute color, color handled not as the medium but as the stuff of art. Sometimes his realization of a theme is complete but only rarely, at least as far as composition and mass go. But his color carries every bit of his work splendidly just as texture and mass carry Manet so that nothing of his is worthless.

Degas,[8] the third of the quartet, is the most distinctively intellectual. Scarcely anything that he has done is unachieved either in part or in whole. He is incomparably the greatest master of composition of our time, the greatest master probably of movement of line, with a colossal feeling for form and superb color. And all his qualities are held together and brought to a focus by a perfection of control that only the finest mentality could give.

---

[4] *Manet:* The son of a magistrate, Édouard Manet (1832–1883) went to sea and returned to study art under Couture; he was influenced by Velázquez, Goya, and Japanese prints. Never accepted by contemporary art critics, his paintings (*Olympia, The Balcony,* portrait of Zola) are appreciated today.

[5] *Renoir:* Pierre Auguste Renoir (1841–1919), French impressionist painter and sculptor and friend of Cézanne and Monet. The last twenty years of his life he spent in Provence, ill with gout but finally able to live off the proceeds from his art.

[6] *Rubens:* Peter Paul Rubens (1577–1640) remains the foremost painter of the Flemish school. Famous in both Italy and France, this prolific painter is known for both landscapes and figures, using lush colors and abundant vitality.

[7] *Fragonard:* Considered one of the eighteenth century's greatest painters, Jean-Honore Fragonard (1732–1806) is important today for his brush stroke technique and his virtuosity in capturing the sensuous and charming aspects of the age of Louis XV.

[8] *Degas:* Edward Degas (1834–1917), French impressionist who introduced strikingly modern elements into his oils, charcoals, and other works.

Fourth comes Cézanne[9] and here again is great mind, a perfect concentration, and great control. Cézanne's essential problem is mass and
he has succeeded in rendering mass with a vital intensity that is unpar-
alleled in the whole history of painting. No matter what his subject
is — the figure, landscape, still life — there is always this remorseless
intensity, this endless unending gripping of the form, the unceasing
effort to force it to reveal its absolute self-existing quality of mass.
There can scarcely be such a thing as a completed Cézanne. Every canvas is a battlefield and victory an unattainable ideal. Cézanne rarely
does more than one thing at a time and when he turns to composition
he brings to bear the same intensity, keying his composition up till it
sings like a harp string. His color also, though as harsh as his forms, is
almost as vibrant. In brief, his is the most robust, the most intense, and
in a fine sense the most ideal of the four.

The work of these four men is exceedingly diverse; in fact there is in
general aspect no resemblance whatever. They do not constitute a
school in any sense in which that word can be reasonably used and yet
they have something in common. Their work is all non-dramatic.
When figures are composed in a group their relations are merely spatial. At most they are relations of movement concurrent or opposite.
This fact is intimately related to another, that the work is done in the
main direct from life. This is in a sense not true of Degas, for though
he makes elaborate studies, his compositions are not painted from life;
yet even with him the model remains dominant. The consequences of
this are enormous and in Cézanne we find their logical working out.
They mean that the path of pictorial accomplishment lies in the reaching to the last drop the virtue which lies in the model. The roots of this
procedure lie far back. Holbein[10] did it with his line, Velasquez[11] did it
with his values, but never till this recent time was the conception fun-
damentally worked out. Velasquez is, in fact, to me a man more of
Monet's[12] type, only ten thousand thousand times finer — for the

---

[9] *Cézanne:* Unappreciated until near his death in 1906, Paul Cézanne (1839–1906)
quickly became the quintessential modern French painter, whose craft of placing oil on
canvas in new spatial patterns influenced hundreds of artists, both in language and
paint.

[10] *Holbein:* Hans Holbein the younger (1497–1543) carried on his German father's
expert craft, but infused technique with a more visible religious spirit.

[11] *Velazquez:* Diego Rodriguez de Silva y Velázquez (1599–1660), the most celebrated painter of the Spanish school, influenced by Tintoretto and still revered for his
accomplishment in achieving color effects.

[12] *Monet's:* Claude Monet (1840–1926) was a French landscape painter, a friend of
Cézanne, Renoir, Sisley, Pissarro, and Berthe Morisot. He lived much of his life in
poverty, until his later work brought him fame.

things that Monet does not badly, nor too well either, Velasquez did magnificently. Both of them are what might fairly be called naturalists with the addition of a sense for composition both in color and form. Renoir, Manet and Cézanne substitute for that the abstraction of the quality of color, and form from the model. Manet pushed the thing through to the bitter end, for he was continually experimenting, but Renoir succeeded with his color and Cézanne with the form or rather with what is the essence of form — mass. Degas most perfectly combined both things, adding composition and movement, but losing perhaps a trifle of the pure virgin force that the other three have. Whistler[13] lost it almost completely, substituting for it an artifice so brilliant in its accomplishment as almost to succeed in disguising the loss.

The loss of dramatic quality was a necessary consequence of this devotion to the model and we have an art that is full of ideas and personality, but which does not attempt to render ideas of personality. This was its great limitation, its element of "materialism" that made it caviare to the general. This other side was ministered to most adequately among contemporaries by Puvis de Chavannes, Gustave Moreau[14] and Fantin-Latour.[15] Moreau, though not a great artist, seems to have been a remarkable teacher and had a wide influence. The other two you know well enough for the purposes in hand.

The influence of all these people, except perhaps Fantin, has been enormous. Practically every young man who counts artistically has undergone it in some place and in all sorts of combinations. Degas has had an overwhelming influence on the satirists. Forain,[16] perhaps the greatest of them, is one of his very rare pupils, and began by doing work entirely in his manner. The same is true of Toulouse-Lautrec,[17] an imp of Satan of great genius, a powerful designer, a sinuous and vigorous draughtsman and a keen satirical wit. The Autumn Salon

[13] *Whistler:* James Whistler (1834–1903), one of the few American painters to figure in the international world of aesthetic modernism.

[14] *Gustave Moreau:* Professor at the École des Beaux-Arts, this French painter (1826–1898) chose subjects that were weird and mystical. Among his students were Matisse and Rouault.

[15] *Fantin-Latour:* French painter and lithographer, Henri Fantin-Latour (1836–1904) was known for his portraits of groups of famous contemporaries. Influenced by Courbet, he depicted his friends with a near-photographic realism.

[16] *Forain:* Jean Louis Forain (1852–1931), French painter, lithographer, and etcher, known primarily today for his etchings.

[17] *Toulouse-Lautrec:* Henri de Toulouse-Lautrec (1864–1901), French painter and lithographer, crippled from a childhood fall that broke both legs. This painter of French brothels and circuses died at the age of thirty-seven.

contained a special exhibition of his pictures (he died some years since at the age of thirty-seven) and it was there that I first came to know his work. He has borrowed a great deal in design from Japanese prints but the spirit of his work is all his own. Gauguin[18] who died two years ago began as a pupil of Pissarro[19] but afterward followed more particularly Degas and Cézanne. He lived in Tahiti and developed a color of his own. He revels in massive golden and keen lemon yellows, his drawing has the firmness and feeling for form of Degas, and through all his work runs Cézanne's splendid elemental feeling which gives his work its monumental character. By these derivative references I don't mean to imply that he is a mosaic of qualities; none of the men I mention are. They are all real artistic personalities. Maurice Denis[20] derives from Degas, Puvis de Chavannes and Cézanne. His decorative feeling is of the finest; in a mother and child that we have there is a distribution of blacks such as I have never seen in European art. He has an almost Fra Angelico[21] daintiness yet firmness of religious feeling with an amplitude of conception that makes him a worthy follower of Puvis, and while reveling in the grays and pale tones of the latter he has yet at command a palette of ringing intensity. There are a number of other men of interest, but what's the use of telling of people whose work you never saw. All I have tried to do in the cases of the last few is to indicate my impression of them for what that might be worth to you. It would be useless to continue the same process with Deparquet,[22] the ablest immediate follower of Renoir, whose last development comes nearer to the quality of Titian,[23] almost of Gior-

---

[18] *Gauguin:* Paul Gaugin (1848–1903), French painter and woodcut artist who traded several careers (as a sailor and a banker) for art. He lived away from civilization at intervals — in Martinique, with Van Gogh at Arles, and in Tahiti. Credited with breaking from naturalism to use nature in his own idiosyncratic configuration and colors, he is one of the fathers of twentieth-century art.

[19] *Pissarro:* Camille Pissarro (1830–1903) was both an impressionist and a follower of Seurat; he was a mentor to Gauguin, Van Gogh, and Cézanne.

[20] *Maurice Denis:* French painter and theorist of art (1870–1943) commonly known for his paintings on religious themes.

[21] *Fra Angelico:* Florentine painter (1400–1455) variously named Guidolino di Pietro and Giovanni da Fiesole. After taking orders with the Dominicans, he became a revered painter of religious subjects, oftentimes in small formats.

[22] *Deparquet:* One of the evanescent French painters who was highly thought of for a time; in his exploratory way, Leo Stein was one of the few collectors who would purchase paintings from the less renowned artists such as Picasso, Denis, Matisse, and Deparquet in the early 1900s.

[23] *Titian:* Venetian painter (c. 1490–1576) whose name was Tiziano Vecellio (Pieve deCadore in the Dolomites). Best known for religious decoration and paintings.

gione,[24] than any modern work I know, or of Vuillard,[25] who paints interiors with speckled carpets and things with a myriad beauty of color, of Bonnard[26] who has much of Manet's fine feeling for color as texture, and so on. The general drift of this letter will be to indicate what I think of modern art and to repeat that it vitally goes marching on. That's enough about art. . . .

. . . Gertrude says to tell you that her lapis lazuli chain is beautchiful. She says that I'm to say that it's beautiful. She says that's not a nice way to act, that I'm to say it. She wants to know whether I don't think it. She says don't be nasty. She repeats that remark four times. She drools some more.

Regards to everybody.

Sincerely,
Leo Stein

[24] *Giorgione:* Venetian painter (c. 1478–1510) also known as Zorgo (Zorgi da Castelfranco) and as Giogio Barbarelli. Credited with being the father of sixteenth-century Venetian painting.

[25] *Vuillard:* French painter and illustrator, Jean-Édouard Vuillard (1868–1940) was best known for his domestic and, later, decorative works.

[26] *Bonnard:* Now enjoying a resurgence, Pierre Bonnard (1867–1947) was a French painter, lithographer, and illustrator who focused on the domestic, the intimate, and the study of light.

PAUL CÉZANNE

Letter to His Son Paul (September 26, 1906)

Although relatively unknown at the time of his death in 1906, Paul Cézanne (b. 1839) had been an outstanding French painter for years. Son of a banker and a friend of Émile Zola, Cézanne went to Paris in 1861 and became friends with Pissarro, who strongly influenced his early development. Known for his heavy use of the palette knife to create a thickly textured scene, he later became an erstwhile impressionist (through his association with Manet) and exhibited at the impressionist show of 1874. Never satisfied with current relationships between painting and nature, Cézanne kept changing techniques to achieve the new, almost moving, effects of landscape as well as portraits. His new types of spatial patterns, sometimes underlying the ostensible surfaces, led to his prominence during the twentieth century.

That Gertrude Stein wrote the stories of *Three Lives* sitting under Cézanne's large portrait of his wife *(Portrait of Mme Cézanne)* speaks to the prominence of the painter in the Steins' lives and imaginations. The prices of Cézanne's paintings soared quickly, and soon this painting was the most valuable of the Steins' large collection. When Leo and Gertrude divided their art collection in 1913, in order to set up separate households, Gertrude kept this painting.

This letter is published in *Paul Cézanne's Letters,* ed. John Rewald (London: Bruno Cassirer, 1976).

Aix, 26 September, 1906

My dear Paul,

I received a notification from the *Salon d'Automne,* signed by Lapigie, who is doubtless one of the big organizers and . . . of the exhibition; from it I see that eight of my pictures[1] are on view. Yesterday I saw that valiant Marseillean Carlos Camoin, who came to show me a pile of paintings and to seek my approval; what he does is good, if anything he seems to be making progress, he is coming to spend a few days at Aix and is going to work on the *petit chemin du Tholonet.*[2] He showed me a photograph of a figure by the unfortunate Émile Bernard; we are agreed on this point, that he is an intellectual constipated by recollections of museums, but who does not look enough at nature, and that is the great thing, to make himself free from the school and indeed from all schools. — So that Pissarro was not mistaken, though he went a little too far, when he said that all the necropoles of art should be burned down.

Certainly one could make a strange menagerie with all the professionals of art and their kindred spirits. — The Secretary-General is himself an artist. In this case, in view of his position, he would be the equal, in the Salon d'Automne of course, of a member of the Institute. It is, therefore, a structure which rises proudly if not victoriously opposite the barracks of the Quai Conti, the library of which was founded by the man Sainte-Beuve described as an 'artful Italian'.[3] — I

Note: The footnotes in this selection are original to the selection.

[1] In fact there were ten.

[2] The road leading to the Château Noir, where a few years before Cézanne had painted several landscapes.

[3] This is an allusion to Cardinal Mazarin.

still go into the country, to the banks of the Arc, and I leave my baggage with a man named Bossy who offered me hospitality for it.
I embrace you and Mamma with all my heart, your father,

Paul Cézanne

THE NATION

Anonymous Review of
Stein's Three Lives (January 20, 1910)

This 1910, unsigned review of Stein's *Three Lives,* describes both the attraction to and the distaste for realism that many readers felt. The clarity of method that the reviewer craves comments, presumably, on Stein's use of repetition, or *insistence* as she termed it, a style she was experimenting with intentionally, and not something she would likely "consent to clarify" (p. 371). In fact, both Stein's use of realism and repetition serve to keep her writing fresh, even now, over half a century after her death.

Three Lives. By Gertrude Stein. New York: The Grafton Press.

These stories of the Good Anna, Melanctha, and the Gentle Lena have a quite extraordinary vitality conveyed in a most eccentric and difficult form. The half-articulated phrases follow unrelentingly the blind mental and temperamental gropings of three humble souls wittingly or unwittingly at odds with life. Whoever can adjust himself to the repetitions, false starts, and general circularity of the manner will find himself very near real people. Too near, possibly. The present writer had an uncomfortable sense of being immured with a girl wife, a spinster, and a woman who is neither, between imprisoning walls which echoed exactly all thoughts and feelings. These stories utterly lack construction and focus, but give that sense of urgent life which one gets more commonly in Russian literature than elsewhere. How the Good Anna spent herself barrenly for everybody in reach, the Gentle Lena for the notion of motherhood, while the mulattress Melanctha perished partly of her own excess of temperament, but more from contact with a life-diminishing prig and emotionally inert surroundings, readers who are willing to pay a stiff entrance fee in patient attention may learn for themselves. From Miss Stein, if she can consent to clarify her method, much may be expected. As it is, she writes quite as a Browning escaped from the bonds of verse might wallow in fiction, only without his antiseptic whimsicality.

MABEL DODGE LUHAN

From *Speculations, or Post-Impressionism in Prose*

One of the more famous expatriate American women of the twentieth century, Mabel Dodge Luhan (1879–1962) was the only child of wealthy New York parents. After a childhood in Buffalo, she located in Florence, Italy, where her Villa Curonia became an extended intellectual salon. (Stein wrote a now-famous portrait of her friend there, "Portrait of Mabel Dodge at the Villa Curonia.") Part of the committee that brought French Impressionist and avant-garde painting to the United States through the 1913 Armory Show collection, Luhan was also one of the first intellectuals to promote Stein's writing. Her 1913 essay, excerpted here, made Stein's literary efforts explicable, as they were linked to the graphic art that so dominated modern European aesthetics.

After the Armory Show, Luhan remained in New York, her Greenwich Village salon becoming a spiritual center. In 1918 she moved to Taos, New Mexico, where she was hostess to a number of internationally known writers and painters, among them D. H. Lawrence, Marsden Hartley, and Leo Stein. There, she married her fourth husband, Antonio Lujan, a Pueblo Indian, and wrote her four-volume memoir, *Intimate Memories*.

This influential essay, which served to promote the avant-garde in the States, appeared in *Arts and Decorations* 3 (March 1913).

Many roads are being broken today, and along these roads consciousness is pursuing truth to eternity. This is an age of communication, and the human being who is not a "communicant" is in the sad plight which the dogmatist defines as being a condition of spiritual non-receptivity.

Some of these newly opened roads lie parallel and almost touch.

In a large studio in Paris, hung with paintings by Renoir,[1] Matisse[2] and Picasso,[3] Gertrude Stein is doing with words what Picasso is doing with paint. She is impelling language to induce new states of consciousness, and in doing so language becomes with her a creative art rather than a mirror of history.

In her impressionistic writing she uses familiar words to create perceptions, conditions, and states of being, never before quite consciously experienced. She does this by using words that appeal to her as having the meaning that they *seem* to have. She has taken the English language and, according to many people, has misused it, or has used it roughly, uncouthly and brutally, or madly, stupidly and hideously, but by her method she is finding the hidden and inner nature of nature.

To present her impressions she chooses words for their inherent quality, rather than for their accepted meaning.

Her habit of working is methodical and deliberate. She always works at night in the silence, and brings all her will power to bear upon the banishing of preconceived images. Concentrating upon the impression she has received and which she wishes to transmit, she suspends her selective faculty, waiting for the word or group of words that will perfectly interpret her meaning, to rise from her subconsciousness to the surface of her mind.

Then and then only does she bring her reason to bear upon them, examining, weighing and gauging their ability to express her meaning. It is a working proof of the Bergson[4] theory of intuition. She does not go after words — she waits and lets them come to her, and they do.

It is only when art thus pursues the artist that his production will bear the mark of inevitability. It is only when the *"élan vital"* drives the artist to the creative overflow that life surges in his production.

[1] *Renoir:* Pierre Auguste Renoir (1841–1919), French impressionist painter and sculptor.

[2] *Matisse:* Henri Matisse (1869–1954), French painter, sculptor, and lithographer who first exhibited in 1896. Although more strenuously collected by the Michael Steins, both Leo and Gertrude also owned the flamboyant, and clearly innovative, oils. Early a key figure among the fauves, known for raw, untempered colors and bold outlines, Matisse went through a number of styles during his long lifetime.

[3] *Picasso:* Pablo Ruiz y Picasso (1881–1973), a Spanish painter who made France his home early in his art career, he was perhaps the most visible artist of the twentieth century — being credited with founding Cubism and other innovative movements.

[4] *Bergson:* Henri Bergson (1859–1941), French philosopher who attributed a number of concepts to the notion of modern art and thought.

Vitality directed into a conscious expression is the modern definition of genius.

It is impossible to define or to describe fully any new manifestation in esthetics or in literature that is as recent, as near to us, as the work of Picasso or of Gertrude Stein; the most that we can do is to suggest a little, draw a comparison, point the way and then withdraw. . . .

Of course, comment is the best of signs. Any comment. One that Gertrude Stein hears oftenest is from conscientious souls who have honestly tried — and who have failed — to get anything out of her work at all. "But why don't you make it simpler?" they cry. "Because this is the only way in which I can express what I want to express," is the invariable reply, which of course is the unanswerable argument of every sincere artist to every critic. Again and again comes the refrain that is so familiar before the canvases of Picasso — "But it is so ugly, so brutal!" But how does one know that it is ugly, after all? How does one know? Each time that beauty has been reborn in the world it has needed complete readjustment of sense perceptions, grown all too accustomed to the blurred outlines, faded colors, the death in life of beauty in decline. It has become jaded from over-familiarity, from long association and from inertia. If one cares for Rembrandt's[5] paintings today, then how could one have cared for them at the time when they were painted, when they were glowing with life. If we like St. Marks in Venice today, then surely it would have offended us a thousand years ago. Perhaps it is not Rembrandt's paintings that one cares for, after all, but merely for the shell, the ghost — the last pale flicker of the artist's intention. Beauty? One thing is certain, that if we must worship beauty as we have known it, we must consent to worship it as a thing dead. "*Une grande, belle chose — morte,*" And ugliness — what is it? Surely, only death is ugly.

In Gertrude Stein's writing every word lives and, apart from the concept, it is so exquisitely rhythmical and cadenced, that when read aloud and received as pure sound, it is like a kind of sensuous music. Just as one may stop, for once in a way, before a canvas of Picasso, and, letting one's reason sleep for an instant, may exclaim: "It *is* a fine pattern!"— so listening to Gertrude Stein's words and forgetting to try to understand what they mean, one submits to their gradual charm. Huntley Carter,[6] of the *New Age,* says that her use of language has a

---

[5] *Rembrandt's:* Rembrandt Harmenszoon van Rijn (1606–1669), Dutch painter and etcher, known for his landscapes, portraits, and his use of striking color.

[6] *Huntley Carter:* Book reviewer for *New Age.*

curious hypnotic effect when read aloud. In one part of her writing she made use of repetition and the rearranging of certain words over and over, so that they became adjusted into a kind of incantation, and in listening one feels that from the combination of repeated sounds, varied ever so little, that there emerges gradually a perception of some meaning quite other than that of the contents of the phrases. Many people have experienced this magical evocation, but have been unable to explain in what way it came to pass, but though they did not know what meaning the words were bearing, nor how they were affected by them, yet they had *begun* to know what it all meant, because they were not indifferent.

In a portrait that she has finished recently, she has produced a coherent totality through a series of impressions which, when taken sentence by sentence, strike most people as particularly incoherent. To illustrate this, the words in the following paragraph are strenuous words — words that weigh and qualify conditions; words that are without softness yet that are not hard words — perilous abstractions they seem, containing agony and movement and conveying a vicarious livingness. "It is a gnarled division, that which is not any obstruction, and the forgotten swelling is certainly attracting. It is attracting the whiter division, it is not sinking to be growing, it is not darkening to be disappearing, it is not aged to be annoying. There cannot be sighing. This is this bliss."

Many roads are being broken — what a wonderful word — "broken"! And out of the shattering and petrifaction of today — up from the cleavage and the disintegration — we will see order emerging tomorrow. Is it so difficult to remember that life at birth is always painful and rarely lovely? How strange it is to think that the rough-hewn trail of today will become tomorrow the path of least resistance, over which the average will drift with all the ease and serenity of custom. All the labor of evolution is condensed into this one fact, of the vitality of the individual making way for the many. We can but praise the high courage of the road breakers, admitting as we infallibly must, in Gertrude Stein's own words, and with true Bergsonism faith — "Something is certainly coming out of them!"

CARL VAN VECHTEN

From *How to Read Gertrude Stein*

Carl Van Vechten (1880–1964) was born to a middle-class, Iowa family. Educated at the University of Chicago, earning the Ph.D. in English in 1903, he was an arts critic (music, dance, paintings) for several United States papers. While the Paris correspondent for the *New York Times,* he became involved with the 1913 Armory Show and met Mabel Dodge Luhan. In 1914, having by this time met Gertrude Stein and Alice B. Toklas, he wrote this essay, which appeared in the *New York Times,* and later arranged for his friend Donald Evans to publish Stein's poetry collection *Tender Buttons.*

Friends with Stein for more than thirty-three years, Van Vechten was loyal to Stein and Toklas through the vicissitudes of their lives; their correspondence has been published in two volumes. Van Vechten was both a novelist and photographer, and made a strong imprint on international modern art.

This essay appeared first in the economics/business section of the *New York Times,* and later in August 1914, in *Trend,* 7, no. 4.

The English language is a language of hypocrisy and evasion. How not to say a thing has been the problem of our writers from the earliest times. The extraordinary fluidity and even naivete of French makes it possible for a writer in that language to babble like a child; de Maupassant[1] is only possible in French, a language in which the phrase, "Je t'aime" means everything. But what does "I love you" mean in English? Donald Evans,[2] of our poets, has realized this peculiar quality of English and he is almost the first of the poets in English to say unsuspected and revolting things, because he so cleverly avoids saying them.

Miss Stein discovered the method before Mr. Evans. In fact his Patagonian Sonnets were an offshoot of her later manner, just as Miss Kenton's[3] superb story, "Nicknames," derives its style from Miss

---

[1] *de Maupassant:* Guy de Maupassant (1850–1893) was a French short story writer whose writing influenced the United States short story form appreciably.

[2] *Donald Evans:* Poet and publisher who was a friend of Van Vechten's, and who published Gertrude Stein's poetry collection *Tender Buttons* in 1914.

[3] *Miss Kenton's:* Edna Kenton (1876–1954), little-known short story writer of the early twentieth century.

Stein's *Three Lives.* She has really turned language into music, really made its sound more important than its sense. And she has suggested to the reader a thousand channels for his mind and sense to drift along, a thousand, instead of a stupid only one.

Miss Stein has no explanations to offer regarding her work. I have often questioned her, but I have met with no satisfaction. She asks you to read. Her intimate connection with the studies of William James[4] have been commented upon; some say that the "fringe of thought," so frequently referred to by that writer, may dominate her working consciousness. Her method of work is unique. She usually writes in the morning, and she sets down the words as they come from her pen; they bubble, they flow; they surge through her brain and she sets them down. You may regard them as nonsense, but the fact remains that effective imitations of her style do not exist. John Reed[5] tells me that, while he finds her stimulating and interesting, an entity, he feels compelled to regard her work as an offshoot, something that will not be concluded by followers. She lives and dies alone, a unique example of a strange art. It may be in place also to set down here the fact that once in answer to a question Miss Stein asserted that her art was for the printed page only; she never expects people to converse or exchange ideas in her style.

As a personality Gertrude Stein is unique. She is massive in physique, a Rabelaisian woman with a splendid thoughtful face; mind dominating her matter. Her velvet robes, mostly brown, and her carpet slippers associate themselves with her indoor appearance. To go out she belts herself, adds a walking staff, and a trim unmodish turban. This garb suffices for a shopping tour or a box party at the *Opéra.*

Paris is her abode. She settled there after Cambridge, and association with William James, Johns Hopkins and a study of medicine. Her orderly mind has captured the scientific facts of both psychology and physiology. And in Paris the early painters of the new era captured her heart and purse. She purchased the best of them, and now such examples as Picasso's *Acrobats* and early Matisses[6] hang on her walls.

---

[4] *William James:* Harvard philosopher and psychologist (1842–1910) whose theories of the mind and thought influenced Gertrude Stein from the years when she was his student at Radcliffe/Harvard Annex to her death in 1946.

[5] *John Reed:* Journalist and radical leader (1887–1920) who graduated from Harvard in 1910. An editor of *Masses,* he covered political stories in Mexico, Europe, and Russia. His *Ten Days That Shook the World* (1919) is the best account of the Bolshevik revolution. Reed died in Moscow in 1920 of typhus.

[6] *Matisses:* Henri Matisse (1869–1954) encouraged the Steins to purchase his paintings and sculpture. Although Michael and Sally Stein were the more avid collectors, Leo and Gertrude also bought a large number, especially from Matisse's fauve period.

There is also the really authoritative portrait of herself, painted by Pablo Picasso.[7]

These two painters she lists among her great friends. And their influence, perhaps, decided her in her present mode of writing. Her pictures are numerous, and to many, who do not know of her as a writer, she is mentioned as the Miss Stein with the collection of post-impressionists. On Saturday nights during the winter one can secure a card of admission to the collection and people wander in and out the studio, while Miss Stein serves her dinner guests unconcernedly with after-dinner coffee. And conversation continues, strangely unhindered by the picture viewers. . . .

In *Three Lives* Miss Stein attained at a bound an amount of literary facility which a writer might strive in vain for years to acquire. Simplicity is a quality one is born with, so far as literary style is concerned, and Miss Stein was born with that. But to it she added, in this work, a vivid note of reiteration, a fascinatingly complete sense of psychology and the workings of minds one on the other, which at least in "Melanctha: Each [One] as She May" reaches a state of perfection which might have satisfied such masters of craft as Turgenev,[8] or Balzac,[9] or Henry James.[10]

[7] *Picasso:* Pablo Ruiz y Picasso (1881–1973) was a Spanish painter who lived most of his life in Paris. A great friend of Gertrude Stein's, he was the most visible painter of the twentieth century. His portrait of Gertrude is reproduced here.

[8] *Turgenev:* Ivan Sergeyevich Turgenev (1818–1883), the leading Russian novelist, dramatist, and short story writer of his time (paired as he often was with Leo Tolstoy). His studies of the simple lives of peasants bore some relation to *Three Lives*.

[9] *Balzac:* Honore de Balzac (1799–1850), French novelist who created his own fictional province.

[10] *Henry James:* Brother of William, Henry (1843–1916) gained even greater fame through his novels and short stories, which not only captured some essential truths about American life, but created an inclusive style that led to modernist writing effects.

Famous for her somewhat idiosyncratic "portraits" of her friends — which ➤ included painters such as Henri Matisse and Pablo Picasso (works that appeared in *Camera Work*) and writers such as Harold Loeb and Ernest Hemingway — Stein's publications from 1912 through the 1930s were often these lyric portraits, some very short and others extremely long, book-length even. Her portrait of Hemingway was written at the beginning of their short-lived friendship, and appeared in *Ex Libris* in December 1923.

DECEMBER 1923

Volume 1   Number 6

He and They, Hemingway :
A Portrait

GERTRUDE STEIN

Among and then young.
Not ninety-three.
Not Lucretia Borgia. [1]
Not in or on a building.
Not a crime not in the time.
Not by this time.
Not in the way.

On their way and to head away. A head any way.
What is a head. A head is what every one not in the
north of Australia returns for that. In English we know.
And it is to their credit that they have nearly finished
and claimed, is there any memorial of the failure of
civilization to cope with extreme and extremely well
begun, to cope with extreme savagedom.

There and we know.

Hemingway.

How do you do and good-by. Good-by and how do
you do. Well and how do you do.

AMERICAN LIBRARY IN PARIS
10 RUE DE L'ELYSEE

[1] *Lucretia Borgia:* Lucretia Borgia (1480–1519), the Spanish-Italian Duchess of
Ferrara and illegitimate daughter of Pope Alexander VI., set up a brilliant court and was
known for her beauty, kindness and criminal vices.

Selected Bibliography

▾

This bibliography is divided into two parts, "Works Cited" and "Suggestions for Further Reading." The first part contains all primary and secondary works quoted or discussed in the general, chapter, and selection introductions. The second part is a selective list of materials that will be useful to students who want to know more about Gertrude Stein's life and culture or who are interested in reading critical studies of *Three Lives*. A book or article that appears in "Works Cited" is not recorded again under "Suggestions for Further Reading." Thus, both lists should be consulted.

WORKS CITED

Anderson, Margaret C. *Forbidden Fires*. Ed. Mathilda M. Hills. Tallahassee, FL: Naiad, 1996. (Excerpted in this volume, pp. 281–85.)

Bergson, Henri. "Brain and Thought: A Philosophical Illusion." (1904). *Mind-Energy, Lectures and Essays*. Trans. H. Wildon Carr. New York: Holt, 1920. (In this volume, pp. 321–25.)

Blackmer, Corrine E. "African Masks and the Arts of Passing in Gertrude Stein's 'Melanctha' and Nella Larsen's *Passing*." *Journal of the History of Sexuality*. 4.2 (October 1993): 230–63.

Bodnar, John. *The Transplanted: A History of Immigrants in Urban America*. Bloomington: Indiana UP, 1985.

Cézanne, Paul. Letter to his son Paul. 26 September 1906. *Paul Cézanne's Letters*. Ed. John Rewald. London: Cassirer, 1976. (In this volume, pp. 369–71.)

Cooper, Anna Julia Haywood. "The Higher Education of Woman." *A Voice from the South: By a Black Woman of the South*. 1892. By Cooper. Ed. Henry Louis Gates, Jr. with an introduction by Mary Helen Washington. New York: Oxford UP (The Schomburg Library of Nineteenth-Century Black Women Writers), 1988. (In this volume, pp. 229–33.)

Crane, Stephen. "An Experiment in Misery." *New York Press* 22 April 1894. (Excerpted in this volume, pp. 341–47.)

Dearborn, Mary V. *Pocahontas's Daughters, Gender and Ethnicity in American Culture*. New York: Oxford UP, 1986.

Dickie, Margaret. "Recovering the Repression in Stein's Erotic Poetry." *Gendered Modernisms: American Women Poets and Their Readers*. Ed. Margaret Dickie and Thomas Travisano. Philadelphia: U of Pennsylvania P, 1996. 3–25.

———. *Stein, Bishop and Rich, Lyrics of Love, War and Place*. Chapel Hill: U of North Carolina P, 1997.

Du Bois, W. E. B. *The Souls of Black Folk: Essays and Sketches*. Chicago: McClurg, 1903. (Excerpted in this volume, pp. 352–59.)

Ellis, Havelock. *Studies in the Psychology of Sex*. Vol. II. New York: Random, 1906. (Excerpted in this volume, pp. 285–90.)

Farber, Lawren. "Fading: A Way. Gertrude Stein's Sources for *Three Lives*." *Journal of Modern Literature* 5.3 (September 1976): 463–80.

Fern, Fanny (Sara Payson [Willis] Parton). "The Working-Girls of New York." *Folly As It Flies*. By Fern. New York: Carleton, 1868. (In this volume, pp. 226–29.)

Flaubert, Gustave. Letter to Émile Zola. 3 June 1874. *The Complete Works of Gustave Flaubert*. New York: M. Walter Dunne, 1904. (In this volume, pp. 332–34.)

Freud, Sigmund. *Three Contributions to the Theory of Sex*. 1905. Trans. A. A. Brill. New York: Dutton, 1962. (Excerpted in this volume, pp. 271–81.)

Gallup, Donald, ed. *The Flowers of Friendship: Letters Written to Gertrude Stein*. New York: Octagon, 1979.

Garland, Hamlin. *Crumbling Idols, Twelve Essays on Art and Literature*. New York: Stone, 1894. (Excerpted in this volume, pp. 338–40.)

Gilman, Charlotte Perkins Stetson. *Women and Economics*. Boston: Small, 1898. (Excerpted in this volume, pp. 236–46.)

Gilman, Sander L. *Difference and Pathology: Stereotypes of Sexuality, Race, and Madness*. Ithaca: Cornell UP, 1985.

James, Henry. "The Art of the Novel." *Partial Portraits*. New York: Macmillan, 1888. (Excerpted in this volume, pp. 335–37.)

James, William. "The Varieties of Attention" and "The Stream of Thought." *The Principles of Psychology*. New York: Holt, 1890. (Excerpted in this volume, pp. 310–19.)

Lemke, Sieglinda. *Primitivist Modernism, Black Culture and the Origins of Transatlantic Modernism*. New York: Oxford UP, 1998.

Luhan, Mabel Dodge. "Speculations, or Post-Impressionists in Prose." *Arts and Decorations* 3 (March 1913): 172, 174. (In this volume, pp. 372–75.)

Matisse, Henri. "Notes of a Painter, 1908." *Matisse on Art*. Ed. Jack D. Flam. London: Phaidon, 1973.

Meyer, Steven. "The New Novel: *A Novel of Thank You* and the Characterization of Thought." *A Novel of Thank You*. By Gertrude Stein. Normal, IL: Dalkey, 1994. vii–xxvii.

Mitchell, S. Weir. *Fat and Blood and How To Make Them*. Philadelphia: J. B. Lippincott, 1877. (Excerpted in this volume, pp. 249–51.)

Morris, Wright. *About Fiction*. New York: Harper, 1975.

Norris, Frank. "Simplicity in Art." *The Responsibilities of the Novelist and Other Literary Essays*. By Norris. New York: Doubleday, 1901. (In this volume, pp. 347–51.)

North, Michael. *The Dialect of Modernism: Race, Language & Twentieth-Century Literature*. New York: Oxford UP, 1994.

"One Farmer's Wife." *The Independent* 58 (February 9, 1905): 294–98. (In this volume, pp. 233–36.)

Osler, William. "Medicine in the Nineteenth Century." 1901. *Aequanimitas: With Other Addresses to Medical Students, Nurses and Practitioners of Medicine*. Philadelphia: Blakiston's, 1932. (Excerpted in this volume, pp. 251–57.)

"Review of Gertrude Stein's *Three Lives*." *The Nation* 90 (January 20, 1910): 65. (In this volume, pp. 371–72.)

Santayana, George. *Persons and Places*. New York: Scribner's, 1963. (Excerpted in this volume, pp. 319–21.)

Stanton, Elizabeth Cady. "Address to the New York State Legislature." 1860. *History of Woman Suffrage*. Vol. I. Ed. Elizabeth Cady Stanton, Susan B. Anthony, and Matilda Joslyn Gage. Rochester, New York, 1881. (In this volume, pp. 223–26.)

Stein, Gertrude. *The Autobiography of Alice B. Toklas*. New York: Random, 1933.

———. "Cultivated Motor Automatism: A Study of Character in Its Relation to Attention." *Harvard Psychological Review* (May 1898): 295–306.

———. *Lectures in America*. New York: Random, 1935.

————. Letter to Henri Pierre Roche. 12 June 1912. Stein and Toklas holdings. Harry Ransom Humanities Research Center, The University of Texas at Austin. (In this volume, pp. 246–48.)

————. "The Modern Jew Who Has Given Up the Faith of His Fathers Can Reasonably and Consistently Believe in Isolation." Radcliffe forensics unpublished essay. Yale American Literature Collection. (In this volume, pp. 291–92.)

————. *Three Lives*. New York: Grafton, 1909. (In this volume, pp. 36–215.)

————. "A Transatlantic Interview." 1946. Ed. Walter Sutton. *A Primer for the Gradual Understanding of Gertrude Stein*. Ed. Robert Bartlett Haas. Los Angeles: Black Sparrow P, 1971. 15–35.

Stein, Gertrude, and Leon Solomons. "Normal Motor Automatism." *Harvard Psychological Review* (September 1896): 495–512.

Stein, Leo. Letter to Mabel Weeks. Undated. *Journey into the Self*. Ed. Edmund Fuller. New York: Crown, 1950. 15–18. (In this volume, pp. 363–69.)

Van Vechten, Carl. "How to Read Gertrude Stein." *Trend* 7. 4 (August 1914): 553–57. (Excerpted in this volume, pp. 375–78.)

Wagner-Martin, Linda. *"Favored Strangers": Gertrude Stein and Her Family*. New Brunswick, NJ: Rutgers UP, 1995.

Weininger, Otto. *Sex and Character*. Authorised translation from the 6th German edition. New York: Putnam's, 1908. (Excerpted in this volume, pp. 293–302.)

Weiss, M. Lynn. *Gertrude Stein and Richard Wright: The Poetics and Politics of Modernism*. Jackson: UP of Mississippi, 1998.

Wells-Barnett, Ida B. "The Case Stated." *Southern Horrors: Lynch Law in all its Phases,* The New York Age Print: 1892. (Excerpted in this volume, pp. 302–07.)

Whitehead, Alfred North. "The Aim of Philosophy." *Modes of Thought*. By Whitehead. New York: Macmillan, 1957. (In this volume, pp. 325–28.)

Wilde, Oscar. "The Ballad of Reading Gaol." *The Works of Oscar Wilde, 1856–1900*. Ed. G. F. Maine. London: Collins, 1948. (Excerpted in this volume, pp. 263–71.)

Williams, J. Whitridge. *Obstetrics*. New York: Appleton, 1923. (Excerpted in this volume, pp. 257–59.)

Williams, William Carlos. "The Work of Gertrude Stein." *Critical Essays on Gertrude Stein*. Ed. Michael J. Hoffman. Boston: G. K. Hall, 1986. 55–58.

Wilson, Edmund. "A Guide to Gertrude Stein." *Vanity Fair,* 21 (September 1923). 60, 80.

Wineapple, Brenda. *Sister Brother, Gertrude and Leo Stein*. New York: Putnam's, 1996.

Woloch, Nancy. *Women and the American Experience*. New York: Knopf, 1984.

SUGGESTIONS FOR FURTHER READING

Benstock, Shari. *Women of the Left Bank, Paris, 1900–1940*. Austin: U of Texas P, 1986.

Berry, Ellen E. *Curved Thought and Textual Writing: Gertrude Stein's Postmodernism*. Ann Arbor: U of Michigan P, 1992.

Bridgman, Richard. *Gertrude Stein in Pieces*. New York: Oxford UP, 1970.

———. "Melanctha," *American Literature* 33 (1961): 350–59.

Brodkin, Karen. *How Jews Became White Folks and What That Says about Race in America*. New Brunswick, NJ: Rutgers UP, 1998.

Brooks, Peter. *Body Work: Objects of Desire in Modern Narrative*. Cambridge, MA: Harvard UP, 1993.

Burns, Edward, Ulla E. Dydo, and William Rice, eds. *The Letters of Gertrude Stein and Thornton Wilder*. 1986. New Haven: Yale UP, 1996.

Castle, Terry. *The Apparitional Lesbian: Female Homosexuality and Modern Culture*. New York: Columbia UP, 1993.

Cohen, Milton A. "Black Brutes and Mulatto Saints: The Racial Hierarchy of Stein's 'Melanctha.'" *Black American Literature Forum* 18.3 (Fall 1984): 119–21.

Curnutt, Kirk. "Parody and Pedagogy: Teaching Style, Voice, and Authorial Intent in the Works of Gertrude Stein." *College Literature* 23.2 (June 1996): 1–24.

DeKoven, Marianne. *A Different Language: Gertrude Stein's Experimental Writing*. Madison: U of Wisconsin P, 1983.

———. *Rich and Strange: Gender, History, Modernism*. Princeton: Princeton UP, 1991.

Everett, Patricia R., ed. *A History of Having of a Great Many Times Not Continued to Be Friends: The Correspondence between Mabel Dodge and Gertrude Stein, 1911–1934*. Albuquerque: U of New Mexico P, 1996.

Faderman, Lillian. *Odd Girls and Twilight Lovers: A History of Lesbian Life in Twentieth-Century America*. New York: Penguin, 1991.

Fahy, Thomas. "Iteration as a Form of Narrative Control in Stein's 'The Good Anna'." *Style* 34.1 (2000).

Garber, Eric. "A Spectacle in Color: The Lesbian and Gay Subculture of Jazz Age Harlem." *Hidden from History: Reclaiming the Gay and Lesbian Past.* Ed. Martin Bauml Duberman, Martha Vicinus, and George Chauncey, Jr. New York: New American, 1989. 318–31.

Gay, Peter. *The Bourgeois Experience, Victoria to Freud.* New York: Oxford UP, 1986.

Gilman, Sander L. "Black Bodies, White Bodies: Toward an Iconography of Female Sexuality in Late Nineteenth-Century Art, Medicine, and Literature." *Critical Inquiry* 12 (Autumn 1985): 204–42.

Grosz, Elizabeth. *Space, Time and Perversion, Essays on the Politics of Bodies.* New York: Routledge, 1995.

Harrowitz, Nancy A., and Barbara Hyams, eds. *Jews and Gender: Responses to Otto Weininger.* Philadelphia: Temple UP, 1995.

Herndl, Diane Price. *Invalid Women: Figuring Feminine Illness in American Fiction and Culture, 1840–1940.* Chapel Hill: U of North Carolina P, 1993.

Katz, Leon. "Weininger and *The Making of Americans.*" *Critical Essays on Gertrude Stein.* Ed. Michael J. Hoffman. Boston: G. K. Hall, 1986. 139–49.

Kellner, Bruce, ed. *A Gertrude Stein Companion, Content with the Example.* New York: Greenwood, 1988.

Kleeblatt, Norman L., ed. *The Dreyfus Affair, Art, Truth, and Justice.* Berkeley: U of California P, 1987.

Mellow, James R. *Charmed Circle: Gertrude Stein & Company.* New York: Praeger, 1974.

Mishkin, Tracy. *The Harlem and Irish Renaissances: Language, Identity, and Representation.* Gainesville: UP of Florida, 1998.

Perelman, Bob. *The Trouble with Genius: Reading Pound, Joyce, Stein and Zukofsky.* Berkeley: U of California P, 1994.

Perkins, Linda. "Black Women and Racial 'Uplift' Prior to Emancipation." *The Black Woman Cross-Culturally.* Ed. Filomina Chioma Steady. Cambridge, MA: Schenkman, 1981. 317–34.

Peterson, Carla L. "The Remaking of Americans: Gertrude Stein's 'Melanctha' and African-American Musical Traditions." *Criticism and the Color Line: Desegregating American Literary Studies.* Ed. Henry B. Wonham. New Brunswick, NJ: Rutgers UP, 1996. 140–57.

Power, Tyrone. *Impressions of America during the Years 1833, 1834, and 1835.* London: R. Bentley, 1836. 2:238–44. Rpt. in *The Way We Lived.* Vol. I. Ed. Frederick M. Binder and David M. Reimers. Lexington, MA: Heath, 1988. 238–40.

Quartermain, Peter. *Disjunctive Poetics: From Gertrude Stein and Louis Zukofsky to Susan Howe.* Cambridge, MA: Cambridge UP, 1992.

Roof, Judith. *Come as You Are: Sexuality and Narrative*. New York: Columbia UP, 1996.

Ruddick, Lisa. "'Melanctha' and the Psychology of William James." *Modern Fiction Studies* 28 (1982–83): 543–56.

Ryan, Judith. *The Vanishing Subject, Early Psychology and Literary Modernism*. Chicago: U of Chicago P, 1991.

Saldivar-Hull, Sonia. "Wrestling Your Ally: Stein, Racism, and Feminist Critical Practice." *Women's Writing in Exile*. Ed. Mary Lynn Broe and Angela Ingram. Chapel Hill: U of North Carolina P, 1989. 181–98.

Smith-Rosenberg, Carroll. *Disorderly Conduct: Visions of Gender in Victorian America*. New York: Oxford UP, 1986.

——. "The Female World of Love and Ritual: Relations between Women in Nineteenth-Century America." *Signs* I. 1 (1975): 1–31.

Stein, Gertrude. "On the Value of a College Education for Women." Unpublished essay. The Cone Collection Archives, The Baltimore Museum of Art.

Steiner, Wendy. *Exact Resemblance to Exact Resemblance: The Literary Portraiture of Gertrude Stein*. New Haven: Yale UP, 1978.

Stimpson, Catharine R. "The Mind, the Body, and Gertrude Stein." *Critical Inquiry* 3 (1977): 491–506.

Thernstrom, Stephan, ed. *Harvard Encyclopedia of American Ethnic Groups*. Cambridge, MA: Harvard UP, 1980.

Torgovnick, Marianne. *Gone Primitive: Savage Intellects, Modern Lives*. Chicago: U of Chicago P, 1990.

Tuttle, William M. "Going into Canaan," *The Way We Lived*, Vol. II. Eds. Frederick M. Binder, and David M. Reimers. Lexington, MA: Heath, 1988. 122–32.

Van Vechten, Carl. Introduction. *Three Lives*. By Gertrude Stein. New York: Modern Library, 1933. v–xi.

Wald, Priscilla. *Constituting Americans: Cultural Anxiety and Narrative Form*. Durham: Duke UP, 1995.

Wood, Carl. "Continuity of Romantic Irony: Stein's Homage to Laforgue in *Three Lives*." *Comparative Literature Studies* 12.2 (June 1975): 147–58.

Zola, Émile. *The Dreyfus Affair*. Ed. Alain Pages. New Haven: Yale UP, 1998.

(*Continued from page iv*)

Reprinted from *Forbidden Fires* by Margaret C. Anderson, edited by Mathilda M. Hills, The Naiad Press, 1997, with the permission of the publisher and the editor.

Paul Cézanne's 1906 letter to his son, reprinted from *Paul Cézanne's Letters*, edited by John Rewald, translation to English Marguerite Kay and G. & E. M. Hill, reprinted with the permission of Bruno Cassirer (Publishers) Ltd.

Unpublished Gertrude Stein materials are used with permission of The Yale Collection of American Literature, Beinecke Rare Book and Manuscript Library, Yale University.

Excerpt from Stein's letter to Henri Pierre Roche, dated June 12, 1912, used by permission of the Harry Ransom Humanities Center, The University of Texas, Austin.

Printed in the United States
By Bookmasters